Past, Present, and Future of Radiochemical Synthesis

Past, Present, and Future of Radiochemical Synthesis

Editor

Svend Borup Jensen

Basel • Beijing • Wuhan • Barcelona • Belgrade • Novi Sad • Cluj • Manchester

Editor
Svend Borup Jensen
Department of Nuclear
Medicine
University Hospital Aalborg
Aalborg C
Denmark

Editorial Office
MDPI
St. Alban-Anlage 66
4052 Basel, Switzerland

This is a reprint of articles from the Special Issue published online in the open access journal *Molecules* (ISSN 1420-3049) (available at: www.mdpi.com/journal/molecules/special_issues/radiochemical_synthesis).

For citation purposes, cite each article independently as indicated on the article page online and as indicated below:

Lastname, A.A.; Lastname, B.B. Article Title. *Journal Name* **Year**, *Volume Number*, Page Range.

ISBN 978-3-03928-614-0 (Hbk)
ISBN 978-3-03928-613-3 (PDF)
doi.org/10.3390/books978-3-03928-613-3

© 2024 by the authors. Articles in this book are Open Access and distributed under the Creative Commons Attribution (CC BY) license. The book as a whole is distributed by MDPI under the terms and conditions of the Creative Commons Attribution-NonCommercial-NoDerivs (CC BY-NC-ND) license.

Contents

About the Editor vii

Preface ix

Pia Afzelius, Aage Kristian Olsen Alstrup, Ole Lerberg Nielsen, Karin Michaelsen Nielsen and Svend Borup Jensen
Attempts to Target *Staphylococcus aureus* Induced Osteomyelitis Bone Lesions in a Juvenile Pig Model by Using Radiotracers
Reprinted from: *Molecules* 2020, 25, 4329, doi:10.3390/molecules25184329 1

Rosalba Mansi, Guillaume Pierre Nicolas, Luigi Del Pozzo, Karim Alexandre Abid, Eric Grouzmann and Melpomeni Fani
Evaluation of a New ^{177}Lu-Labeled Somatostatin Analog for the Treatment of Tumors Expressing Somatostatin Receptor Subtypes 2 and 5
Reprinted from: *Molecules* 2020, 25, 4155, doi:10.3390/molecules25184155 16

Panagiotis Kanellopoulos, Emmanouil Lymperis, Aikaterini Kaloudi, Marion de Jong, Eric P. Krenning, Berthold A. Nock and Theodosia Maina
Tc]Tc-DB1 Mimics with Different-Length PEG Spacers: Preclinical Comparison in GRPR-Positive Models
Reprinted from: *Molecules* 2020, 25, 3418, doi:10.3390/molecules25153418 32

Kristina Søborg Pedersen, Christina Baun, Karin Michaelsen Nielsen, Helge Thisgaard, Andreas Ingemann Jensen and Fedor Zhuravlev
Design, Synthesis, Computational, and Preclinical Evaluation of ^{nat}Ti/^{45}Ti-Labeled Urea-Based Glutamate PSMA Ligand
Reprinted from: *Molecules* 2020, 25, 1104, doi:10.3390/molecules25051104 45

Cassis Varlow, Emily Murrell, Jason P. Holland, Alina Kassenbrock, Whitney Shannon, Steven H. Liang, et al.
Revisiting the Radiosynthesis of [^{18}F]FPEB and Preliminary PET Imaging in a Mouse Model of Alzheimer's Disease
Reprinted from: *Molecules* 2020, 25, 982, doi:10.3390/molecules25040982 64

Alessandra Cavaliere, Katrin C. Probst, Stephen J. Paisey, Christopher Marshall, Abdul K. H. Dheere, Franklin Aigbirhio, et al.
Radiosynthesis of [^{18}F]-Labelled Pro-Nucleotides (ProTides)
Reprinted from: *Molecules* 2020, 25, 704, doi:10.3390/molecules25030704 73

Andrei Molotkov, John W. Castrillon, Sreevidya Santha, Paul E. Harris, David K. Leung, Akiva Mintz and Patrick Carberry
The Radiolabeling of a Gly-Sar Dipeptide Derivative with Flourine-18 and Its Use as a Potential Peptide Transporter PET Imaging Agent
Reprinted from: *Molecules* 2020, 25, 643, doi:10.3390/molecules25030643 91

Klaudia Cybulska, Lars Perk, Jan Booij, Peter Laverman and Mark Rijpkema
Huntington's Disease: A Review of the Known PET Imaging Biomarkers and Targeting Radiotracers
Reprinted from: *Molecules* 2020, 25, 482, doi:10.3390/molecules25030482 100

Gonçalo S. Clemente, Tryfon Zarganes-Tzitzikas, Alexander Dömling and Philip H. Elsinga
Late-Stage Copper-Catalyzed Radiofluorination of an Arylboronic Ester Derivative of Atorvastatin
Reprinted from: *Molecules* **2019**, *24*, 4210, doi:10.3390/molecules24244210 . . . . **121**

Sajid Mushtaq, Seong-Jae Yun and Jongho Jeon
Recent Advances in Bioorthogonal Click Chemistry for Efficient Synthesis of Radiotracers and Radiopharmaceuticals
Reprinted from: *Molecules* **2019**, *24*, 3567, doi:10.3390/molecules24193567 . . . . **130**

Falguni Basuli, Xiang Zhang, Burchelle Blackman, Margaret E. White, Elaine M. Jagoda, Peter L. Choyke and Rolf E. Swenson
Fluorine-18 Labeled Fluorofuranylnorprogesterone ([^{18}F]FFNP) and Dihydrotestosterone ([^{18}F]FDHT) Prepared by "Fluorination on Sep-Pak" Method
Reprinted from: *Molecules* **2019**, *24*, 2389, doi:10.3390/molecules24132389 . . . . **160**

About the Editor

Svend Borup Jensen

Svend Borup Jensen received his BSc in Chemistry and Biochemistry from Aarhus University, Denmark (1992), a MPhil in Chemistry from University of Strathclyde, Scotland (1994), and a PhD in Chemistry from University of Paisley, Scotland (1998).

Following graduation, Svend worked as an organic chemist and as an analysis chemist for private companies. However, his main career has been in radiochemistry, first at Aarhus University Hospital as head of production, and, since 2009, as the head of radiochemistry/QP (Qualified Person) at Aalborg University Hospital. Svend's responsibility as QP is the production and release of radiopharmaceutical drugs used in humans for diagnosis. He has also been strongly involved in designing and qualifying the new and heavily enlarged cleanroom facilities for the Dept. of Nuclear Medicine at "Nyt Aalborg Universitetshospital" (NAU) to ensure that the facilities are in compliance with the legislation for production of sterile pharmaceuticals and contains the suitable apparatus for radioactive drug production.

Research in the field of radioactive drugs is very versatile. Most of Svend research has the synthesis of new radiopharmaceutical drug or the optimization of known syntheses as the focal point, but the analysis and identification of the radioactive drugs and the by-products has also received a lot of attention.

As a researcher, Svend has authored or co-authored 49 research papers in peer-reviewed journals, 3 book chapters, and 11 articles in Danish professional journals with a more educational angle.

Preface

The purpose of this Special Issue was to host research and review papers on the past, present, and future of radiochemical synthesis. I wished for this Issue to be about radiochemistry and related topics. I came up with a title, "Past, Present, and Future of Radiochemical Synthesis", in the hopes of inspiring authors.

The Special Issue contains nine original research papers and two reviews within the area of radiochemistry. These address how and why a certain radioactive compound has been prepared, how a radioactive compound has taken our understanding of a biological system to a different level, how a radiopharmaceutical can be used for treatment, and/or how a radioactive compound has improved our understanding of a certain drug (binding pattern and/or metabolization of the drug).

I hope you enjoy reading this Special Issue. Best regards.

Svend Borup Jensen
Editor

Article

Attempts to Target *Staphylococcus aureus* Induced Osteomyelitis Bone Lesions in a Juvenile Pig Model by Using Radiotracers

Pia Afzelius [1,2,*], **Aage Kristian Olsen Alstrup** [3,4], **Ole Lerberg Nielsen** [5], **Karin Michaelsen Nielsen** [1,5] **and Svend Borup Jensen** [1,6]

1 Department of Nuclear Medicine, Aalborg University Hospital, 9100 Aalborg, Denmark; karinmn007@hotmail.com (K.M.N.); svbj@rn.dk (S.B.J.)
2 North Zealand Hospital, Copenhagen University Hospital, 3400 Hillerød, Denmark
3 Department of Nuclear Medicine and PET, Aarhus University Hospital, 8200 Aarhus, Denmark; aagealst@rm.dk
4 Department of Clinical Medicine, Aarhus University, 8200 Aarhus, Denmark
5 Department of Veterinary and Animal Sciences, Faculty of Health and Medical Sciences, University of Copenhagen, 2000 Copenhagen F, Denmark; olelerbergnielsen@gmail.com
6 Department of Chemistry and Biochemistry, Aalborg University, 9100 Aalborg, Denmark
* Correspondence: Pia.Afzelius@dadlnet.dk; Tel.: +45-25732501

Received: 12 August 2020; Accepted: 16 September 2020; Published: 21 September 2020

Abstract: Background [^{18}F]FDG Positron Emission Tomography cannot differentiate between sterile inflammation and infection. Therefore, we, aimed to develop more specific radiotracers fitted for differentiation between sterile and septic infection to improve the diagnostic accuracy. Consequently, the clinicians can refine the treatment of, for example, prosthesis-related infection. Methods: We examined different target points; *Staphylococcus aureus* biofilm (^{68}Ga-labeled DOTA-K-A9 and DOTA-GSGK-A11), bone remodeling ([^{18}F]NaF), bacterial cell membranes ([^{68}Ga]Ga-Ubiquicidin), and leukocyte trafficking ([^{68}Ga]Ga-DOTA-Siglec-9). We compared them to the well-known glucose metabolism marker [^{18}F]FDG, in a well-established juvenile *S. aureus* induced osteomyelitis (OM) pig model. Results: [^{18}F]FDG accumulated in the OM lesions seven days after bacterial inoculation, but disappointingly we were not able to identify any tracer accumulation in OM with any of the supposedly more specific tracers. Conclusion: These negative results are, however, relevant to report as they may save other research groups from conducting the same animal experiments and provide a platform for developing and evaluating other new potential tracers or protocol instead.

Keywords: osteomyelitis; animal experimentation; pigs; PET/CT; [^{18}F]FDG; [^{18}F]NaF; [^{68}Ga]Ga-DOTA-K-A9; [^{68}Ga]Ga-DOTA-GSGK-A11 [^{68}Ga]Ga-DOTA-Siglec-9; [^{68}Ga]Ga-ubiquicidin

1. Introduction

Staphylococcus aureus (*S. aureus*) causes skin, soft tissue, bone, pleuropulmonary infections, infective endocarditis, sepsis, osteoarticular, and device-related infections [1]. To prevent disease progression and to reduce the number of complications, it is essential to diagnose bacterial infections at an early stage and initiate the best therapy. As stated by the WHO and others, the annual consumption of antibiotics is increasing, and the support for its use is often insignificant [2,3]. The prevalence of specific Multi-Drug Resistant (MDR) bacteria is associated with the usage of broad-spectrum antibiotics, both for empiric as well as for definite therapy [4]. According to a recent report, more than 2.8 million antibiotic-resistant infections occur in the U.S. each year, and more than 35,000 people die as a result [5]. To use the correct treatment, clinicians need to identify the location of inflammation and discriminate between sterile lesions and infections by pathogenic microorganisms. To fit this purpose, nuclear medicine imaging techniques in combination with Computed Tomography (CT) provide valuable information on functional and anatomical data. These imaging techniques visualize the extent, the activity, the size, and the location of the disease. The molecular imaging techniques of nuclear medicine, such as scintigraphy and Positron Emission Tomography (PET), are characterized by visualizing the whole body.

However, the commonly used tracer [^{18}F]-fluorodeoxyglucose ([^{18}F]FDG) and [^{111}In]-labeled leukocytes cannot distinguish between sterile inflammation and infection. Osteomyelitis is a severe bone infection, and the overall purpose of our project was to identify radiotracers that may improve the scanning of the bacterial infection. We have, therefore, searched for other tracers; tracers made for targeting the *S. aureus* biofilms or the infected bone lesion itself.

^{68}Ga-labeled PET tracers are of particular interest. This radiopharmaceutical is nontoxic. The half-life of 68 minutes makes it suitable for peptide pharmacology. ^{68}Ga is available from ^{68}Ge/^{68}Ga generators and independent of reliance on local cyclotron production [6]. In vitro specificity for bacterial binding and in vivo stability with favorable kinetics makes these radiotracers candidates for infection imaging with PET [7].

S. aureus tends to form biofilms, especially on catheters, artificial heart valves, bone, joint prostheses, and bone sequesters [8]. Biofilms reduce the effect of the immune response and antibiotics. As previously described, we thus studied the applicability of two ^{68}Ga-labeled phage-display selected dodecapeptides with an affinity towards *S. aureus* biofilm: DOTA-K-A9 and DOTA-GSGK-A11 in our juvenile pig osteomyelitis model [9].

We have previously used ^{99m}Tc-label methylene diphosphate ([^{99m}Tc]Tc-DPD) to examine bone remodeling in pigs. We did not observe any tracer uptake in the osteomyelitis lesions [10]. We, therefore, moved on to use [^{18}F]NaF in the present study because [^{18}F]NaF though acting through similar uptake mechanisms as [^{99m}Tc]Tc-DPD, [^{18}F]NaF is supposed to have a two-fold higher uptake in bones and a better sensitivity for metastatic osteoblastic metastases [11].

Antimicrobial peptides, such as ubiquicidin, can distinguish between mammalian and bacterial or fungal cells and may be targeting vector-candidates for molecular imaging due to their selectivity for bacterial cell membranes in the innate immune system response [12]. We used a [^{68}Ga]Ga-ubiquicidin as a bacteria-specific imaging probe.

We also examined the vascular adhesion protein 1 (VAP-1)-targeted PET tracer [^{68}Ga]Ga-DOTA-Siglec-9, as an alternative to the traditional tracers for detecting infections. Sialic acid-binding immunoglobulin-like lectin 9 (Siglec-9) is a natural ligand for VAP-1 and is expressed on monocytes and neutrophils and is involved in leukocyte trafficking [13].

2. Results

2.1. Animal Model

Forty-five OM lesions developed in 15 pigs and none developed in the non-inoculated left control hind limb (Table 1). Bacterial culturing and histopathology confirmed the presence of the inoculated *S. aureus* strain S54F9 in the OMs.

2.2. Tracers

[^{68}Ga]Ga-DOTA-K-A9 and [^{68}Ga]GaDOTA-GSGK-A11

While previous evaluation in murine subcutaneous *S. aureus* infections showed uptake of [^{68}Ga]Ga-DOTA-K-A9 [9,14], we saw no increased tracer activity of the two *S. aureus* phage displayed selected peptides, [^{68}Ga]Ga-DOTA-K-A9 (Figure 1) and [^{68}Ga]Ga-GSG-KA-11 (Figure 2) in 9 and 5, respectively, porcine OM lesions. Figure 3 shows the bio-distribution of the peptides in pigs. Both peptides were excreted by the liver and kidneys.

Table 1. Standard uptake values (SUV) of different radiotracers in OM lesions.

Pig Tracer	1	2	3	4	5	6	7	8	9	10	11	12	13	14	15
[68Ga]Ga-DOTA-Siglec-9 Leukocyte trafficking	0.4–0.4	0.4–0.5	0.2–0.5									1.3	0.9–1	0.7–0.9	1
[68Ga]Ga-DOTA-K-A9 S. aureus (biofilm)				0.7–0.7	0.9–1.2	0.5–0.7									
[68Ga]Ga-DOTA-GSGK-A11 S. aureus (biofilm)							0.7–0.8	0.2–0.3							
[68Ga]Ga-ubiquicidin Bacterial cell membranes									0.6–0.8	0.6–0.9	0.5–0.6				
[18F]FDG Glucose metabolism	4–7.3	2.3–5.9	2.9–4.5	10.5–16.3	3.4–9	4–9.2	3–6.4	2.4–8	5.4–5.9	4.4–5.1	1–3.9	6.1	7.8–10.1	4–7.8	3.6
[18F]NaF Bone remodeling									9.2–13.6	15.2–17	5.7–19.2	19.8	7.8–17.6		
Number of lesions	3	4	5	2	3	4	3	2	4	3	5	1	2	3	1

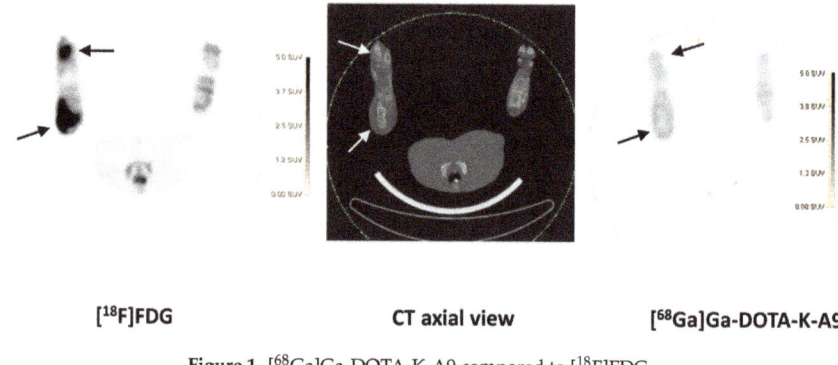

Figure 1. [^{68}Ga]Ga-DOTA-K-A9 compared to [^{18}F]FDG.

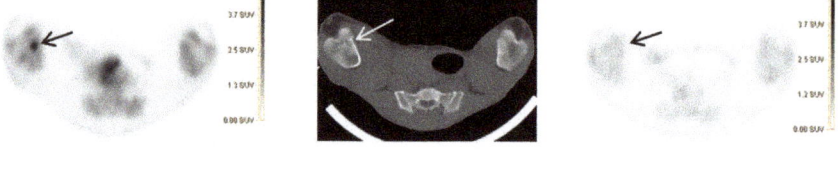

Figure 2. [^{68}Ga]Ga-DOTA-GSGK-A11 compared to [^{18}F]FDG.

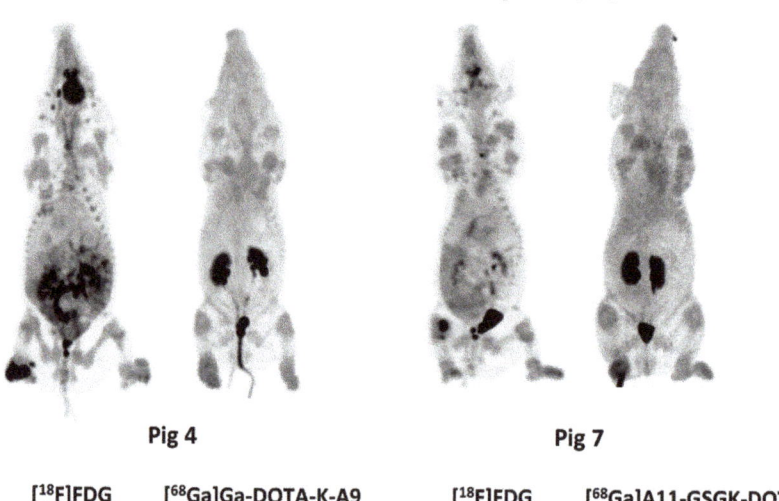

Figure 3. Maximum intensity projections (MIPs) of [^{18}F]FDG, [^{68}Ga]Ga-DOTA-K-A9, and [^{68}Ga]Ga-DOTA-GSGK-A11.

[^{18}F]FDG (left) and Ga-DOTA-K-A9 (right) accumulation is shown in an OM lesion in the right calcaneus and distal II metatarsus of pig 4 (indicated by arrows). The lesions show sequester formation and lysis of the cortical bone on CT in the axial view (middle). Comparable SUV scales are shown to the right of the PET images.

[^{18}F]FDG (left) and [^{68}Ga]Ga-DOTA-GSGK-A11 (right) accumulation in the right distal femur of pig 7. An OM lesion close to the medial part of the growth zone of the distal right femur with sequester formation and lysis of the cortical bone was revealed on the axial CT image (middle). Comparable SUV scales are shown to the right of the PET images.

MIPs of the [^{18}F]FDG, [^{68}Ga]Ga-DOTA-K-A9, and [^{68}Ga]Ga-DOTA-GSGK-A11.

CT images of two OM in the distal right femur in pig 9 in the axial view. [^{68}Ga]Ga-ubiquicidin and [^{18}F]NaF compared to [^{18}F]FDG accumulation in the same lesions. Comparable SUV scales are shown to the right of the PET images. For a demonstration of the uptake of [^{18}F]NaF in the growth zones of long bones, a wider SUV-scale is used in the last picture. The CT image shows sequester formations and lysis of the cortical bones.

2.3. [^{18}F]NaF

We saw no accumulation of [^{18}F]NaF in 15 OM lesions (Figure 4). [^{18}F]NaF was distributed to the growth zones of the bones (Figure 4) and excreted by the kidneys (Figure 5).

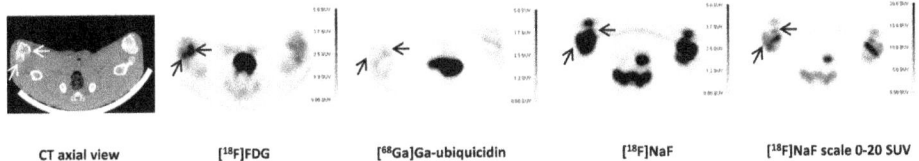

Figure 4. [^{68}Ga]Ga-ubiquicidin and [^{18}F]NaF compared to [^{18}F]FDG.

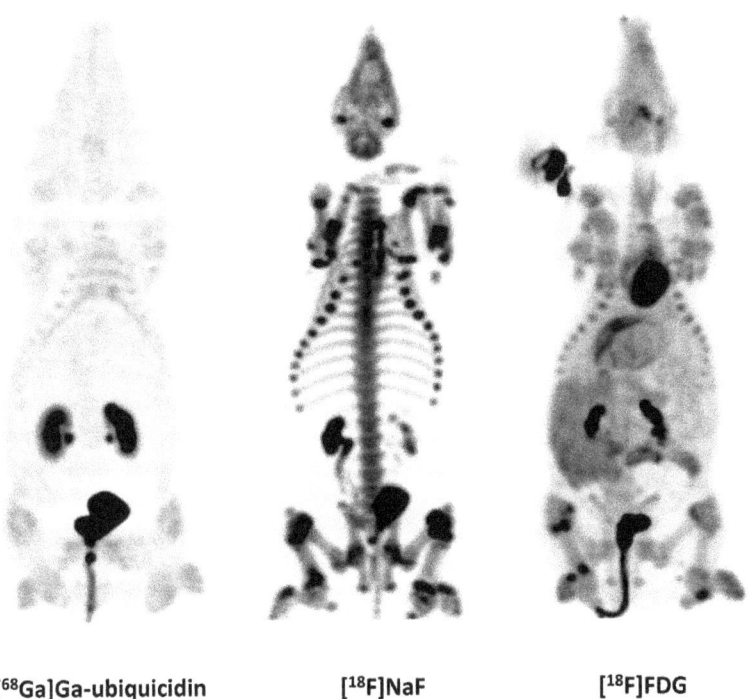

Figure 5. MIPs of [^{68}Ga]Ga-ubiquicidin, [^{18}F]NaF, and [^{18}F]FDG.

2.4. [^{68}Ga]Ga-Ubiquicidin

We saw no accumulation of [^{68}Ga]Ga-ubiquicidin in 12 OM lesions. [^{68}Ga]Ga-ubiquicidin was excreted by the kidneys (Figure 5).

MIPs of the [^{68}Ga]Ga-ubiquicidin, [^{18}F]NaF, and [^{18}F]FDG distribution in pig 9.

2.5. [^{68}Ga]Ga-DOTA-Siglec-9

We saw no tracer activity neither in the early nor the late images of [^{68}Ga]Ga-DOTA-Siglec-9 in 19 osteomyelitis lesions (Figure 6). Instead, the tracer accumulated in dorsocaudal parts of both lungs of, for example, pig 1 (Figure 7). The CT image showed signs of infectious foci in dorsocaudal parts of both lungs. We noticed, however, a general tendency of [^{68}Ga]Ga-DOTA-Siglec-9 accumulation in the dorsocaudal parts of the lungs. We also observed this in pigs that did not develop OM or pulmonary infections (data not shown). Figure 8 demonstrates in images acquired 10 to 30 min after tracer injection, a marked ^{68}Ga]Ga-DOTA-Siglec-9 uptake in the margin of an inguinal abscess adjacent to the *S. aureus* inoculation site. In the static images acquired after 60 min, the activity of [^{68}Ga]Ga-DOTA-Siglec-9 decreased. Whereas the activity of [^{18}F]FDG increased slightly in the margin of the abscess (Figure 8).

[^{18}F]FDG and [^{68}Ga]Ga-DOTA-Siglec-9] (early and late acquisition; 1 and 2 h) uptake in osteomyelitis lesion in the right medial condyle of the right distal femur (indicated by an arrow) of pig 1 and the corresponding CT (axial view). Comparable SUV scales are shown to the right of the PET images.

The CT image (axial view) of the dorsocaudal parts of the lungs of pig 1 showed signs of infection and partial atelectasis (arrow) and the corresponding [^{68}Ga]Ga-DOTA-Siglec-9 uptake in the corresponding anatomical area. Axial views of CT and PET and a MIP (side view) of [^{68}Ga]Ga-DOTA-Siglec-9 distribution in the pig's body. The SUV scale bar is shown in the axial PET image.

Pig 12 demonstrated [^{68}Ga]Ga-DOTA-Siglec-9 uptake in 5 × 3 × 2 cm chronic inoculation abscess (arrow) in the right inguinal region (10 to 30 min after tracer injection; this is an average image of the frames covering 10 to 30 min, simulating a static image 10 to 30 min) and the uptake of [^{18}F]FDG in the same region. Note the similar uptake of [^{68}Ga]Ga-DOTA-Siglec-9 in the growth zones of the distal femur and proximal tibia bones and the intense uptake of [^{18}F]FDG in the OM lesions of these bones (with sequesters and fistulas formations) in the right hind limb. A hyperplasic medial iliacus lymph node (broken arrow) on the right site had increased [^{18}F]FDG uptake compared to the lymph node in the non-infected left hind limb. In the second image, the acquisition was at 60 min post-injection of tracers. The activity in the abscess had decreased for [^{68}Ga]Ga-DOTA-Siglec-9 and had increased for [^{18}F]FDG.

2.6. [^{18}F]FDG

[^{18}F]FDG accumulated distinguishably in 45 of the 45 OM lesions.

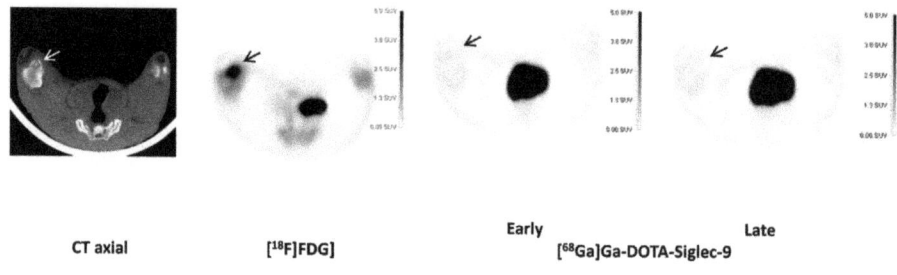

Figure 6. [^{68}Ga]Ga-DOTA-Siglec-9 compared to [^{18}F]FDG].

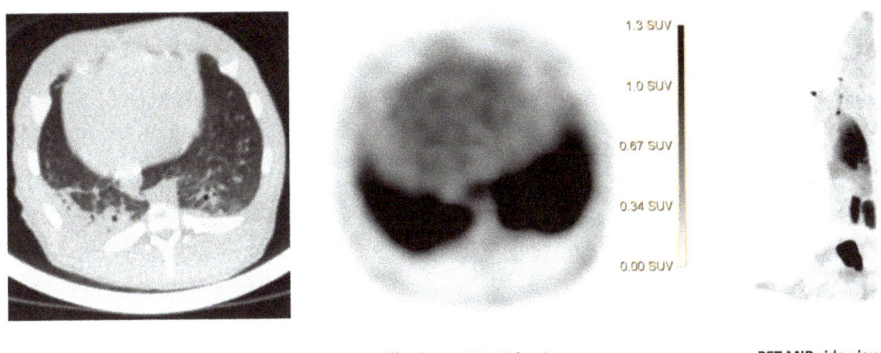

Figure 7. [^{68}Ga]Ga-DOTA-Siglec-9 uptake in the lungs.

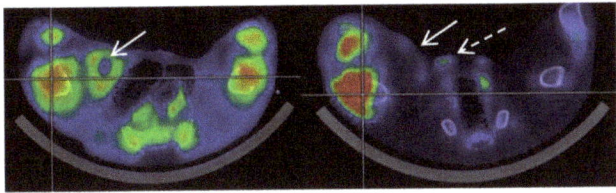

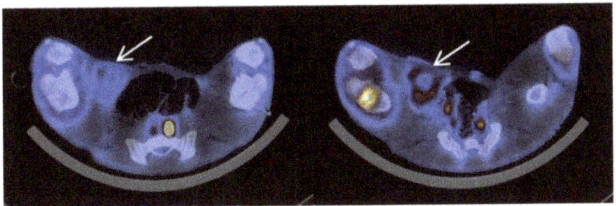

Figure 8. [^{68}Ga]Ga-DOTA-Siglec-9 compared to [^{18}F]FDG.

3. Discussion

A concern for the health authorities is the increasing MDR in society and its negative effect on morbidity and mortality. Correct antibiotic use in both humans and animals reduces the threat of antibiotic resistance. Timely and specific diagnosis of infectious diseases is essential for the patient's outcome. Blood cultures can detect invading microorganisms but cannot discriminate between sterile inflammatory and bacterial infections in tissues. Imaging can reveal infection in the body, and early imaging may help to direct the choice of antibiotic treatment and justify continued treatment by antibiotics. Nuclear medicine techniques can visualize the whole body but the available radiopharmaceuticals, such as [^{18}F]FDG, are, however, not capable of distinguishing between sterile inflammation and bacterial or fungal infections. Osteomyelitis is a severe bone infection, and the overall purpose of our project was to identify radiotracers that may improve the scanning of bone infection. For diagnosing OM, [^{18}F]FDG is usable but unspecific, especially in adults. Therefore, we searched for radioactive tracers for nuclear imaging fitted for differentiation between sterile and septic inflammation to improve the diagnostic accuracy and thus the treatment of, for example, prosthesis-related infections.

We used tracers prepared for targeting the *S. aureus* biofilms or the infected bone lesion itself. We used a well-characterized juvenile porcine model, where OM was selectively induced by an also well-characterized *S. aureus* strain in one hind limb leaving the contralateral hind limb for comparison. Worldwide, *S. aureus* is involved in most cases of hematogenously spread OM. Hematogenously spread OM most often affects children and elderly patients. Our juvenile porcine model has previously demonstrated to mimic osteomyelitis in children [15].

When *S. aureus* enters the skin, neutrophils and macrophages migrate to the site of infection. *S. aureus* avoids this imminent attack by the immune system of the host, for example, by blocking chemotaxis of leukocytes, sequestering host antibodies, hiding from detection via polysaccharide capsule, or biofilm formation, and resisting destruction after ingestion by phagocytes. Bacteriophages are viruses that show no specificity for mammalian cells and infect bacteria exclusively [16]. Most bacteriophages have specificity toward a single bacterial strain. The binding mechanism consists of the attachment of the phages to specific surface receptors or domains located on the surface of the bacterium, subsequently transferring their genetic material into the host cell dedicated for phage replication/reproduction. We have previously examined two ^{68}Ga-labeled phage-display selected peptides with affinity for *S. aureus* biofilm and evaluated their potential as bacteria-specific PET imaging agents [9,14]. However, in the present study, we saw no increased accumulation of neither [^{68}Ga]Ga-DOTA-K-A9 nor [^{68}Ga]Ga-DOTA-GSGK-A11 in the porcine OM lesions, indicating that either no biofilm had formed one week after the inoculation of *S. aureus* or the tracer did not bind to the biofilm in vivo, or was metabolized before the static scan.

The static imaging was for logistic and financial purposes acquired in the same set up as the dynamic scans. We have not analyzed the dynamic scans yet. We, therefore, do not know the kinetics of these two tracers in juvenile pigs. The static scan 60 min after tracer-injection may have been obsolete.

The PET tracer [^{68}Ga]Ga-DOTA-Siglec-9 binds to a protein involved in leukocyte extravasations (vascular adhesion protein 1, VAP-1). Only recently, we have analyzed the dynamic data [13]. The uptake of [^{68}Ga]Ga-DOTA-Siglec-9 had reversible kinetics and could be modeled with the rev2TCM (4 k-parameters). We demonstrated that [^{68}Ga]Ga-DOTA-Siglec-9 was metabolized very quickly and no increased [^{68}Ga]Ga-DOTA-Siglec-9 uptake was seen in OM [13]. However, the distribution volume for the uptake of soft tissue infections was elevated, concluding the tracer has a role for soft tissue infections, but not for bone infections (osteomyelitis). In the present study, we also observed an affinity of [^{68}Ga]Ga-DOTA-Siglec-9 for infected soft tissue (Figures 7 and 8), although the tracer was present for a short period in this infectious compartment (Figure 8). We can conclude that static scanning should have been earlier when visualizing infections in soft tissues, as the tracer turned out to be reversible [13]. The reason for uptake in the lungs (Figure 7) may be due to an infection, or it may be a result of tracer-trapping in the pulmonary tissue, as the whole-body biodistribution of [^{68}Ga]Ga-DOTA-Siglec-9 showed, in general, a significant uptake or accumulation in the lungs, also in non-infected lungs. Jaakkola and coworkers demonstrated that VAP-1 was present in porcine lung endothelium [17]. The tracer accumulation in the lungs may be an effect of high constitutional expression of the receptor, perhaps as a consequence of the lungs being the "excretion organ" for neutrophils. The lungs of pigs have much of the same filter-like function as the spleen in humans. Or it may be an effect of the pulmonary intravascular macrophages in pigs [18].

Theoretically, highly infection-specific radiopharmaceuticals targeting antimicrobial peptides such as the human antimicrobial peptide ubiquicidin seemed to be a worthy approach. The peptide is in mammals, birds, amphibians, and insects, and it protects these animals against infection [19]. We used the small UBI 29-41 synthetic peptide, derived from humans. This UBI binds preferentially to bacteria in vitro and not to activated leukocytes, and can distinguish bacterial infections from inflammation with greater specificity than other UBI peptides [19–22]. We saw, however, no increase in [^{68}Ga]Ga-ubiquicidin accumulation in the *S. aureus* induced OM lesions in our study.

The [^{68}Ga]Ga-ubiquicidin, though human-derived, should not be species-specific. Also, the *S. aureus* strain we used for inoculation in our animal model was isolated from human pneumonia. Perhaps the mice or rabbit models, inoculated with a human-derived *S. aureus* strain, is not comparable to a pig model. Healthy pigs may eliminate bacterial infections quickly and, thus, bacteria in the pigs may not be detectable with ^{68}Ga-labeled UBI 29-41. But it may not exclude the tracer for usage in humans. We have not analyzed the dynamic scans yet. We, therefore, do not know the kinetics of this tracer.

It would have been optimal to analyze the tracer-kinetics before performing the static scans. The kinetic analyses can indicate the optimal times to scan after the tracer-injections. The dynamic scans were performed first and lasted one hour, and consequently delaying the static scans, but the delay may not have been significant as we used [^{18}F]- and [^{68}Ga]-marked tracers.

S. aureus can also invade osteoblasts and can form small-colony variants in the intracellular compartment, where they can survive in a metabolically inactive state while preserving the integrity of the host cell [23,24]. We do not know if this was the case. And if it was, if this was the explanation for the missing sign of osteoblast activity by both [^{99m}Tc]Tc-DPD and [^{18}F]NaF. Another reason could be that the infection was in its early phase before osteoblast recruitment and reparative processes took place. [^{18}F]NaF diffuses into the extravascular fluids of bones and is incorporated in hydroxyapatite. Thus, the uptake of [^{18}F]NaF reflects both blood flow and bone remodeling. We have recently demonstrated only slightly increased blood perfusion in the infected limb compared to the not infected limb [25]. This amount of blood perfusion may not have been sufficient to increase the [^{18}F]NaF accumulation. It was, however, not ethically justifiable for us to keep the pigs alive for more than seven days after they had established the *S. aureus* infection as human endpoints were reached. This model, therefore, reflected acute/subacute OM.

[^{18}F]FDG accumulated very distinguishably in the OM in our juvenile pig model, and even small lesions were recognizable on CT images alone. The same may be the case in children as juveniles have fewer concurring diseases and seldom degenerative changes causing inflammation of the bone and joint tissues. [^{18}F]FDG secures that no OM is unnoticed. We have previously addressed a possible reduction of [^{18}F]FDG activity in children [26].

The overall osteomyelitis project used an ambitious protocol as we scanned each pig several times, using a combination of scanning techniques (SPECT/CT and PET/CT) and series of tracers from 18 to 20 hours. We scheduled the scan protocol so that a new scan was not initiated until at least five half-lives after the latest PET-tracer had passed. The exceptions occurred in some of the pigs when the ^{68}Ga-labeled tracer followed by the ^{18}F-labeled tracers. It was impractical to wait the full 5 × 67.7 min from the injection of ^{68}Ga-labeled tracer to [^{18}F]FDG scan. Instead, close to four half-lives passed, and the injected activity of [^{18}F]FDG was at least double the injected activity of the ^{68}Ga-based tracers. The pigs were euthanized after the last scan and transported for necropsy at noon the following day; therefore, it was possible to raise the injected activity for the final scan. Due to the expected shorter half-lives and slightly slow uptake in OM, we gave more [^{68}Ga]Ga-Ubiquicidin. We would have liked to administer the same amount as of [^{68}Ga]Ga-DOTA-Siglec-9 but due to the fractionate and usage of only half the activity, we were only able to synthesize half as much [^{68}Ga]Ga-Ubiquicidin as [^{68}Ga]Ga-DOTA-Siglec-9. Regarding the radio-labeled phage displayed selected peptides, [^{68}Ga]Ga-DOTA-K-A9 and [^{68}Ga]Ga-DOTA-GSGK-A11, we injected the maximum amount of achieved radiotracer. The main reasons for varying injected activities were decreasing yield from ^{68}Ga-generator, changing ^{68}Ga-labeling yield (especially for A11 peptide [9]), and different delays between the time of production to the time of injection. In general, it is not easy to guess the adequate and optimal injectable radioactivity, so we tried variable activities.

4. Materials and Methods

4.1. Animal Model

Fifteen domestic pigs, all clinically healthy, specific pathogen-free Danish Landrace Yorkshire cross-bred female pigs aged 8–9 weeks weighing 18.3–24.5 kg, were purchased from local commercial pig farmers. After at least one week of acclimatization, the pigs fasted over-night. Then they were premedicated with s-ketamine (Pfizer, Ballerup, Denmark) and midazolam (B. Braun Medical, Frederiksberg, Denmark) intramuscularly and anesthetized with propofol (B. Braun Medical, Frederiksberg, Denmark) intravenously. Finally, they were inoculated with a well-characterized porcine strain of *S. aureus* S54F9 (approximately 10,000 CFU/kg body weight) into the femoral artery of the right hind limb to induce OM in that limb, as described elsewhere [18]. After the onset of clinical signs, for example, limping of the right hind limb, redness, and local swelling of the leg, starting typically on day 3–4 after inoculation, the pigs were supplied with intramuscular procaine benzylpenicillin (10,000 IE)/kg procaine benzylpenicillin (Penovet, Boehringer Ingelheim, Copenhagen, Denmark) once daily until 48 h before scanning [27]. The pigs were all treated with 30–45 µg/kg buprenorphine (Indivior, Berkshire, United Kingdom) intramuscularly thrice daily from the time before inoculation and until scanning at day 6 and 7 after inoculation [27]. After scanning, pigs were euthanized with an overdose of pentobarbitone (Scan Vet Animal Health, Fredensborg, Denmark) while still under anesthesia. Humane endpoints were: anorexia for more than 24 h, superficial respiration, the inability to stand up, or signs of systemic infection. After the inoculation, pigs were housed in separate boxes with soft bedding, fed twice daily with a restricted standard pellet diet (DIA plus FI, DLG, Denmark), and had ad libitum access to tap water. The environment was characterized by a room temperature of 20 °C, relative humidity of 51–55%, 12 h in light/12 h in dark cycles, an exchange of air at least eight times per hour. Pigs were fasted overnight before anesthesia.

Osteomyelitis was verified by CT imaging, necropsy, histology, and/or bacterial cultivation.

The study was approved by the Danish Animal Experimentation Board, license no. 2012-15-2934-000123 and 2017-15-0201-01239 in accordance with 2010/63/EU. All facilities were approved by the Danish Occupational Health Surveillance.

4.2. Tracers

4.2.1. *S. aureus* Biofilm (Phage Selected Peptides ^{68}Ga-labeled DOTA-K-A9 and DOTA-GSGK-A11)

The selection, radio-synthesis, the in vitro binding, and preliminary in vivo evaluation of the two phage display selected DOTA-peptide derivatives with an affinity towards *S. aureus* has previously been described [9,14]. Briefly, in the ^{68}Ga-labeling of DOTA-K-A9, as described in [14], the ^{68}GaCl$_3$ from the generator was trapped on a Varian SCX cartridge and eluted with a 5 M NaCl/5.5 M HCl (2.5%) solution (700 µL). The following ^{68}Ga-labeling was achieved by mixing the pre-purified ^{68}GaCl$_3$ with the precursor DOTA-K-A9 (70 µg, 37 nmol) in 0.1 M HEPES solution (5 mL) pH 6.5 adjusted with 30% ultrapure HCl and heated at 95 °C for 3.5 min. The post-purification of [^{68}Ga]Ga-DOTA-K-A9 was performed using a C18-light cartridge, preconditioned with 50% EtOH in water solution (5 mL) and 0.1 M HEPES (5 mL) pH 5.2 adjusted with 30% ultrapure HCl. The [^{68}Ga]Ga-DOTA-K-A9 reaction mixture was immediately transferred and trapped on a C18-light cartridge and rinsed with the 0.1 M HEPES pH 5.2 (2 mL) before it was eluted with 50% EtOH in water solution (1 mL). The final formulation (9 mL) of the product was achieved by adding isotonic saline (8 mL). The product formulation was further diluted prior to in vivo testing, to achieve an EtOH content <5% (*v/v*). Synthesis time: around 30 min. RCY: 56 ± 4%. RCP: 93 ± 1% EOS and 90 ± 1% at 2.5 h. pH: 5.1–5.7.

The ^{68}Ga-labeling of DOTA-GSGK-A11, as described in [9], was achieved by mixing the precursor DOTA-GSGK-A11 (70 µg, 34 nmol) in 0.1 M HEPES (5 mL) pH 6.5 adjusted with 30% ultrapure HCl and adding the pre-purified ^{68}Ga-eluate trapped on a Varian SCX cartridge and eluted with a 5 M NaCl/5.5 M HCl (2.5%) solution (700 µL). The mixture was heated at 95 °C for 7 min before purification was performed using a C18-light cartridge preconditioned with a 50% EtOH in water/sodium phosphate (2%) solution (5 mL) and isotonic saline (5 mL). After end heating, [^{68}Ga]Ga-DOTA-GSGK-A11 was immediately trapped on the C18-light cartridge, while non-chelated ^{68}Ga passed through. The cartridge was rinsed with isotonic saline (2 mL) before the purified [^{68}Ga]Ga-DOTA-GSGK-A11 was eluted by employing a 50% EtOH in water/sodium phosphate (2%) solution (1 mL). The final formulation of the product for biological testing was achieved by adding isotonic saline (5 mL). Synthesis time: around 30 min. RCY: 69 ± 2%. RCP: 96 ± 3% EOS and 96 ± 2% at 2 h. pH: 6.5–7.0.

The average injected activity was a median of 760 MBq (range 348–848 MBq) of [^{68}Ga]Ga-DOTA-K-A9, corresponding to ~17–46 MBq/kg and median 475 MBq (range 65–500 MBq) of [^{68}Ga]Ga-DOTA-GSGK-A11, corresponding to ~3–23 MBq/kg. PET was performed median ~60 min (60–61min) after injection of [^{68}Ga]Ga-DOTA-K-A9 (pigs 4–6, Table 1) and ~62 min (62–62 min) after injection of [^{68}Ga]Ga-DOTA-GSGK-A11 (pigs 7–8, Table 1) in the jugular vein.

4.2.2. Bone Remodeling ([^{18}F]NaF)

Fluoride-18 was produced via the ^{18}O (p,n) ^{18}F nuclear reaction using a cyclotron. The target content was passed through a QMA-light Sep-Pak (Waters, Hedehusene, Denmark) to trap F-18. The solution was washed with 10 mL sterile water for injection (10 mL) to remove any residual [^{18}O]H$_2$O and then dried with an argon stream. [^{18}F]NaF was set free by eluting the QMA Sep-Pak with 10 mL NaHCO$_3$ and 10–20 mL air into the product vial. The average injected activity was as median 184 MBq (range 165–196 MBq), corresponding to ~8–10 MBq/kg. The PET was performed ~57 min (33–78 min) after injection of [^{18}F]NaF (pigs 9–13, Table 1).

4.2.3. Bacterial Cell Membranes ([^{68}Ga]Ga-Ubiquicidin)

NOTA-Ubiquicidine 29-41 acetate was obtained from ABX, Radeberg, Germany. Ultrapure Water (metal-free water) and ultra-pure hydrochloric acid (30%) was obtained from Merck (Merck KGaA, Darmstadt Germany) and cation exchange cartridge Strata XC 10, Phenomenex Inc., Værløse, Denmark (Strong cation exchange) was used. Sterile water and sterile isotonic saline solution (9 mg/L) were purchased from the Hospital Pharmacy in Aalborg. Sodium acetate was of lab-grade from Sigma Aldrich (St. Louis, MI, USA). The synthesis was inspired by the synthesis published by Ebenhan et al. [20] and Vilche et al. [28]. Briefly, Ga-68 was obtained by eluting a ^{68}Ge/^{68}Ga generator (IGG100, Eckert & Ziegler AG Eurotope GmbH, Berlin, Germany). The generator eluate was fractionated and the 0.7 mL fraction with the highest radioactivity was collected and used for the synthesis. In our case, this was the fraction from 1. 6–2.3 mL (7 −/+ 2% of the radioactivity came in the early fraction 0–1.6 mL, 50 −/+ 2% came in the fraction we used, and 41 −/+ 2% came after (2.3–10 mL)). A total of 625 µL of the 0.7 mL eluate was mixed with 30 µL sodium buffer solution (5 M, pH 5.5) and 600 µL of this mixture was transferred into a reactor containing NOTA-Ubiquicidin 29-41 acetate (400 µg, 202 nmol, dissolved in 64 µL metal-free water) resulting in an overall concentration of NOTA-Ubiquicidin 29-41 of 0.3045 mM. The reaction vial was placed in a water bath at a temperature of 90 °C for 12 min. Sterile water (6 mL) was added. The mixture was run through a cation exchange cartridge Strata XC 10. The reaction vial was washed with water (2 mL), and the wash was also run through the cation exchange cartridge. Approximately 80% of the reaction mixture was trapped on the cation exchange cartridge; approximately 17% stayed in the reactor, and 3% went through the cation exchange cartridge. The product was released from the cation exchange cartridge by 0.5 mL 60% EtOH in sterile water, followed by 5 mL sterile saline, leaving 8% of the radioactivity on the cation exchange cartridge. The synthesis took 18–20 min, and the labeling yield was 162–203 MBq depended on how strong the Ge/Ga generator was (72–74% d.c). The specific radioactivity was ~1 MBq/nmol, assuming that all the NOTA-Ubiquicidin 29-41 applied in the synthesis ends up in the product vial. The radiochemical purity of [^{68}Ga]Ga-Ubiquicidin was examined using the setup described by Vilche et al. [17]. Briefly, the radio-HPLC system consisted of Dionex Ultimate 3000 HPLC system (Dionex Denmark) and a NaI radio detector from LabLogic (LabLogic, UK) and C-18 columns (4.6 × 150 mm, 5 µm). The HPLC conditions were as follows: flow rate = 1 mL/min; Channel A = 0.1% TFA/water; Channel B = 0.1% TFA/acetonitrile. Gradient: during 0–6 min, from 100% A to 40% A; Cfrom 6–10 min isocratic 40% A. The radiochemical purity was higher than 96% in all our syntheses.

The average injected activity was median 117 MBq (range 109–120 MBq), corresponding to ~5–6 MBq/kg. The PET was performed ~67 min (66–69 min) after injection of [^{68}Ga]Ga-Ubiquicidin (pigs 9–11, Table 1).

4.2.4. Leukocyte Trafficking ([^{68}Ga]Ga-DOTA-Siglec-9)

The radioactive labeling of DOTA-Siglec-9 was previously discussed and described in detail [13,29]. Briefly, ^{68}Ga was eluted from a *GalliaPharm®* ^{68}Ge/^{68}Ga generator (Eckert & Ziegler, Radiopharma GmbH, Berlin, Germany) using aq HCl (10 mL, 0.1 M). Gallium was trapped on a cation exchange cartridge (Strata-X-C 33 u Polymeric Strong, Phenomenex, Værløse Denmark), and released from the cartridge with an acidified acetone solution (0.6 mL, made from ultrapure HCl (30%, 0.21 mL) and sterile water (2.23 mL) in a total volume of 100 mL with acetone) from the SCX cartridge into a preheated reaction vial (100 °C). The acetone contents were reduced by azeotrope evaporation (100 °C, the addition of HCl (0.5 mL, 0.1 M) after 1.5 min, metal-free water (0.5 mL) after 2.5 min and HEPES buffer (124.0 ± 2.0 mg in 500 µL metal-free water) after 5.5 min). Followed by an addition of the DOTA-Siglec-9 peptide (40.0 µg, 16.5 nmol, dissolved in 200 µL of metal-free water. The reaction mixture was heated for 5 min for the ^{68}Ga incorporation to take place before it was quenched by adding sterile water (10 mL). The product mixture was then run through a preconditioned C-18 Sep-Pak cartridge (Waters, Hedehusene, Denmark) to trap [^{68}Ga]Ga-DOTA-Siglec-9 (preconditioned with 67% EtOH/33% sterile water (10 mL) followed by sterile saline (10 mL)).

The product ready for injection, [^{68}Ga]Ga-DOTA-Siglec-9, was released from the cartridge with ethanol (67%, 0.5 mL) followed by saline (9 mL) directly into the product vial through a sterile filter.

The radiochemical purity of [^{68}Ga]Ga-DOTA-Siglec-9 was determined using a radio detector coupled with a reversed-phase high-performance liquid chromatography (radio-HPLC) (Jupiter C18 column, 4.6 × 150 mm, 300 Å, 5 µm; Phenomenex, Torrance, CA, USA). The HPLC conditions were as follows: flow rate = 1 mL/minute; λ = 274 nm; Channel A = 0.1% TFA/water; Channel B = 0.1% TFA/acetonitrile; gradient: during 0–2 min, 82% A and 18% B; during 2–11 min, from 82% A and 18% B to 40% A and 60% B; during 11–15 min, from 40% A and 60% B to 82% A and 18% B; during 15–20 min, 82% A and 18% B. The radio-HPLC system consisted of Dionex Ultimate 3000 HPLC system equipped with a 4-wavelength simultaneous data collector (Dionex Denmark) and of a NaI radiodetector from LabLogic (LabLogic, UK). The overall yield of the reaction was approximately 62% ndc with a reaction time of approximately 25 min. The molar activity was 35 ± 10 MBq/nmol.

The average injected activity was as median 178 MBq (range 126–253 MBq), corresponding to ~8–12 MBq/kg. The PET was performed ~72 min (60–126 min, early scan) and 139 min (118–186 min, late scan) after injection of [^{68}Ga]Ga-DOTA-Siglec-9 (pigs 1–3 and 12–15, Table 1).

4.2.5. Glucose Metabolism ([^{18}F]FDG)

^{18}F were produced at the PET Center Aarhus using either a PETtrace 800 series cyclotron (GE Healthcare, Uppsala, Sweden) or a Cyclone 18/18 cyclotron (IBA, Louvain La Neuve, Belgium). [^{18}F]FDG was produced by a standard procedure applying a GE Healthcare MX Tracerlab synthesizer, Mx cassettes supplied by Rotem Industries (Arava, Israel), and chemical kits supplied by ABX GmbH (Radeberg, Germany). Radiochemical purity was higher than 99%. The median injected activity was 125 MBq (range 94–497 MBq), corresponding to ~5–24 MBq/kg. The PET was performed ~61 min (50–76 min) after injection of [^{18}F]FDG (pigs 1–15, Table 1).

4.3. CT and PET

The pigs were scanned at locations; Aalborg and Aarhus PET Centers. The pigs were carefully placed in dorsal recumbency to make sure the infected and not infected limb would be comparable in CT images. They were in propofol anesthesia and mechanically ventilated throughout the scan. Reflexes, pulse, oxygen saturation, and body temperature were monitored. After unenhanced CT for attenuation correction and anatomical co-registration, whole-body static PET imaging was performed with 5 min per bed position in three-dimensional mode. Pigs are not designed for erect walking, and their limbs cannot be stretched in the same manner as human extremities. Despite that, we kept the standards for imaging of humans; views in axial, coronal, and sagittal planes. But in the thorax, we changed the directions to cranial and caudal.

In Aalborg, to evaluate if any lytic lesions had developed, the pigs had a diagnostic CT scan (GE VCT Discovery True 64 PET/CT scanner 2006, GE Healthcare, Chicago, Illinois, USA).

The scan fields covered 15 cm in the axial direction. PET images were reconstructed using an ordered subset expectation maximization (OSEM) iterative algorithm (3D ViewPoint algorithm, GE Healthcare). The reconstruction parameters were 2 iterations, 28 subsets, 128 × 128 matrix in 47 slices, (5.5 × 5.5 × 3.3) mm^3 voxel size, and a 6 mm Gaussian filter.

Due to scanner replacement, later pigs were scanned on a different scanner than the previous pigs. They were scanned on a Siemens Biograph mCT (Siemens, Erlangen, Germany) with time-of-flight (TOF) detection. The scan field covered 22 cm in the axial direction. The images were reconstructed with an OSEM algorithm without using the resolution recovery option (setting "Iterative + TOF"). The reconstruction parameters were 3 iterations, 21 subsets, 400 × 400 matrix in (1.02 × 1.02 × 2.03) mm^3 voxels, and a 3 mm Gaussian filter. On both scanners, image reconstruction included attenuation-correction based on the CT scan.

In Aarhus, all examinations were performed with Siemens Biograph TruePoint™ 64 PET/CT, Siemens Healthineers, Erlangen, Germany. The pigs were placed in recumbency. Initially, a scout view was obtained to secure body coverage from snout to tail. We used visually comparable PSF reconstruction protocols (TrueX) (four iterations, 21 subsets, 3-mm Gaussian post-processing filter, matrix size 336 × 336, voxel size (2 mm × 2 mm × 2 mm).

In the present study, only the whole-body static PET/CT scans are presented.

5. Conclusions

We report these negative results as we find them valuable to other research groups developing new tracers for osteomyelitis. They may save other research groups from doing the same experiments and as a platform for the development of new potential tracers, or to use other scanning protocols. We also have an ethical responsibility to report negative results to prevent unnecessary animal studies from being repeated by other researchers.

[^{18}F]FDG was the best tracer for the detection of OM in peripheral bones in our juvenile pig model seven days after *S. aureus* induced OM.

Author Contributions: Conceptualization, P.A., S.B.J., and A.K.O.A.; methodology, P.A., S.B.J., A.K.O.A., and K.M.N.; formal analysis: P.A.; resources: P.A., S.B.J., A.K.O.A., and O.L.N.; data curation: P.A.; writing original draft preparation: P.A.; writing review and editing: P.A., S.B.J., A.K.O.A., K.M.N., and O.L.N.; founding acquisition: S.B.J., A.K.O.A., and O.L.N. All authors have read and agreed to the published version of the manuscript.

Funding: This work was supported by the Danish Council for Independent Research, Technology, and Production Sciences grant number 0602-01911B (11-107077).

Conflicts of Interest: The authors declare no conflict of interest.

References

1. Tong, S.Y.C.; Davis, J.S.; Eichenberger, E.; Holland, T.L.; Fowler, V.G., Jr. Staphylococcus aureus Infections: Epidemiology, Pathophysiology, Clinical Manifestations, and Management. *Clin. Microbiol. Rev.* **2015**, *28*, 603–661. [CrossRef]
2. Klein, E.Y.; Van Boeckel, T.P.; Martinez, E.; Pant, S.; Gandra, S.; Levin, S.A.; Goossens, H.; Laxminarayan, R. Global increase and geographic convergence in antibiotic consumption between 2000 and 2015. *Proc. Natl. Acad. Sci. USA* **2018**, *115*, E3463–E3470. [CrossRef] [PubMed]

3. Romo, A.L.; Quirós, R. Appropriate use of antibiotics: An unmet need. *Ther. Adv. Urol.* **2019**, *11*, 9–17. [CrossRef]
4. Ena, J. The epidemiology of intravenous vancomycin usage in a university hospital. A 10-year study. *JAMA* **1993**, *269*, 598–602. [CrossRef] [PubMed]
5. Centers for Disease Control and Prevention. *Antibiotic Resistance Threats in the United States 2019*; Centers for Disease Control and Prevention: Atlanta, GA, USA, 2020; pp. 1–150. Available online: https://www.cdc.gov/drugresistance/Biggest-Threats.html (accessed on 8 June 2020).
6. Decristoforo, C.; Pickett, R.D.; Verbruggen, A. Feasibility and availability of ^{68}Ga-labelled peptides. *Eur. J. Nucl. Med. Mol. Imaging* **2012**, *39*, 31–40. [CrossRef] [PubMed]
7. Bhatt, J.; Mukherjee, A.; Korde, A.; Kumar, M.; Sarma, H.D.; Dash, A. Radiolabeling and Preliminary Evaluation of Ga-68 Labeled NODAGA-Ubiquicidin Fragments for Prospective Infection Imaging. *Mol. Imaging Biol.* **2016**, *19*, 59–67. [CrossRef] [PubMed]
8. Otto, M. Staphylococcal Biofilms. *Microbiol. Spectr.* **2018**, *6*, 699–711. [CrossRef]
9. Nielsen, K.M.; Kyneb, K.H.; Alstrup, A.K.O.; Jensen, J.J.; Bender, D.; Afzelius, P.; Nielsen, O.L.; Jensen, S.B. ^{68}Ga-labelled phage-display selected peptides as tracers for positron emission tomography imaging of Staphylococcus aureus biofilm forming infections: Selection, radiolabelling and preliminary biological evaluation. *Nucl. Med. Biol.* **2016**, *43*, 593–605. [CrossRef]
10. Afzelius, P.; Alstrup, A.K.; Schønheyder, H.C.; Borghammer, P.; Jensen, S.B.; Bender, D.; Nielsen, O.L. Utility of ^{11}C-methionine and ^{11}C-donepezil for imaging of Staphylococcus aureus induced osteomyelitis in a juvenile porcine model: Comparison to autologous 111In-labelled leukocytes, ^{99m}Tc-DPD, and ^{18}F-FDG. *Am. J. Nucl. Med. Mol. Imaging* **2016**, *6*, 286–300.
11. Beheshti, M.; Mottaghy, F.M.; Payche, F.; Behrendt, F.F.F.; Wyngaert, T.V.D.; Fogelman, I.; Strobel, K.; Celli, M.; Fanti, S.; Giammarile, F.; et al. ^{18}F-NaF PET/CT: EANM procedure guidelines for bone imaging. *Eur. J. Nucl. Med. Mol. Imaging* **2015**, *42*, 1767–1777. [CrossRef]
12. Bhusari, P.; Bhatt, J.; Sood, A.; Kaur, R.; Vatsa, R.; Rastogi, A.; Mukherjee, A.; Dash, A.; Mittal, B.; Shukla, J. Evaluating the potential of kir-based ^{68}Ga-ubiquicidin formulation of infection: A pilot study ^{68}Ga. *Nucl. Med. Commun.* **2019**, *40*, 228–234. [CrossRef] [PubMed]
13. Jødal, L.; Roivainen, A.; Oikonen, V.; Jalkanen, S.; Hansen, S.B.; Afzelius, P.; Alstrup, A.K.O.; Nielsen, O.L.; Jensen, S.B. Kinetic Modelling of [^{68}Ga]Ga-DOTA-Siglec-9 in Porcine Osteomyelitis and Soft Tissue Infections. *Molecules* **2019**, *24*, 4094. [CrossRef] [PubMed]
14. Nielsen, K.M.; Jørgensen, N.P.; Kyneb, M.H.; Borghammer, P.; Meyer, R.L.; Thomsen, T.R.; Bender, D.; Jensen, S.B.; Nielsen, O.L.; Alstrup, A.K.O. Preclinical evaluation of potential infection-imaging probe [^{68}Ga]Ga-DOTA-K-A9 in sterile and infectious inflammation. *J. Label. Compd.* **2018**, *61*, 780–795. [CrossRef]
15. Johansen, L.K.; Koch, J.; Kirketerp-Møller, K.; Wamsler, O.J.; Nielsen, O.L.; Leifsson, P.S.; Frees, D.; Aalbæk, B.; Jensen, H.E. Therapy of haematogenous osteomyelitis—A comparative study in a porcine model and Angolan children. *In Vivo* **2013**, *27*, 305–312.
16. Rusckowski, M.; Gupta, S.; Liu, G.; Shuping, D.; Hnatowich, J.D. Investigations of a (99m) Tc labeled bacteriophage as a potential infection-specific imaging agent. *J. Nucl. Med.* **2004**, *45*, 1201–1208. [PubMed]
17. Jaakkola, K.; Nikula, T.; Holopainen, R.; Vahasilta, T.; Matikainen, M.T.; Laukkanen, M.L.; Huupponen, R.; Halkola, L.; Nieminen, L.; Hiltunen, J.; et al. In vivo detection of vascular adhesion protein-1 in experimental inflammation. *Am. J. Pathol.* **2000**, *157*, 463–471. [CrossRef]
18. Nielsen, O.L.; Afzelius, P.; Bender, D.; Schønheyder, H.C.; Leifsson, P.S.; Nielsen, K.M.; Larsen, J.O.; Jensen, S.B.; Alstrup, A.K.O. Comparison of 111In-leucocyte single-photon emission computed tomography (SPECT) and Positron Emission Tomography (PET) with four different tracers to diagnose osteomyelitis in a juvenile porcine experimental haematogenous Staphylococcus aureus model. *Am. J. Nucl. Med. Mol. Imaging* **2015**, *5*, 169–182.
19. Akhtar, M.S.; Qaisar, A.; Irfanullah, J.; Iqbal, J.; Khan, B.; Jehangir, M.; Nadeem, M.A.; Khan, M.A.; Afzal, M.S.; Ul-Haq, I.; et al. Antimicrobial peptide ^{99m}Tcubiquicidin 29-41 as human infection-imaging agent: Clinical trial. *J. Nucl. Med.* **2005**, *46*, 567–573.
20. Ebenhan, T.; Chadwick, N.; Sathekge, M.; Govender, P.; Govender, T.; Kruger, H.G.; Marjanovic-Painter, B.; Zeevaart, J.R. Peptide synthesis, characterization and ^{68}Ga-radiolabeling of NOTA-conjugated ubiquicidin fragments for prospective infection imaging with PET/CT. *Nucl. Med. Biol.* **2014**, *41*, 390–400. [CrossRef]

21. Ferro-Flores, G.; De Murphy, C.A.; Pedraza-López, M.; Alafort, L.; Zhang, Y.-M.; Rusckowski, M.; Hnatowich, N.J. In vitro and in vivo assessment of ^{99m}Tc-UBI specificity for bacteria. *Nucl. Med. Biol.* **2003**, *30*, 597–603. [CrossRef]
22. Ebenhan, T.; Zeevaart, J.R.; Venter, J.D.; Govender, T.; Kruger, H.G.; Jarvis, N.V.; Sathekge, M. Preclinical Evaluation of ^{68}Ga-Labeled 1,4,7-Triazacyclononane-1,4,7-Triacetic Acid-Ubiquicidin as a Radioligand for PET Infection Imaging. *J. Nucl. Med.* **2014**, *55*, 308–314. [CrossRef] [PubMed]
23. Shi, S.; Zhang, X. Interaction of Staphylococcus aureus with osteoblasts (Review). *Exp. Ther. Med.* **2011**, *3*, 367–370. [CrossRef] [PubMed]
24. Proctor, R.A.; Von Eiff, C.; Kahl, B.C.; Becker, K.; McNamara, P.; Herrmann, M.; Peters, G. Small colony variants: A pathogenic form of bacteria that facilitates persistent and recurrent infections. *Nat. Rev. Genet.* **2006**, *4*, 295–305. [CrossRef] [PubMed]
25. Jødal, L.; Nielsen, O.L.; Afzelius, P.; Alstrup, A.K.O.; Hansen, S.B. Blood perfusion in osteomyelitis studied with [^{15}O]water PET in a juvenile porcine model. *EJNMMI Res.* **2017**, *7*, 4. [CrossRef]
26. Afzelius, P.; Nielsen, O.L.; Schønheyder, H.C.; Alstrup, A.K.O.; Hansen, S.B. An untapped potential for imaging of peripheral osteomyelitis in paediatrics using [^{18}F] FDG PET/CT—The inference from a juvenile porcine model. *EJNMMI Res.* **2019**, *9*, 29. [CrossRef]
27. Alstrup, A.K.O.; Nielsen, K.M.; Jensen, S.B.; Afzelius, P.; Schønheyder, H.C.; Nielsen, O.L. Refinement of a haematogenous, localized osteomyelitis model in pigs. *Scand. J. Lab. Anim. Sci.* **2015**, *41*, 1–4.
28. Vilche, M.B.; Reyes, A.L.; Vasilskis, E.; Oliver, P.; Balter, H.S.; Engler, H.W. ^{68}Ga-NOTA-UBI-29-41 as a PET Tracer for Detection of Bacterial Infection. *J. Nucl. Med.* **2016**, *57*, 622–627. [CrossRef]
29. Jensen, S.B.; Käkelä, M.; Jødal, L.; Moisio, O.; Jalkanen, S.; Roivainen, A.; Alstrup, A.K.O. Exploring the radiosynthesis and in vitro characteristics of [^{68}Ga]Ga-DOTA-Siglec-9. *J. Label. Compd. Radiopharm.* **2017**, *60*, 439–449. [CrossRef]

Sample Availability: Samples of the compounds available from Ole Lerberg Nielsen.

© 2020 by the authors. Licensee MDPI, Basel, Switzerland. This article is an open access article distributed under the terms and conditions of the Creative Commons Attribution (CC BY) license (http://creativecommons.org/licenses/by/4.0/).

Article

Evaluation of a New ^{177}Lu-Labeled Somatostatin Analog for the Treatment of Tumors Expressing Somatostatin Receptor Subtypes 2 and 5

Rosalba Mansi [1], Guillaume Pierre Nicolas [2], Luigi Del Pozzo [1], Karim Alexandre Abid [3], Eric Grouzmann [3] and Melpomeni Fani [1,*]

1. Division of Radiopharmaceutical Chemistry, Clinic of Radiology and Nuclear Medicine, University Hospital Basel, 4031 Basel, Switzerland; rosalba.mansi@usb.ch (R.M.); Luigi.DelPozzo@usb.ch (L.D.P.)
2. Division of Nuclear Medicine, Clinic of Radiology and Nuclear Medicine, University Hospital Basel, 4031 Basel, Switzerland; guillaume.nicolas@usb.ch
3. Catecholamine and Peptides Laboratory, Department of Laboratories, University Hospital of Lausanne, 1011 Lausanne, Switzerland; Karim-Alexandre.Abid@chuv.ch (K.A.A.); eric.grouzmann@chuv.ch (E.G.)
* Correspondence: melpomeni.fani@usb.ch; Tel.: +41-615565891

Academic Editor: Svend Borup Jensen
Received: 24 July 2020; Accepted: 9 September 2020; Published: 11 September 2020

Abstract: Targeted radionuclide therapy of somatostatin receptor (SST)-expressing tumors is only partially addressed by the established somatostatin analogs having an affinity for the SST subtype 2 (SST2). Aiming to target a broader spectrum of tumors, we evaluated the bis-iodo-substituted somatostatin analog ST8950 ((4-amino-3-iodo)-D-Phe-c[Cys-(3-iodo)-Tyr-D-Trp-Lys-Val-Cys]-Thr-NH$_2$), having subnanomolar affinity for SST2 and SST5, labeled with [^{177}Lu]Lu^{3+} via the chelator DOTA (1,4,7,10-tetraazacyclododecane-1,4,7,10-tetraacetic acid). Human Embryonic Kidney (HEK) cells stably transfected with the human SST2 (HEK-SST2) and SST5 (HEK-SST5) were used for in vitro and in vivo evaluation on a dual SST2- and SST5-expressing xenografted mouse model. ^{nat}Lu-DOTA-ST8950 showed nanomolar affinity for both subtypes (IC$_{50}$ (95% confidence interval): 0.37 (0.22–0.65) nM for SST2 and 3.4 (2.3–5.2) for SST5). The biodistribution of [^{177}Lu]Lu-DOTA-ST8950 was influenced by the injected mass, with 100 pmol demonstrating lower background activity than 10 pmol. [^{177}Lu]Lu-DOTA-ST8950 reached its maximal uptake on SST2- and SST5-tumors at 1 h p.i. (14.17 ± 1.78 and 1.78 ± 0.35%IA/g, respectively), remaining unchanged 4 h p.i., with a mean residence time of 8.6 and 0.79 h, respectively. Overall, [^{177}Lu]Lu-DOTA-ST8950 targets SST2-, SST5-expressing tumors in vivo to a lower extent, and has an effective dose similar to clinically used radiolabeled somatostatin analogs. Its main drawbacks are the low uptake in SST5-tumors and the persistent kidney uptake.

Keywords: radiolabeled somatostatin analogs; SST2; SST5; Lu-177; targeted radionuclide therapy; neuroendocrine tumors; dual-tumor mouse model

1. Introduction

Targeted radionuclide therapy of neuroendocrine tumors (NET) via somatostatin receptor (SST) is proven very effective. The NETTER-1 phase III study showed that treatment with the ^{177}Lu-labeled somatostatin analog DOTA-TATE ([1,4,7,10-tetraazacyclododecane-1,4,7,10-tetraacetic acid^0, Tyr^3,Thr^8]-octreotate), in combination with single-dose nonradiolabeled octreotide (Sandostatin® LAR®), significantly improved objective response, progression-free survival and quality of life versus treatment with double-dose octreotide [1,2]. Earlier studies involving large cohorts treated with the ^{90}Y- or ^{177}Lu-labeled analog DOTA-TOC ([DOTA^0, Tyr^3]-octreotide) support these findings [3,4].

Molecules 2020, 25, 4155; doi:10.3390/molecules25184155 www.mdpi.com/journal/molecules

[^{177}Lu]Lu-DOTA-TOC is under evaluation in the phase III trial COMPETE (NCT03049189) versus the mTOR inhibitor, everolimus, while [^{177}Lu]Lu-DOTA-TATE (Lutathera®) is approved by the U.S. Food and Drug Administration (FDA) and the European Medicines Agency (EMA). As a companion to these analogs for radionuclide therapy, two ^{68}Ga-labeled radio-diagnostics, [^{68}Ga]Ga-DOTA-TATE (NETSPOT®) and [^{68}Ga]Ga-DOTA-TOC (SOMAKIT TOC®), received FDA approval for positron emission tomography (PET) imaging. The molecular basis of these successful approvals rests on the high affinity of the (radio)metallated DOTA-TATE and DOTA-TOC for the somatostatin receptor subtype 2 (SST2), which is known to be overexpressed by NET cells. However, poorly differentiated neuroendocrine carcinoma (NEC), high-grade NETs, and to a certain extent, well-differentiated NETs, may show low and/or heterogeneous SST2 expression [5,6], leading to suboptimal tumor targeting with these analogs. On the other hand, these tumors may express or co-express other SST subtypes among the five known ones (SST1-5). SST5 is the second highly expressed subtype in gastroenteropancreatic neuroendocrine tumors (GEP-NETs) [7], behind the predominant expression of SST2 (at least in primary tumors) and is concomitantly expressed with SST2 in 70–100% of GEP-NETs, in breast cancer, growth hormone (GH)-secreting pituitary adenomas and in 20–50% of intestinal or bronchial NETs [5,8–10]. NETs from G1/2 to G3 show a downregulation of SST2, while SST5 is constantly present [5,11]. SST5 is predominantly expressed, compared to SST2, in other tumors such as glioblastomas [12], tumor capillaries of pancreatic adenocarcinomas [13] or in lung cancer [14]. There are also cases of NETs where SST2 is absent while SST5 is present [6,15]. There are other tumors where SST5 is expressed, while SST2 is absent, including ACTH pituitary adenoma, cervix carcinoma and ovarian carcinoma [16]. Therefore, analogs targeting SST2 and SST5 (but also other subtypes), may potentially target a broader spectrum of various tumors and/or increase the radiation dose in a given tumor.

Most of the known analogs are synthetic cyclic octapeptides with a disulfide six-membered ring and SST2-selectivity. We are interested in developing radiolabeled somatostatin analogs for multireceptor subtype targeting. Replacement of key amino acids on the octreotide motif resulted in the analog [DOTA, 1-Nal^3]-octreotide (DOTA-NOC) with a high affinity for the SST2 and SST5 and lower for the SST3 [17]. Other designs include, highly constrained bicyclic octreotide analogs, consisting of a head-to-tail cyclization and an inner disulfide six-membered ring [18], the head-to-tail cyclo-hexapeptide pasireotide (Signifor or SOM230) [19], cyclic nonapeptides with nondisulfide eight-membered ring [20], and the 14mer and pseudo-14mer cyclic somatostatin-14 (SS-14) mimics, with ring-size of 12, 9, 8 and 6 amino acids—with the higher number ring favoring multireceptor subtype recognition [21]. All the above-mentioned analogs showed certain limitations, with [^{68}Ga]Ga-DOTA-NOC being, so far, the only well-established analog for SST2, SST3 and SST5 targeting [22]. Clinical data with [^{68}Ga]Ga-DOTA-NOC indicate that multi-receptor subtype targeting is relevant for improving the diagnostic accuracy and sensitivity of PET imaging of SST-expressing tumors [23,24]. Such clinical data on therapy are lacking. In fact, the therapeutic equivalent [^{177}Lu]Lu-DOTA-NOC has been evaluated in only 69 NET patients, compared to [^{177}Lu]Lu-DOTA-TATE, showing higher uptake in normal tissues with subsequently higher effective dose [25]. None of the other analogs has been evaluated for targeted radionuclide therapy.

In this work, we evaluate the ^{177}Lu-labeled somatostatin analog ST8950 ((4-amino-3-iodo)-D-Phe-c[Cys-(3-iodo)-Tyr-D-Trp-Lys-Val-Cys]-Thr-NH$_2$) for potential treatment of SST2- and SST5-expressing tumors. ST8950 (identified as peptide #9 in [26] and as AP102 in [27,28] is a disulfide-bridged bis-iodo-substituted octapeptide that exhibits sub-nanomolar affinity to SST2 and SST5 [26]. ST8950 is as potent as the natural SS-14 in its ability to inhibit growth hormone and prolactin release [26], it has intermediate agonistic potency between octreotide and pasireotide at both subtypes [27], and it acutely reduces growth hormone secretion without causing hyperglycemia (a known undesirable effect of pasireotide) in a healthy rat model [28]. Our previous work demonstrated that coupling of the chelator DOTA and complexation of Ga^{3+} does not alter the affinity to SST2, and while it reduces its affinity to SST5, this still remains in a low nanomolar range [29]. [^{68}Ga]Ga-DOTA-ST8950 showed high and specific accumulation in SST2 and SST5-expressing tumors in vivo, comparable to [^{68}Ga]Ga-DOTA-NOC [29]. Herein we report the comprehensive evaluation of the therapeutic equivalent [^{177}Lu]Lu-DOTA-ST8950. The influence of Lu-complexation on the affinity for SST2 and SST5 was assessed, together with the internalization and efflux rate of [^{177}Lu]Lu-DOTA-ST8950 vs. [^{177}Lu]Lu-DOTA-NOC on intact cells. A series of in vivo characteristics, including the influence of the injected peptide masses on the biodistribution, the specificity, the role of nephroprotective agents on the kidney uptake, the pharmacokinetics over 168 h, the residence time on the tumors and critical organs and finally the dosimetry of [^{177}Lu]Lu-DOTA-ST8950 were assessed on a dual SST2- and SST5-expressing xenografted model.

2. Results and Discussion

2.1. Synthesis of the (Radio)Metallated Peptide Conjugates, Stability and Lipophilicity

The ^{nat}Lu complexes of DOTA-ST8950 and DOTA-NOC (Figure 1) were obtained with a yield of 70–80%, based on the initial amount of peptide used.

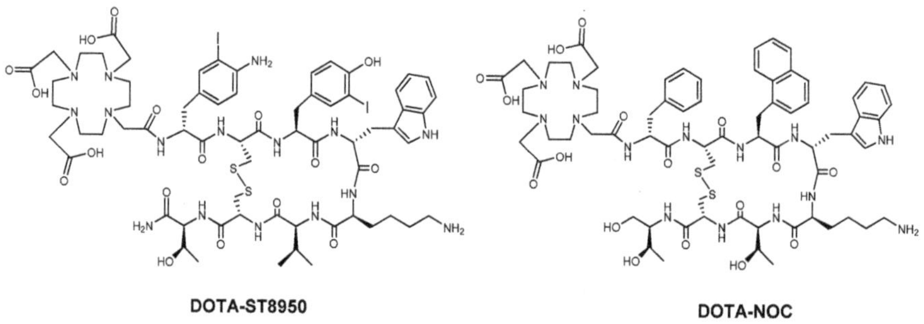

Figure 1. The structural formulae of DOTA-ST8950 (DOTA-(4-amino-3-iodo)-D-Phe-c[Cys-(3-iodo)-Tyr-D-Trp-Lys-Val-Cys]-Thr-NH$_2$) and DOTA-NOC (DOTA-D-Phe-c[Cys-1-NaI-D-Trp-Lys-Thr-Cys]Thr(ol)).

The purity and identity were confirmed by reverse-phase high-performance liquid chromatography (RP-HPLC) and electrospray ionization mass spectrometry (ESI-MS). The analytical data of the peptide conjugates and their ^{nat}Lu complexes are reported in Table 1.

Table 1. Analytical data of the DOTA-conjugates and of their corresponding ^{nat}Lu complexes.

Compounds	Purity (%)	MW (Calculated)	MW (Observed)	HPLC (t_r min)
DOTA-ST8950	100	1699.5	1700.1	10.51
DOTA-NOC	96	1454.6	1456.2	11.00
^{nat}Lu-DOTA-ST8950	100	1870.5	1870.8	11.03
^{nat}Lu-DOTA-NOC	97	1625.6	1627.3	11.52

MW: molecular weight; t_r: retention time.

^{177}Lu-labeling did not require any purification step. Radiochemical yield (non-isolated product, estimated by radio-HPLC) was > 95% and radiochemical purity > 93%. The radiochemical purity of [^{177}Lu]Lu-DOTA-ST8950 without any formulation or addition of scavengers was decreased over a period of 24 h to 74%, when the radiotracer was stored at 4 °C, and to 67% when it was stored at room temperature (RT).

[^{177}Lu]Lu-DOTA-ST8950 was found to be more lipophilic than the reference compound [^{177}Lu]Lu-DOTA-NOC, with log D = −1.2 ± 0.1 vs. −1.6 ± 0.1, respectively.

2.2. Binding Affinity Studies

The results of the binding affinity for the human SST2 and SST5 are summarized in Table 2. Conjugation of the chelate ^{nat}Lu-DOTA to ST8950 did not alter its binding affinity to SST2 (IC$_{50}$: 0.37 vs. 0.28 nM for ^{nat}Lu-DOTA-ST8950 and ST8950, respectively) but reduced by more than a factor of four its affinity to SST5 (IC$_{50}$: 3.4 vs. 0.77 nM, respectively). This observation is in agreement with the affinity data of ^{nat}Ga-DOTA-ST8950 having the same affinity as ST8950 for SST2, but reduced affinity for SST5 [29]. The loss of affinity for SST5 after chelate conjugation is more prominent with ^{nat}Lu-DOTA (factor of four) than with ^{nat}Ga-DOTA (factor of two); IC$_{50}$: 3.4 nM for ^{nat}Lu-DOTA-ST8950 vs. 1.9 nM for ^{nat}Ga-DOTA-ST8950 vs. 0.77 nM for ST8950.

Table 2. Binding affinities of the ^{nat}Lu-DOTA-conjugates for somatostatin receptor subtype 2 (SST2) and SST5, compared to reference compounds.

Compounds	SST2		SST5	
	IC$_{50}$ (nM)	95% Confidence Intervals (nM)	IC$_{50}$ (nM)	95% Confidence Intervals (nM)
Somatostatin-14 [¥,*]	0.11	0.08 to 0.15	0.35	0.22 to 0.55
ST8950 [¥]	0.28	0.19 to 0.42	0.77	0.48 to 1.2
^{nat}Lu-DOTA-ST8950	0.37	0.22 to 0.65	3.4	2.3 to 5.2
^{nat}Lu-DOTA-NOC	0.51	0.33 to 0.78	4.8	3.1 to 7.6
^{nat}Ga-DOTA-ST8950 [¥]	0.32	0.20 to 0.50	1.9	1.1 to 3.1
^{nat}Ga-DOTA-NOC [¥]	0.70	0.50 to 0.96	3.4	1.8 to 6.2

IC$_{50}$: half maximal inhibitory concentration; Experiments were performed in 3 to 4 separate sessions in duplicate; * somatostatin-14 is the natural ligand and was used as control; [¥] from [29].

The affinity of ^{nat}Lu-DOTA-ST8950 was compared favorably with ^{nat}Lu-DOTA-NOC in both receptor subtypes (IC$_{50}$: 0.37 vs. 0.51 nM in SST2 and 3.4 vs. 4.8 nM in SST5, respectively). When compared with the native hormone SS-14, ^{nat}Lu-DOTA-ST8950 and ^{nat}Lu-DOTA-NOC had slightly lower affinities for the SST2 and significantly lower for the SST5.

Overall, the conjugation of a chelate affects the affinity of ST8950 to SST5, but not to SST2, with the ^{nat}Lu-DOTA-ST8950 having lower affinity than its ^{nat}Ga-equivalent. This is in agreement with the observation made previously by Antunes et al. [30] regarding the lower affinity of ^{nat}Lu-DOTA-NOC, compared to its ^{nat}Ga-equivalent, which was attributed to the different coordination number and geometry between the ^{nat}Ga and ^{nat}Lu complexes.

2.3. In Vitro Characterization

2.3.1. Internalization Studies

The internalization rate in Human Embryonic Kidney (HEK)-somatostatin receptor subtype 2 (SST2) cells is reported in Figure 2. [^{177}Lu]Lu-DOTA-ST8950 and [^{177}Lu]Lu-DOTA-NOC showed specific and time-dependent cellular uptake. [^{177}Lu]Lu-DOTA-ST8950 showed lower internalization compared with [^{177}Lu]Lu-DOTA-NOC (18.1 ± 0.7 vs. 26.8 ± 0.1% at 4 h, respectively). The percentage of the cell-surface-bound fraction was very low (about 1%) in all cases, demonstrating that the cell-surface-bound radiotracer is rapidly internalized inside the cells. This confirms their agonistic nature.

Neither of the two radiotracers had substantial internalization on HEK-SST5 cells (0.5–1.2%, at 4 h). The lack of internalization on HEK-SST5 cells of radiolabeled analogs with an affinity for this subtype, like [^{177}Lu]Lu-DOTA-ST8950 and [^{177}Lu]Lu-DOTA-NOC, has been observed by others [30,31], while we and others confirmed these results specifically for [^{67}Ga]Ga-DOTA-NOC [29,30] and for [^{67}Ga]Ga-DOTA-ST8950 [29]. Despite the lack of internalization in vitro, these analogs are able to bind to SST5-expressing tumors in vivo [29].

Unfortunately, no in vitro data are available for [^{177}Lu]Lu-DOTA-NOC to allow direct comparison. Nevertheless, our internalization results on [^{177}Lu]Lu-DOTA-NOC are in agreement with the data reported on [^{111}In]In-DOTA-NOC in HEK-SST2 [30]. This is also in line with the similar affinities found for ^{nat}Lu-DOTA-NOC and ^{nat}In-DOTA-NOC, measured in the same assay [30].

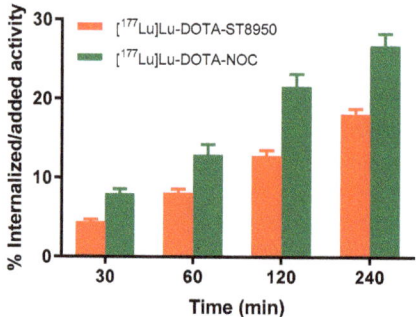

Figure 2. Internalization of [^{177}Lu]Lu-DOTA-ST8950 and [^{177}Lu]Lu-DOTA-NOC in Human Embryonic Kidney (HEK)-SST2 at 37 °C. The results are expressed as % (mean ± SD) of applied activity in the cells and normalized per million cells. All values refer to specific internalization after subtracting the nonspecific (measured in the presence of 1000-fold excess of SS-14) from the total internalized fraction, at each time point.

2.3.2. Efflux Studies

The cellular retention of [^{177}Lu]Lu-DOTA-ST8950 and [^{177}Lu]Lu-DOTA-NOC in HEK-SST2 is presented in Figure 3. Both radiotracers showed the same efflux rate, with more than 50% remaining inside the cells (internalized) after 4 h at 37 °C.

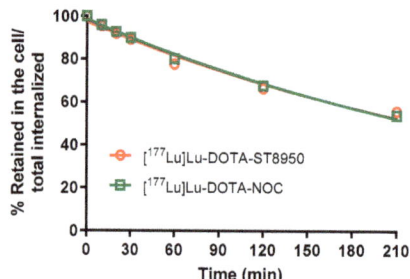

Figure 3. Cellular retention of [^{177}Lu]Lu-DOTA-ST8950 and [^{177}Lu]Lu-DOTA-NOC in HEK-SST2 after 2 h of incubation at 37 °C and acid wash to remove the cell-surface-bound fraction. The results are expressed as % (mean ± SD) of the internalized activity.

2.4. In Vivo Evaluation of [^{177}Lu]Lu-DOTA-ST8950 in Tumor-Bearing Mice

2.4.1. Influence of the Injected Peptide Mass

Biodistribution studies at two different peptide masses, 10 pmol and 100 pmol, were performed at 1 and 4 h p.i. to determine whether the amount of the injected peptide influences the distribution of [^{177}Lu]Lu-DOTA-ST8950 and its uptake in the tumors and organs. The results are reported in Table 3 and refer to a dual SST2- and SST5-tumor mouse model.

Table 3. Biodistribution results of 10 vs. 100 pmol [^{177}Lu]Lu-DOTA-ST8950 on dual SST2- and SST5-expressing xenografts at 1 and 4 h p.i. The results are expressed as mean of the % injected activity per gram of tissue (%IA/g) ± standard deviation (SD).

Organ	1 h			4 h		
	10 pmol *	100 pmol [&]	p	10 pmol *	100 pmol [#]	p
Blood	1.30 ± 0.24	0.96 ± 0.23	0.034	0.15 ± 0.02	0.09 ± 0.03	<0.001
Heart	0.84 ± 0.18	0.69 ± 0.09	*0.075*	0.20 ± 0.04	0.17 ± 0.03	*0.140*
Lung	7.12 ± 2.59	3.90 ± 0.84	0.011	3.81 ± 0.98	1.69 ± 0.40	<0.001
Liver	2.73 ± 0.59	2.60 ± 0.42	*0.660*	2.20 ± 0.27	2.19 ± 0.48	*0.998*
Pancreas	5.99 ± 1.75	6.36 ± 1.39	*0.693*	3.21 ± 0.50	4.38 ± 1.02	0.029
Spleen	1.82 ± 0.26	1.29 ± 0.19	0.002	0.96 ± 0.18	0.68 ± 0.11	0.001
Stomach	9.68 ± 2.34	5.63 ± 0.79	0.002	6.03 ± 1.28	4.94 ± 0.90	*0.063*
Intestine	3.08 ± 0.38	1.95 ± 0.65	0.006	2.29 ± 0.42	1.14 ± 0.42	<0.001
Adrenal	6.01 ± 1.15	4.23 ± 0.68	0.007	5.37 ± 1.17	5.21 ± 1.92	*0.952*
Kidney	11.72 ± 1.79	9.88 ± 0.99	0.045	10.84 ± 1.27	9.75 ± 1.87	*0.254*
Muscle	0.40 ± 0.07	0.30 ± 0.07	0.031	0.13 ± 0.02	0.10 ± 0.02	0.005
Bone	1.35 ± 0.25	0.74 ± 0.11	<0.001	1.12 ± 0.38	0.49 ± 0.11	<0.001
SST2-tumor	12.12 ± 3.94	14.17 ± 1.78	*0.322*	10.34 ± 2.78	15.50 ± 3.63	0.016
SST5-tumor	1.94 ± 0.46	1.78 ± 0.35	*0.561*	1.52 ± 0.75	1.87 ± 0.49	*0.240*

* $n = 5$, [&] $n = 7$; [#] $n = 12$; $p < 0.05$ statistically significant (black), $p > 0.05$ statistically not significant (italics).

[^{177}Lu]Lu-DOTA-ST8950 performed better when injected in higher amounts (100 vs. 10 pmol), as this led to a more desirable lower uptake in the blood, in the blood-rich organs, such as the spleen, lungs, bone marrow, and in other organs, such as the intestine and stomach. The uptake in SST2- and SST5-expressing tumors remained unchanged at 1 h p.i., but it was greater for the higher peptide mass on SST2-tumor at 4 h p.i. For all other organs, the uptake was at the same level for both peptide masses.

The "suppressed" uptake by increasing the injected peptide mass in certain organs expressing SST and in the blood resulted in improved tumor-to-background ratios. This has been reported in the literature [32] and we hypothesize that it is partially attributed to receptor saturation in vivo. Even though the binding of [^{177}Lu]Lu-DOTA-ST8950 was only tested for the human SST2 and SST5, which are expressed on the tumors, we contemplate that this is very similar to the mouse SST2 and SST5, which are expressed in certain organs through the mouse body. This is because the somatostatin receptors are highly conserved with 82–99% amino acid homology between humans and rodents, depending on the subtype [33–35]. More specifically, 93–96% sequence identity between the human, rat, mouse, porcine and bovine SST2 subtype and 82–83% sequence homology between human and rodent SST5. The fact that tumor uptake was not reduced might be explained by a higher receptor density in the tumors, compared to non-tumor-bearing organs, where only further increases in the injected peptide mass can produce saturation. Knowing the effect of the peptide mass is essential when therapy is being planned as it may reduce the radiation exposure of certain non-targeted organs and possibly whole-body radiation dose, without influencing the tumor dose.

The tumor uptake of [^{177}Lu]Lu-DOTA-ST8950 was found to be 1.8 (14.17 ± 1.78 vs. 26.27 ± 8.20%IA/g, $p = 0.0084$) and 8.3 (1.78 ± 0.35 vs. 14.87 ± 5.90%IA/g, $p = 0.0005$) times lower in SST2- and SST5-expressing tumors, respectively, compared with [^{68}Ga]Ga-DOTA-ST8950 at the same time point (1 h p.i.) and with the same peptide mass (100 pmol) [29]. Unfortunately, ex vivo histology to confirm the in situ expression of SST2 and SST5 and compare it with the expression on the [^{68}Ga]Ga-DOTA-ST8950 study was not feasible. Nevertheless, the SST2 and SST5 expression on the HEK transfected cells used on the xenograft model had been confirmed by western blot [27]. On the other hand, [^{177}Lu]Lu-DOTA-ST8950 had significantly lower uptake in the blood (0.96 ± 0.23 vs. 1.86 ± 0.56%IA/g, $p = 0.001$), the liver (2.60 ± 0.42 vs. 6.39 ± 1.93%IA/g, $p = 0.0001$) and the kidneys (9.88 ± 0.99 vs. 14.13 ± 3.69%IA/g, $p = 0.0095$), compared with [^{68}Ga]Ga-DOTA-ST8950, leading to better or similar tumor-to-non tumor ratios, based on the SST2-tumor uptake.

The discordance between ^{177}Lu-labeled and ^{68}Ga-labeled conjugates, such as [^{68}Ga]Ga-/[^{177}Lu]Lu-DOTA-ST8950, needs to be considered for all theranostic pairs. This affects for example dosimetry studies performed with the diagnostic companion for therapy planning. It is known that tumor-targeting properties, pharmacokinetics and body distribution of radiotracers may be affected by selecting a different radiometal. Therefore, each radiotracer needs to be evaluated independently and modifications implemented accordingly to produce a "matching" pair. There are examples in the literature where the theranostic pair consists of two radiotracers that differ not only by the radiometal, but also by the conjugate, e.g., [^{68}Ga]Ga-OPS202/[^{177}Lu]Lu-OPS201 ([^{68}Ga]Ga-NODAGA-JR11/[^{177}Lu]Lu-DOTA-JR11) for SST2 [36] or ^{68}Ga-pentixafor/^{177}Lu-pentixather for CXCR4 [37].

2.4.2. Metabolic Stability

In vivo metabolic stability of [^{177}Lu]Lu-DOTA-ST8950 was assessed by radio-HPLC on blood samples of mice collected at 30 and 60 min after injection of the radiotracer. [^{177}Lu]Lu-DOTA-ST8950 showed high in vivo stability, with 90% remaining intact in the blood 60 min after injection.

2.4.3. In Vivo Specificity and Kidney Protection

The uptake of [^{177}Lu]Lu-DOTA-ST8950 in SST-negative tumors (Table 4) was very low (0.70 ± 0.08 vs. 15.50 ± 3.63%IA/g, $p < 0.0001$ vs. 1.87 ± 0.49%IA/g, $p = 0.0014$, in SST2- and SST5-tumors, respectively), confirming the in vivo receptor-mediated uptake (specificity).

Table 4. Biodistribution results of [^{177}Lu]Lu-DOTA-ST8950, 4 h p.i. in SST-negative (SST(−)) tumor and after 5 min pre-injection of lysine (20 mg/100 µL) on dual SST2- and SST5-expressing xenografts. The results are expressed as mean of the % injected activity per gram of tissue (%IA/g) ± standard deviation (SD).

Organ	Control [¶]	SST(−) xenograft [¥]	Lysine [§]
Blood	0.09 ± 0.03	0.10 ± 0.02	0.09 ± 0.02
Heart	0.17 ± 0.03	0.18 ± 0.03	0.18 ± 0.03
Lung	1.69 ± 0.40	2.43 ± 1.21	1.78 ± 0.67
Liver	2.19 ± 0.48	2.28 ± 0.20	2.25 ± 0.37
Pancreas	4.38 ± 1.02	n.d.	4.10 ± 1.29
Spleen	0.68 ± 0.11	0.84 ± 0.06	0.73 ± 0.17
Stomach	4.94 ± 0.90	4.89 ± 0.94	2.87 ± 0.63
Intestine	1.14 ± 0.42	1.35 ± 0.14	1.05 ± 0.37
Adrenal	5.21 ± 1.92	5.45 ± 2.26	4.87 ± 1.49
Kidney	9.75 ± 1.87	8.96 ± 1.39	5.93 ± 0.56
Muscle	0.10 ± 0.02	0.15 ± 0.02	0.12 ± 0.02
Femur	0.49 ± 0.11	0.56 ± 0.24	0.55 ± 0.16
SST2-tumor	15.50 ± 3.63	-	15.07 ± 2.32
SST5-tumor	1.87 ± 0.49	-	1.69 ± 0.52
SST(−)-tumor	-	0.70 ± 0.08	-

n.d. = not determined, [¶] $n = 12$ (see Table 3); [¥] $n = 3$; [§] $n = 7$.

The pre-injection of the lysine reduced the uptake of the radiotracer in the kidneys by 40% (from 9.75 ± 1.87 to 5.93 ± 0.56%IA/g, $p < 0.0001$), without influencing the total-body biodistribution and the uptake in the tumors (Table 4).

The results indicate that the cationic amino acids such as lysine and arginine that are used for kidney protection in neuroendocrine tumor patients treated with [^{177}Lu]Lu-DOTA-TATE or [^{177}Lu]Lu-DOTA-TOC [38] can also be used in combination with [^{177}Lu]Lu-DOTA-ST8950, with similar effects.

2.4.4. Pharmacokinetics of [^{177}Lu]Lu-DOTA-ST8950

The biodistribution of [^{177}Lu]Lu-DOTA-ST8950 was studied at 1, 4, 48, 72 and 168 h p.i. The results are presented in Table 5. [^{177}Lu]Lu-DOTA-ST8950 was predominantly accumulated in the SST2-expressing tumors, while its accumulation in the SST5-expressing tumors was significantly lower. Normal distribution was also seen in the SST-expressing organs, such as the pancreas, stomach and adrenals. The maximum tumor uptake was observed already at 1 h p.i. remaining essentially unchanged at 4 h p.i. (SST2-tumor: 14.17 ± 1.78 and 15.50 ± 3.63%IA/g (p = 0.459), respectively and SST5-tumor: 1.78 ± 0.35 and 1.87 ± 0.49%IA/g (p = 0.483), respectively).

Table 5. Results of the pharmacokinetics studies of [^{177}Lu]Lu-DOTA-ST8950 (100 pmol) on dual SST2- and SST5-expressing xenografts. The results are expressed as mean of the % injected activity per gram of tissue (%IA/g) ± standard deviation (SD).

Organ	1 h *	4 h #	24 h	72 h	168 h
Blood	0.96 ± 0.23	0.09 ± 0.03	0.02 ± 0.00	0.01 ± 0.00	0.00 ± 0.00
Heart	0.69 ± 0.09	0.17 ± 0.03	0.08 ± 0.01	0.06 ± 0.02	0.04 ± 0.01
Lung	3.90 ± 0.84	1.69 ± 0.40	1.16 ± 0.10	0.46 ± 0.08	0.23 ± 0.09
Liver	2.60 ± 0.42	2.19 ± 0.48	1.23 ± 0.16	0.85 ± 0.21	0.54 ± 0.09
Pancreas	6.36 ± 1.39	4.38 ± 1.02	1.68 ± 0.27	0.72 ± 0.14	0.37 ± 0.05
Spleen	1.29 ± 0.19	0.68 ± 0.11	0.36 ± 0.24	0.38 ± 0.12	0.27 ± 0.05
Stomach	5.63 ± 0.79	4.94 ± 0.90	2.61 ± 0.44	1.66 ± 0.49	0.99 ± 0.04
Intestine	1.95 ± 0.65	1.14 ± 0.42	0.51 ± 0.09	0.28 ± 0.06	0.12 ± 0.02
Adrenal	4.23 ± 0.68	5.21 ± 1.92	3.81 ± 1.08	2.86 ± 1.30	1.84 ± 0.49
Kidney	9.88 ± 0.99	9.75 ± 1.87	5.99 ± 0.41	3.20 ± 0.84	1.42 ± 0.38
Muscle	0.30 ± 0.07	0.10 ± 0.02	0.06 ± 0.00	0.04 ± 0.01	0.01 ± 0.00
Femur	0.74 ± 0.11	0.49 ± 0.11	0.30 ± 0.05	0.19 ± 0.05	0.11 ± 0.01
SST2-tumor	14.17 ± 1.78	15.50 ± 3.63	9.32 ± 2.02	4.22 ± 1.43	0.79 ± 0.13
SST5-tumor	1.78 ± 0.35	1.87 ± 0.49	0.88 ± 0.16	0.35 ± 0.08	0.08 ± 0.01

* n = 7, # n = 12, n = 4 for all other groups.

The radiotracer was cleared rapidly from the blood; only 0.02 %IA/g remained in the blood at 24 h p.i. The kidney was the second organ accumulating activity after the SST2-tumors. The kidney uptake was high (approx. 10%IA/g) at the initial time points of the study, i.e., 1 and 4 h p.i., and was washed out over time, without, however, being negligible 168 h later (1.42 ± 0.38%IA/g). Lung and liver also showed considerable uptake. The uptake in the lungs was washed out quickly, while the uptake in the liver was more persistent and mainly attributed to the lipophilicity of the radiotracer.

Figure 4 shows the area under the time–activity curve (AUC) in SST2- and SST5-tumors and also in the liver and kidneys. The mean residence time for the SST2-tumor was 8.6 h, for the SST5-tumor 0.79 h, for the kidneys 6.3 h and for the liver 1.6 h, based on nondecay corrected biodistribution data and normalized per gram of tumor.

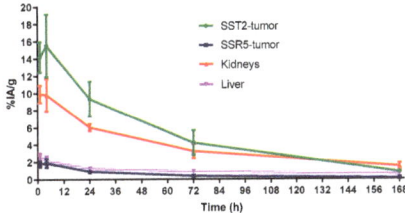

Figure 4. The area under the time–activity curve (AUC) in the SST2- and SST5-tumors, in the kidneys and the liver. These pharmacokinetic data were generated from serial independent biodistribution experiments performed 1, 4, 24, 72 and 168 h post injection.

The low accumulation in SST5-tumors is in contrast with the uptake of the [^{68}Ga]Ga-DOTA-ST8950 in the same tumor model. Unfortunately, very limited biodistribution data of somatostatin analogs in SST5-expressing tumors are available for comparison and a better understanding of these findings. One such case is the pan-somatostatin analog [^{111}In]In-DOTA-LLT-SS28 [39]. [^{111}In]In-DOTA-LLT-SS28 showed 1.7 times higher accumulation in SST5-tumors (2.61 ± 0.39 vs. 1.52 ± 0.75 %IA/g, respectively), when compared with [^{177}Lu]Lu-DOTA-ST8950 in the same animal model and experimental conditions (4 h after injection of 10 pmol of the radiotracer). However, [^{111}In]In-DOTA-LLT-SS28 had 2.3 times lower uptake in the SST2-tumor than [^{177}Lu]Lu-DOTA-ST8950 (4.43 ± 1.52 vs. 10.34 ± 2.78%IA/g, respectively) under the same conditions. These data are in agreement with the higher affinity of ^{nat}Lu-DOTA-ST8950 vs. ^{nat}In-DOTA-LLT-SS28 for SST2 (IC$_{50}$ = 0.35 vs. 1.8 nM) and the lower affinity for SST5 (IC$_{50}$ = 3.4 vs. 1.4 nM) [39].

Overall, the biodistribution profile and pharmacokinetics of [^{177}Lu]Lu-DOTA-ST8950 follow the profile of known radiolabeled somatostatin analogs, nevertheless, its high and persistent accumulation in the kidneys is identified as the main drawback for targeted radionuclide therapy.

2.4.5. SPECT/CT Imaging

SPECT/CT image 4 h after injection of [^{177}Lu]Lu-DOTA-ST8950 is shown in Figure 5. [^{177}Lu]Lu-DOTA-ST8950 clearly visualized the SST2-expressing tumors, but had faint uptake on the SST5 tumors, which confirm the low uptake in SST5 xenografts in the biodistribution studies. The background activity of [^{177}Lu]Lu-DOTA-ST8950 was low at 4 h after injection, with abdominal uptake and kidney uptake indicative of its renal excretion.

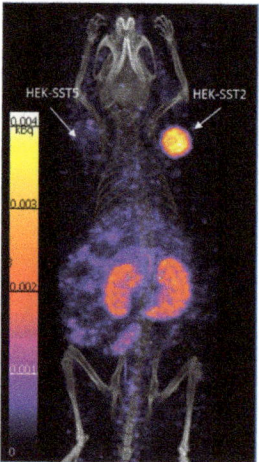

Figure 5. Maximum intensity projection (MIP) SPECT/CT images of [^{177}Lu]Lu-DOTA-ST8950 in a dual SST2- and SST5-expressing tumor mouse model, 4 h after injection.

2.4.6. Radiation Dosimetry Data Extrapolated to Humans

Table 6 shows the radiation dose estimate for human organs and tumors after injection of 100 pmol [^{177}Lu]Lu-DOTA-ST8950, based on a female phantom. The estimated whole-body radiation dose (effective dose) of [^{177}Lu]Lu-DOTA-ST8950 was 0.0252 mSv/MBq, which is within the expected range for ^{177}Lu-labeled somatostatin analogs. The radiation dose delivered to SST2-tumors was higher by a factor of seven than the dose delivered to SST5-tumors. A higher radiation dose targeting the SST5-expressing tumors is desirable to produce effective treatment outcomes. However, in many cases where SST5 is co-expressed with SST2, such as the indications mentioned in the introduction, this dual-targeting brings additive value, even if the uptake on SST5 is not at the same level as on SST2.

Table 6. Radiation dose estimation, extrapolated from mice to humans and expressed as Mean Absorbed Dose (mGy/MBq) for [^{177}Lu]Lu-DOTA-ST8950. [^{177}Lu]Lu-DOTA-TATE values estimated with the same methodology [32] are reported only for comparison.

Organ/tissue	[^{177}Lu]Lu-DOTA-ST8950 (mGy/MBq)	[^{177}Lu]Lu-DOTA-TATE (mGy/MBq)[¥]
Adrenals	2.46×10^{-1}	2.37×10^{-1}
Intestine	5.57×10^{-2}	3.38×10^{-1}
Stomach	7.55×10^{-2}	1.27×10^{-1}
Heart	6.06×10^{-3}	6.44×10^{-3}
Kidneys	5.44×10^{-1}	2.13×10^{-1}
Liver	9.16×10^{-2}	1.16×10^{-2}
Lungs	1.27×10^{-2}	4.07×10^{-2}
Muscle	5.43×10^{-4}	6.94×10^{-4}
Pancreas	2.95×10^{-1}	3.10×10^{-1}
Red marrow	1.59×10^{-3}	1.25×10^{-3}
Spleen	4.66×10^{-2}	3.13×10^{-2}
Total body	7.01×10^{-3}	7.04×10^{-3}
SST2-tumor	5.07×10^{-1}	3.33×10^{-1}
SST5-tumor	6.89×10^{-2}	
Effective dose (mSv/MBq)	2.52×10^{-2}	3.17×10^{-2}

[¥] The results of [^{177}Lu]Lu-DOTA-TATE refer to the injected peptide mass of 10 pmol [32].

Comparison between [^{177}Lu]Lu-DOTA-ST8950 and the FDA approved [^{177}Lu]Lu-DOTA-TATE is not easy since the results of [^{177}Lu]Lu-DOTA-TATE refer to an injected peptide mass of 10 pmol [32], though calculated with the same methodology. However, the following comparison can be made: (a) Given that the renal uptake and the kinetic of washout are not affected by the injected amount of peptide, the comparison between the two radiotracers is valid. [^{177}Lu]Lu-DOTA-ST8950 delivers a 2.5-fold higher radiation dose to the kidneys than [^{177}Lu]Lu-DOTA-TATE. This is a potential limitation, due to the known nephrotoxicity of this type of treatment. Unfortunately, the higher [^{177}Lu]Lu-DOTA-ST8950 uptake in the kidneys, compared to [^{177}Lu]Lu-DOTA-TATE, persists even with the use of basic amino acids that enable kidney uptake reduction by 40% with both radiolabeled analogs. On the other hand, SST5 co-targeting increases the radiation dose to the tumor and enhances the therapeutic effect, compared to SST2-selective targeting. This might mitigate the additional kidney dose by balancing the tumor-to-kidney ratio. (b) Regarding hematotoxicity, the other limiting factor in this type of treatment, the red marrow dosimetry of [^{177}Lu]Lu-DOTA-ST8950 is at about the same level as with [^{177}Lu]Lu-DOTA-TATE, therefore, there are no additional concerns than what is already known. (c) Last, but not least, [^{177}Lu]Lu-DOTA-ST8950 delivers a much higher radiation dose to the liver than [^{177}Lu]Lu-DOTA-TATE, presumably as a consequence of its lipophilicity. Nevertheless, the liver uptake is of greater concern for the diagnostic radiotracer [^{68}Ga]Ga-DOTA-ST8950 considering that the liver is the first site of metastasis of NETs and low background activity is needed for good image contrast and diagnostic accuracy.

3. Materials and Methods

All chemicals were obtained from commercial sources and used without additional purification. ESI-MS was carried out with ESI Bruker Esquire 3000 plus (Bruker Daltonics, Billerica, MA, USA). RP-HPLC was performed on a Bischoff instrument consisting of a LC-CaDi 22–14 interface, a UV-vis Lambda 1010 detector and a flow-through Berthold LB509 γ-detector (BISCHOFF chromatography, Leonberg, Germany), using a Phenomenex Jupiter Proteo 90 Å C12 (250 × 4.6 mm) column (Phenomenex Inc., California, USA). (Eluents: A = H$_2$O (0.1% TFA), B = Acetonitrile (0.1% TFA), Gradient: 95–50% solvent A in 15 min, Flow rate: 1.5 mL/min). Quantitative gamma counting was performed on a COBRA 5003 γ-system well counter from Packard Instruments (Meriden, CT, USA). SPECT/CT images were acquired using a dedicated nanoSPECT/CT system (Bioscan, Mediso, Budapest, Hungary).

Human Embryonic Kidney (HEK) cells were stably transfected with plasmids encoding the human SST2 and SST5 (HEK-SST2 and HEK-SST5) and cultivated as previously described [27]. The SST2 and SST5 expression was confirmed by western blot and has been previously reported [27]. Nontransfected HEK cells were used as negative control.

3.1. Synthesis of the (Radio)Metallated Peptide Conjugates and Stability

DOTA-ST8950 (Figure 1) was custom-made by PolyPeptide (San Diego, CA, USA). DOTA-NOC (Figure 1) was synthesized by standard Fmoc-solid-phase peptide synthesis, purified by preparative RP-HPLC and characterized by ESI-MS and analytical RP-HPLC (Figure 1). ^{nat}Lu-DOTA-ST8950 and ^{nat}Lu-DOTA-NOC were synthesized after incubation of the DOTA-conjugates with a 2.5-fold excess of ^{nat}LuCl$_3$ × 6H$_2$O (Sigma Aldrich, St. Louis, MO, USA) in ammonium acetate buffer (Sigma Aldrich, St. Louis, MO, USA), 0.4 M, pH 5 at 95 °C for 30 min. Free metal ions were eliminated via SepPak C-18 cartridge (Waters), preconditioned with methanol (Merck, Darmstadt, Germany) and water. The reaction mixture was loaded and the free ^{nat}Lu was eluted with water while the metallated peptides were eluted with ethanol, evaporated to dryness, redissolved in water and lyophilized. The ^{177}Lu-labeled conjugates were synthesized by dissolving 5–10 µg (3–6 nmol) of the DOTA-conjugates in 250 µL of sodium acetate buffer (0.4 M, pH 5.0) followed by incubation with [^{177}Lu]LuCl$_3$ (10–200 MBq, depending on the planned experiment) for 30 min at 95 °C. The stability of [^{177}Lu]Lu-DOTA-ST8950 under two different storage conditions, room temperature (RT) and at 4 °C, was evaluated over 24 h after synthesis.

3.2. Log D Measurements

The determination of log D (pH = 7.4) was performed by the "shake-flask" method. To a pre-saturated mixture of 500 µL 1-octanol and 500 µL of phosphate-buffered saline (PBS) (pH 7.4), 10 µL of 1 µM of the ^{177}Lu-labeled conjugates were added. The solution was vortexed for 1 h to reach the equilibrium and then centrifuged (3000 rpm) for 10 min. From each phase, 100 µL of the aliquot was removed and measured in a γ-counter. Each measurement was repeated three times. Care was taken to avoid cross-contamination between the phases. The partition coefficient was calculated as the average of the logarithmic values (n = 3) of the ratio between the radioactivity in the organic and the PBS phase.

3.3. Affinity Studies

The binding affinities of the ^{nat}Lu-DOTA-ST8950, in comparison to ^{nat}Lu-DOTA-NOC, were determined on membranes of HEK-SST2 and HEK-SST5 cells. The natural hormone Somatostatin-14 (SS-14) was used as reference compounds. ^{125}I-labeled SS-14 (^{125}I-SS-14, 81.4 TBq/mmol, Perkin Elmer, Waltham, MA, USA) was used as a radioligand for the competition binding assays. Binding assays were performed as described previously [27].

3.4. In Vitro Characterization of [^{177}Lu]Lu-DOTA-ST8950

For all cell experiments, HEK-SST2 and HEK-SST5 were seeded at a density of 1 × 10^6 cells/well in 6-well plates and incubated overnight with Dulbecco's modified Eagle's medium (DMEM) with 1% Fetal Bovine Serum (FBS, Biochrom GmbH, Merck Millipore, Darmstadt, Germany) to obtain a good cell adherence. For plating HEK-SST2 and HEK-SST5, the plates were pre-treated with a solution of 10% poly-lysine to promote the cell attachment.

3.4.1. Internalization Assay

The cells were washed with PBS and were incubated with fresh medium (DMEM with 1% FBS) for 1 h at 37 °C. [^{177}Lu]Lu-DOTA-ST8950 and [^{177}Lu]Lu-DOTA-NOC (100 µL, 2.5 nM) were added to the medium (0.9 mL) and the cells incubated (in triplicates) for 0.5, 1, 2 and 4 h at 37 °C,

5% CO_2. The internalization was stopped by removing the medium and washing the cells with ice-cold PBS. The cells were then treated twice for 5 min with ice-cold glycine solution (0.05 mol/L, pH 2.8), to distinguish between cell-surface-bound (acid releasable) and internalized (acid resistant) radio-peptide. Finally, the cells were detached with 1 M NaOH at 37 °C. To determine nonspecific cellular uptake, selected wells were incubated with the radio-peptide in the presence of 1000-fold excess of SS-14. Internalization and bound rate are expressed as a percentage of the applied radioactivity.

3.4.2. Cellular Retention (Efflux) Assay

For the cellular retention studies, HEK-SST2 cells were incubated with both radio-peptides (2.5 nM) for 120 min. The medium was then removed and the wells were washed twice with 1 mL ice-cold PBS. The acid wash with a glycine buffer of pH 2.8 was performed twice (5 min each time) on ice to remove the receptor-bound radio-peptide. Cells were then incubated again at 37 °C with fresh buffer (DMEM with 1% FBS). After preselected time points (10, 20, 30, 60, 120 and 210 min) the external medium was removed for quantification of radioactivity and replaced with fresh 37 °C medium. The cells were solubilized in 1 M NaOH and collected for quantification.

3.5. In Vivo Evaluation of [^{177}Lu]Lu-DOTA-ST8950 in Tumor-Bearing Mice

3.5.1. Tumor Implantation

The Veterinary Office (Department of Health) of the Cantonal Basel-Stadt approved the animal experiments (approval no. 2799) in accordance with the Swiss regulations for animal treatment. Female athymic Nude-$Foxn1^{nu}/Foxn1^{+}$ mice (Envigo, The Netherlands), 4–6 weeks old, were injected subcutaneously with 10^7 HEK-SST2 cells in the right shoulder and 10^7 HEK-SST5 cells in the left shoulder, freshly suspended in 100 µL sterile phosphate-buffered saline. The tumors were allowed to grow for 2–3 weeks.

3.5.2. Investigation of the Influence of the Injected Peptide Mass

Groups of mice bearing dual SST2- and SST5-expressing xenografts were injected with two different peptide doses of [^{177}Lu]Lu-DOTA-ST8950: 10 pmol/100 µL/0.5–0.6 MBq and 100 pmol/100 µL/0.6–1.5 MBq. The biodistribution was evaluated at 1 and 4 h post injection. At the selected time points, the mice were sacrificed and the organs of interest were collected, rinsed, blotted, weighed and counted in a γ-counter. The results are expressed as a percentage of injected activity per gram (%IA/g) obtained by extrapolation from counts of an aliquot taken from the injected solution as a standard.

3.5.3. In Vivo Metabolic Stability Studies

The in vivo stability of [^{177}Lu]Lu-DOTA-ST8950 was evaluated after intravenous injection of (100 pmol/100 µL/9.3 MBq) into the tail vein of non-tumor-bearing mice. Blood samples were collected at 30 and 60 min after injection in polypropylene tubes containing ethylenediaminetetraacetic acid (EDTA). After centrifugation at 4 °C, the plasma was collected and 95% ethanol was added in equal volume (1:1 v/v). The mixture was stirred, centrifuged and the supernatant was separated from the precipitated proteins. The solution was treated with acetonitrile in equal volume (1:1, v/v) to promote further precipitation of proteins. After centrifugation, the supernatant was filtered, diluted with water (1:1, v/v) and analyzed by radio-RP-HPLC for the identification and quantification of the intact peptide and possible metabolites.

3.5.4. Specificity and Kidney Protection

The in vivo SST2- and SST5-mediated uptake of [^{177}Lu]Lu-DOTA-ST8950 was assessed in mice bearing HEK-SST(−) negative xenografts after injection of 100 pmol/100 µL/0.3 MBq. The mice were sacrificed 4 h p.i. and accumulation of the radiotracer in the tumors and in all organs of interest was quantified in a γ-counter.

The basic amino acid lysine was evaluated as a nephroprotective agent, in an attempt to reduce the kidney uptake of [^{177}Lu]Lu-DOTA-ST8950. Dual SST2- and SST5-expressing xenografts were treated intravenously with lysine (20 mg/100 µL in PBS) 5 min before the administration of [^{177}Lu]Lu-DOTA-ST8950 (100 pmol/100 µL/0.3 MBq). Animals were sacrificed 4 h p.i. and accumulation of the radiotracer in the tumors and in all organs of interest was quantified in a γ-counter.

3.5.5. Pharmacokinetics of [^{177}Lu]Lu-DOTA-ST8950

Quantitative biodistribution studies of [^{177}Lu]Lu-DOTA-ST8950 (100 pmol/100 µL/0.6–1.5 MBq) was followed from 1 up to 168 h p.i. At the preselected time points, the mice were sacrificed and the organs of interest were collected, rinsed, blotted, weighed and counted in a γ-counter. The results are expressed as a percentage of injected activity per gram (%IA/g) obtained by extrapolation from counts of an aliquot taken from the injected solution as a standard.

3.5.6. SPECT/CT Imaging

Mice bearing dual SST2- and SST5-expressing xenografts were imaged using a nano SPECT/CT system (Bioscan, Mediso, Budapest, Hungary) 4 h after administration of [^{177}Lu]Lu-DOTA-ST8950 (100 pmol/100 µL/6 MBq). A helical CT scan was acquired with the following parameters: current, 177 mA; voltage, 45 kVp; pitch, A helical SPECT scan was acquired using multipurpose pinhole collimators (APT1), 20% energy window width centered symmetrically over the 208 and 113 keV g-peaks of ^{177}Lu, 24 projections, and 1200 s per projection. CT and SPECT images were reconstructed and filtered using the manufacturer's algorithm, resulting in a pixel size of 0.3 mm for the SPECT and of 0.2 mm for the CT.

3.5.7. Dosimetry

Mice biodistribution data were used to generate time–activity curves for each radiotracer. Because of the absence of a specific radioactivity accumulation in bones and red marrow, a linear relationship between the blood and the red marrow residence times was assumed for estimating the red marrow radiation dose [40]. The proportionality factor was the ratio between the red marrow mass and the blood mass in humans. OLINDA/EXM 1.0 (OLINDA/EXM®, Vanderbilt University, USA) was used to integrate the fitted time–activity curves and to estimate the organ and effective doses using the whole-body adult female model. For all calculations, the assumption was made that the mouse biodistribution, determined as the %IA/organ, was the same as the human biodistribution.

3.6. Statistics

Comparison of data was performed using unpaired two-tailed *t*-test with GraphPad Prism 7 software (GraphPad Software, Inc., San Diego, CA, USA). *p* values < 0.05 were considered significant.

4. Conclusions

The 2-iodo-substituted somatostatin analog [^{177}Lu]Lu-DOTA-ST8950 has an affinity for the SST2 and SST5, similarly to [^{177}Lu]Lu-DOTA-NOC. The in vivo uptake and residence time of [^{177}Lu]Lu-DOTA-ST8950 is high on SST2-expressing tumors, but significantly lower on SST5-expressing tumors. Nevertheless, SST5-targeting may provide additive value in the case of SST2 and SST5 coexpression in the same tumor, like gastroenteropancreatic neuroendocrine tumors, pituitary tumors and gastric cancers. [^{177}Lu]Lu-DOTA-ST8950 has an effective dose similar to [^{177}Lu]Lu-DOTA-TATE, however, its persistent kidney uptake is a drawback since nephrotoxicity is of concern in targeted radionuclide therapy with radiolabeled somatostatin analogs.

Author Contributions: Conceptualization, R.M. and M.F.; methodology, M.F., E.G; validation, R.M., E.G., G.P.N. and M.F.; formal analysis, R.M., K.A.A. and G.P.N.; investigation, R.M., G.P.N., L.D.P. and K.A.A.; resources, E.G. and M.F.; data curation, R.M. and K.A.A.; writing—original draft preparation, R.M.; writing—review and editing, M.F. and E.G.; visualization, R.M., K.A.A., and M.F.; supervision, M.F.; project administration, M.F.; funding acquisition, E.G. and M.F. All authors have read and agreed to the published version of the manuscript.

Funding: This research was financially supported by the Swiss Confederation Commission for Technology and Innovation (CTI), Project Nr 18282.2 PFLS-LS.

Acknowledgments: We acknowledge Sandra Zanger and Dominique Santoianni-Klauer for their assistance with the animal experiments.

Conflicts of Interest: The authors declare no conflict of interest.

References

1. Hofman, M.S.; Michael, M.; Hicks, R.J.; Ozdemir, N.Y.; Arslan, S.; Sendur, M.; Matuchansky, C.; Strosberg, J.; Krenning, E. ^{177}Lu-Dotatate for Midgut Neuroendocrine Tumors. *New Engl. J. Med.* **2017**, *376*, 1390–1392. [PubMed]
2. Strosberg, J.; Wolin, E.; Chasen, B.; Kulke, M.; Bushnell, D.; Caplin, M.; Baum, R.P.; Kunz, P.; Hobday, T.; Hendifar, A.; et al. Health-Related Quality of Life in Patients With Progressive Midgut Neuroendocrine Tumors Treated With ^{177}Lu-Dotatate in the Phase III NETTER-1 Trial. *J. Clin. Oncol.* **2018**, *36*, 2578–2584. [CrossRef] [PubMed]
3. Imhof, A.; Brunner, P.; Marincek, N.; Briel, M.; Schindler, C.; Rasch, H.; Mäcke, H.R.; Rochlitz, C.; Müller-Brand, J.; Walter, M.A. Response, Survival, and Long-Term Toxicity after Therapy with the Radiolabeled Somatostatin Analogue [^{90}Y-DOTA]-TOC in Metastasized Neuroendocrine Cancers. *J. Clin. Oncol.* **2011**, *29*, 2416–2423. [CrossRef]
4. Villard, L.; Romer, A.; Marincek, N.; Brunner, P.; Koller, M.T.; Schindler, C.; Ng, Q.K.; Mäcke, H.R.; Müller-Brand, J.; Rochlitz, C.; et al. Cohort Study of Somatostatin-Based Radiopeptide Therapy With [^{90}Y-DOTA]-TOC Versus [^{90}Y-DOTA]-TOC Plus [^{177}Lu-DOTA]-TOC in Neuroendocrine Cancers. *J. Clin. Oncol.* **2012**, *30*, 1100–1106. [CrossRef] [PubMed]
5. Kulaksiz, H.; Eissele, R.; Rössler, D.; Schulz, S.; Höllt, V.; Cetin, Y.; Arnold, R. Identification of somatostatin receptor subtypes 1, 2A, 3, and 5 in neuroendocrine tumours with subtype specific antibodies. *Gut* **2002**, *50*, 52–60. [CrossRef]
6. Asnacios, A.; Courbon, F.; Rochaix, P.; Bauvin, E.; Cances-Lauwers, V.; Susini, C.; Schulz, S.; Boneu, A.; Guimbaud, R.; Buscail, L. Indium-111–Pentetreotide Scintigraphy and Somatostatin Receptor Subtype 2 Expression: New Prognostic Factors for Malignant Well-Differentiated Endocrine Tumors. *J. Clin. Oncol.* **2008**, *26*, 963–970. [CrossRef]
7. Mai, R.; Kaemmerer, D.; Träger, T.; Neubauer, E.; Sänger, J.; Baum, R.P.; Schulz, S.; Lupp, A. Different somatostatin and CXCR4 chemokine receptor expression in gastroenteropancreatic neuroendocrine neoplasms depending on their origin. *Sci. Rep.* **2019**, *9*, 4339. [CrossRef]
8. Forssell-Aronsson, E.B.; Nilsson, O.; A Bejegård, S.; Kölby, L.; Bernhardt, P.; Mölne, J.; Hashemi, S.H.; Wangberg, B.; E Tisell, L.; Ahlman, H. ^{111}In-DTPA-D-Phe1-octreotide binding and somatostatin receptor subtypes in thyroid tumors. *J. Nucl. Med.* **2000**, *41*, 636–642.
9. Reubi, J.C.; Waser, B.; Schaer, J.C.; Laissue, J.A. Somatostatin receptor sst1–sst5 expression in normal and neoplastic human tissues using receptor autoradiography with subtype-selective ligands. *Eur. J. Nucl. Med. Mol. Imaging* **2001**, *28*, 836–846. [CrossRef]
10. Traub, T.; Petkov, V.; Ofluoglu, S.; Pangerl, T.; Raderer, M.; Fueger, B.J.; Schima, W.; Kurtaran, A.; Dudczak, R.; Virgolini, I. ^{111}In-DOTA-lanreotide scintigraphy in patients with tumors of the lung. *J. Nucl. Med.* **2001**, *42*, 1305–1315.
11. Kaemmerer, D.; Träger, T.; Hoffmeister, M.; Sipos, B.; Hommann, M.; Sänger, J.; Schulz, S.; Lupp, A. Inverse expression of somatostatin and CXCR4 chemokine receptors in gastroenteropancreatic neuroendocrine neoplasms of different malignancy. *Oncotarget* **2015**, *6*, 27566–27579. [CrossRef] [PubMed]
12. Lange, F.; Kaemmerer, D.; Behnke-Mursch, J.; Bruck, W.; Schulz, S.; Lupp, A. Differential somatostatin, CXCR4 chemokine and endothelin A receptor expression in WHO grade I–IV astrocytic brain tumors. *J. Cancer Res. Clin. Oncol.* **2018**, *144*, 1227–1237. [CrossRef] [PubMed]

13. Kajtazi, Y.; Kaemmerer, D.; Sänger, J.; Schulz, S.; Lupp, A. Somatostatin and chemokine CXCR4 receptor expression in pancreatic adenocarcinoma relative to pancreatic neuroendocrine tumours. *J. Cancer Res. Clin. Oncol.* **2019**, *145*, 2481–2493. [CrossRef] [PubMed]
14. Stumpf, C.; Kaemmerer, D.; Neubauer, E.; Sänger, J.; Schulz, S.; Lupp, A. Somatostatin and CXCR4 expression patterns in adenocarcinoma and squamous cell carcinoma of the lung relative to small cell lung cancer. *J. Cancer Res. Clin. Oncol.* **2018**, *144*, 1921–1932. [CrossRef] [PubMed]
15. Buscail, L.; Saint-Laurent, N.; Chastre, E.; Vaillant, J.C.; Gespach, C.; Capella, G.; Kalthoff, H.; Lluis, F.; Vaysse, N.; Susini, C. Loss of sst2 somatostatin receptor gene expression in human pancreatic and colorectal cancer. *Cancer Res.* **1996**, *56*, 1823–1827.
16. Lupp, A.; Hunder, A.; Petrich, A.; Nagel, F.; Doll, C.; Schulz, S. Reassessment of sst5 Somatostatin Receptor Expression in Normal and Neoplastic Human Tissues Using the Novel Rabbit Monoclonal Antibody UMB-4. *Neuroendocrinology* **2011**, *94*, 255–264. [CrossRef]
17. Wild, D.; Schmitt, J.S.; Ginj, M.; Maecke, H.R.; Bernard, B.F.; Krenning, E.; De Jong, M.; Wenger, S.; Reubi, J.C. DOTA-NOC a high-affinity ligand of somatostatin receptor subtypes 2, 3 and 5 for labelling with various radiometals. *Eur. J. Nucl. Med. Mol. Imaging* **2003**, *30*, 1338–1347. [CrossRef]
18. Fani, M.; Mueller, A.; Tamma, M.L.; Nicolas, G.; Rink, H.R.; Cescato, R.; Reubi, J.C.; Maecke, H.R. Radiolabeled Bicyclic Somatostatin-Based Analogs: A Novel Class of Potential Radiotracers for SPECT/PET of Neuroendocrine Tumors. *J. Nucl. Med.* **2010**, *51*, 1771–1779. [CrossRef]
19. Liu, F.; Liu, T.; Xu, X.; Guo, X.; Li, N.; Xiong, C.; Li, C.; Zhu, H.; Yang, Z. Design, Synthesis, and Biological Evaluation of 68Ga-DOTA-PA1 for Lung Cancer: A Novel PET Tracer for Multiple Somatostatin Receptor Imaging. *Mol. Pharm.* **2018**, *15*, 619–628. [CrossRef]
20. Ginj, M.; Zhang, H.; Eisenwiener, K.P.; Wild, D.; Schulz, S.; Rink, H.; Cescato, R.; Reubi, J.C.; Maecke, H.R. New Pansomatostatin Ligands and Their Chelated Versions: Affinity Profile, Agonist Activity, Internalization, and Tumor Targeting. *Clin. Cancer Res.* **2008**, *14*, 2019–2027. [CrossRef]
21. Tatsi, A.; Maina, T.; Cescato, R.; Waser, B.; Krenning, E.P.; de Jong, M.; Cordopatis, P.; Reubi, J.-C.; Nock, B.A. [DOTA]Somatostatin-14 analogs and their ^{111}In-radioligands: Effects of decreasing ring-size on sst1-5 profile, stability and tumor targeting. *Eur. J. Med. Chem.* **2014**, *73*, 30–37. [CrossRef] [PubMed]
22. Ambrosini, V.; Campana, D.; Bodei, L.; Nanni, C.; Castellucci, P.; Allegri, V.; Montini, G.C.; Tomassetti, P.; Paganelli, G.; Fanti, S. ^{68}Ga-DOTANOC PET/CT Clinical Impact in Patients with Neuroendocrine Tumors. *J. Nucl. Med.* **2010**, *51*, 669–673. [CrossRef] [PubMed]
23. Wild, D.; Bomanji, J.B.; Benkert, P.; Maecke, H.; Ell, P.J.; Reubi, J.C.; Caplin, M.E. Comparison of ^{68}Ga-DOTANOC and ^{68}Ga-DOTATATE PET/CT Within Patients with Gastroenteropancreatic Neuroendocrine Tumors. *J. Nucl. Med.* **2013**, *54*, 364–372. [CrossRef]
24. Lamarca, A.; Pritchard, D.M.; Westwood, T.; Papaxoinis, G.; Nonaka, D.; Vinjamuri, S.; Valle, J.W.; Manoharan, P.; Mansoor, W. ^{68}Gallium DOTANOC-PET Imaging in Lung Carcinoids: Impact on Patients' Management. *Neuroendocrinology* **2017**, *106*, 128–138. [CrossRef]
25. Wehrmann, C.; Senftleben, S.; Zachert, C.; Müller, D.; Baum, R.P. Results of Individual Patient Dosimetry in Peptide Receptor Radionuclide Therapy with 177Lu DOTA-TATE and 177Lu DOTA-NOC. *Cancer Biother. Radiopharm.* **2007**, *22*, 406–416. [CrossRef]
26. Moore, S.B.; Van Der Hoek, J.; De Capua, A.; Van Koetsveld, P.M.; Hofland, L.J.; Lamberts, A.S.W.J.; Goodman, M. Discovery of Iodinated Somatostatin Analogues Selective for hsst2 and hsst5 with Excellent Inhibition of Growth Hormone and Prolactin Release from Rat Pituitary Cells. *J. Med. Chem.* **2005**, *48*, 6643–6652. [CrossRef] [PubMed]
27. Streuli, J.; Harris, A.; Cottiny, C.; Allagnat, F.; Daly, A.F.; Grouzmann, E.; Abid, K. Cellular effects of AP102, a somatostatin analog with balanced affinities for the hSSTR2 and hSSTR5 receptors. *Neuropeptides* **2018**, *68*, 84–89. [CrossRef]
28. Tarasco, E.; Seebeck, P.; Pfundstein, S.; Daly, A.F.; Eugster, P.J.; Harris, A.; Grouzmann, E.; Lutz, T.; Boyle, C.N. Effect of AP102, a subtype 2 and 5 specific somatostatin analog, on glucose metabolism in rats. *Endocrine* **2017**, *58*, 124–133. [CrossRef]
29. Mansi, R.; Abid, K.; Nicolas, G.P.; Del Pozzo, L.; Grouzmann, E.; Fani, M. A new ^{68}Ga-labeled somatostatin analog containing two iodo-amino acids for dual somatostatin receptor subtype 2 and 5 targeting. *Ejnmmi Res.* **2020**, *10*, 1–10. [CrossRef]

30. Antunes, P.; Ginj, M.; Zhang, H.; Waser, B.; Baum, R.P.; Reubi, J.C.; Maecke, H. Are radiogallium-labelled DOTA-conjugated somatostatin analogues superior to those labelled with other radiometals? *Eur. J. Nucl. Med. Mol. Imaging* **2007**, *34*, 982–993. [CrossRef]
31. Cescato, R.; Schulz, S.; Waser, B.; Eltschinger, V.; E Rivier, J.; Wester, H.J.; Culler, M.; Ginj, M.; Liu, Q.; Schonbrunn, A.; et al. Internalization of sst2, sst3, and sst5 receptors: Effects of somatostatin agonists and antagonists. *J. Nucl. Med.* **2006**, *47*, 502–511.
32. Nicolas, G.P.; Mansi, R.; McDougall, L.; Kaufmann, J.; Bouterfa, H.; Wild, D.; Fani, M. Biodistribution, Pharmacokinetics, and Dosimetry of ^{177}Lu-, ^{90}Y-, and ^{111}In-Labeled Somatostatin Receptor Antagonist OPS201 in Comparison to the Agonist ^{177}Lu-DOTATATE: The Mass Effect. *J. Nucl. Med.* **2017**, *58*, 1435–1441. [CrossRef] [PubMed]
33. Reisine, T.; Bell, G.I. Molecular biology of somatostatin receptors. *Endocr. Rev.* **1995**, *16*, 427–442.
34. Patel, Y.C. Somatostatin and Its Receptor Family. *Front. Neuroendocr.* **1999**, *20*, 157–198. [CrossRef] [PubMed]
35. Patel, Y.; Greenwood, M.T.; Panetta, R.; Demchyshyn, L.; Niznik, H.; Srikant, C. The somatostatin receptor family. *Life Sci.* **1995**, *57*, 1249–1265. [CrossRef]
36. Mansi, R.; Fani, M. Design and development of the theranostic pair ^{177}Lu-OPS201/^{68}Ga-OPS202 for targeting somatostatin receptor expressing tumors. *J. Labelled Comp. Radiopharm.* **2019**, *62*, 635–645. [CrossRef] [PubMed]
37. Schottelius, M.; Osl, T.; Poschenrieder, A.; Hoffmann, F.; Beykan, S.; Hänscheid, H.; Schirbel, A.; Buck, A.K.; Kropf, S.; Schwaiger, M.; et al. ^{177}Lu pentixather: Comprehensive Preclinical Characterization of a First CXCR4-directed Endoradiotherapeutic Agent. *Theranostics* **2017**, *7*, 2350–2362. [CrossRef]
38. Valkema, R.; Pauwels, S.A.; Kvols, L.K.; Kwekkeboom, D.J.; Jamar, F.; De Jong, M.; Barone, R.; Walrand, S.; Kooij, P.P.M.; Bakker, W.H.; et al. Long-term follow-up of renal function after peptide receptor radiation therapy with ^{90}Y-DOTA^{0},Tyr^{3}-octreotide and ^{177}Lu-DOTA^{0}, Tyr^{3}-octreotate. Available online: http://jnm.snmjournals.org/content/46/1_suppl/83S.full.pdf (accessed on 24 July 2020).
39. Maina, T.; Cescato, R.; Waser, B.; Tatsi, A.; Kaloudi, A.; Krenning, E.P.; De Jong, M.; Nock, B.A.; Reubi, J.C. [^{111}In-DOTA]LTT-SS28, a First Pansomatostatin Radioligand for in vivo Targeting of Somatostatin Receptor-Positive Tumors. *J. Med. Chem.* **2014**, *57*, 6564–6571. [CrossRef]
40. Sgouros, G. Bone marrow dosimetry for radioimmunotherapy: Theoretical considerations. *J. Nucl. Med.* **1993**, *34*, 689–694.

Sample Availability: Samples of the compounds are not available from the authors.

© 2020 by the authors. Licensee MDPI, Basel, Switzerland. This article is an open access article distributed under the terms and conditions of the Creative Commons Attribution (CC BY) license (http://creativecommons.org/licenses/by/4.0/).

Article

[^{99m}Tc]Tc-DB1 Mimics with Different-Length PEG Spacers: Preclinical Comparison in GRPR-Positive Models

Panagiotis Kanellopoulos [1,2], Emmanouil Lymperis [1], Aikaterini Kaloudi [1], Marion de Jong [3], Eric P. Krenning [4], Berthold A. Nock [1] and Theodosia Maina [1,*]

1. Molecular Radiopharmacy, INRASTES, NCSR "Demokritos", 15341 Athens, Greece; kanelospan@gmail.com (P.K.); mlymperis@hotmail.com (E.L.); katerinakaloudi@yahoo.gr (A.K.); nock_berthold.a@hotmail.com (B.A.N.)
2. Molecular Pharmacology, School of Medicine, University of Crete, Heraklion, 70013 Crete, Greece
3. Department of Radiology & Nuclear Medicine Erasmus MC, 3015 CN Rotterdam, The Netherlands; m.hendriks-dejong@erasmusmc.nl
4. Cyclotron Rotterdam BV, Erasmus MC, 3015 CE Rotterdam, The Netherlands; erickrenning@gmail.com
* Correspondence: maina_thea@hotmail.com; Tel.: +30-210-650-3908

Academic Editor: Svend Borup Jensen
Received: 5 July 2020; Accepted: 25 July 2020; Published: 28 July 2020

Abstract: *Background*: The frequent overexpression of gastrin-releasing peptide receptors (GRPRs) in human cancers provides the rationale for delivering clinically useful radionuclides to tumor sites using peptide carriers. Radiolabeled GRPR antagonists, besides being safer for human use, have often shown higher tumor uptake and faster background clearance than agonists. We herein compared the biological profiles of the GRPR-antagonist-based radiotracers [^{99m}Tc]Tc-[N$_4$-PEGx-DPhe^6,Leu-NHEt^{13}]BBN(6-13) (N$_4$: 6-(carboxy)-1,4,8,11-tetraazaundecane; PEG: polyethyleneglycol): (i) [^{99m}Tc]Tc-DB7 (x = 2), (ii) [^{99m}Tc]Tc-DB13 (x = 3), and (iii) [^{99m}Tc]Tc-DB14 (x = 4), in GRPR-positive cells and animal models. The impact of in situ neprilysin (NEP)-inhibition on in vivo stability and tumor uptake was also assessed by treatment of mice with phosphoramidon (PA). *Methods*: The GRPR affinity of DB7/DB13/DB14 was determined in PC-3 cell membranes, and cell binding of the respective [^{99m}Tc]Tc-radioligands was assessed in PC-3 cells. Each of [^{99m}Tc]Tc-DB7, [^{99m}Tc]Tc-DB13, and [^{99m}Tc]Tc-DB14 was injected into mice without or with PA coinjection and 5 min blood samples were analyzed by HPLC. Biodistribution was conducted at 4 h postinjection (pi) in severe combined immunodeficiency disease (SCID) mice bearing PC-3 xenografts without or with PA coinjection. *Results*: DB7, -13, and -14 displayed single-digit nanomolar affinities for GRPR. The uptake rates of [^{99m}Tc]Tc-DB7, [^{99m}Tc]Tc-DB13, and [^{99m}Tc]Tc-DB14 in PC-3 cells was comparable and consistent with a radioantagonist profile. The radiotracers were found to be ≈70% intact in mouse blood and >94% intact after coinjection of PA. Treatment of mice with PA enhanced tumor uptake. *Conclusions*: The present study showed that increase of PEG-spacer length in the [^{99m}Tc]Tc-DB7–[^{99m}Tc]Tc-DB13–[^{99m}Tc]Tc-DB14 series had little effect on GRPR affinity, specific uptake in PC-3 cells, in vivo stability, or tumor uptake. A significant change in in vivo stability and tumor uptake was observed only after treatment of mice with PA, without compromising the favorably low background radioactivity levels.

Keywords: gastrin-releasing peptide receptor targeting; [^{99m}Tc]Tc-radiotracer; tumor targeting; [^{99m}Tc]Tc-DB1 mimic; PEG$_x$-spacer; neprilysin-inhibition; phosphoramidon

1. Introduction

The gastrin-releasing peptide receptor (GRPR) has attracted much attention in nuclear oncology owing to its high-density expression in frequently occurring human cancers, such as prostate cancer, mammary carcinoma, and others [1–7]. This finding can be elegantly exploited to direct diagnostic and therapeutic radionuclides to tumor sites by means of suitably designed peptide carriers that specifically interact with the GRPR on tumor cells [8–11]. Diagnostic imaging with gamma emitters (e.g., ^{99m}Tc, ^{111}In) for single-photon-emission computed tomography (SPECT) or positron emitters (e.g., ^{68}Ga, ^{64}Cu) for positron-emission tomography (PET) will allow for initial diagnosis, assessment of disease spread and progression, and selection of patients eligible for subsequent radionuclide therapy. Molecular imaging is likewise essential for dosimetry, therapy planning, and follow-up, enabling a patient-tailored, "theranostic" approach. Therapy per se is conducted with the respective peptide analog carrying a suitable particle emitter (beta, alpha, or Auger electron emitter).

Several analogs of the frog 14 peptide bombesin (BBN; Pyr–Gln–Arg–Leu–Gly–Asn–Gln–Trp–Ala–Val–Gly–His–Leu–Met–NH$_2$) and its C-terminal BBN(6–14) fragment, showing high GRPR affinity, have been derivatized with the appropriate chelator for stable binding of the selected medically relevant radiometal and have been evaluated in animal models and in humans [8,11]. It should be noted that such analogs internalize into target cells and display agonistic profiles at the GRPR. Agonism at the GRPR, however, translates into adverse effects elicited in patients after intravenous injection of BBN analogs and the GRPR activation that follows [12–14]. Such effects intensify at the higher peptide doses administered during radionuclide therapy, thereby restricting the broader clinical use of GRPR agonists. A subsequent shift of paradigm from GRPR radioagonists to antagonists revealed unexpected benefits in their use beyond the anticipated inherent biosafety. The clearance GRPR radioantagonists, in contrast to agonists, turned out to be much faster from physiological tissues than from tumor sites [15]. The basis for this clinically appealing feature has not been elucidated yet, although it has been observed for other receptor radioantagonists as well [16].

We previously reported on a series of radiolabeled analogs of the potent GRPR antagonist [H-DPhe^6,Leu-NHEt^{13}]BBN(6–13) [17,18], generated by coupling suitable chelators to the N-terminus via different linkers, which showed attractive pharmacokinetic profiles [19–23]. Thus, 1,4,8,11-tetraazaundecane has been used for labeling with the pre-eminent SPECT radionuclide [^{99m}Tc]Tc, forming octahedral monocationic *trans*-dioxo Tc-complexes [24]. The resulting radiotracers, [^{99m}Tc]Tc-DB1 mimics, have displayed high GRPR affinity, fair metabolic stability in peripheral mouse blood, and rapid localization in experimental xenografts in mice, whereas background clearance rates varied. [^{99m}Tc]Tc-DB7, whereby 6-(carboxy)-1,4,8,11-tetraazaundecane (N$_4$) is coupled to the peptide N-terminus via a polyethyleneglycol (PEG)$_2$ spacer (Figure 1), showed the highest in vivo metabolic stability and tumor-to-pancreas ratio in mouse models [23].

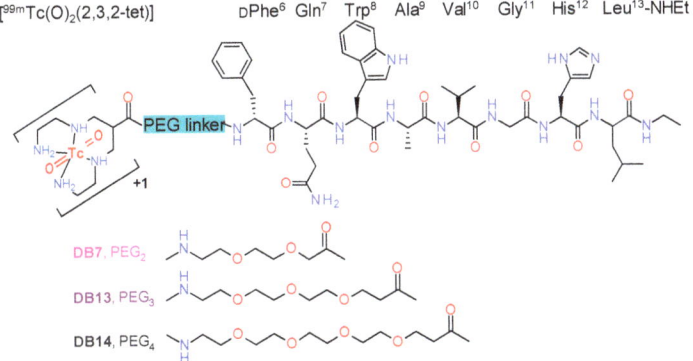

Figure 1. Chemical structure of [^{99m}Tc]Tc-DB7 (PEG$_2$), [^{99m}Tc]Tc-DB13 (PEG$_3$), and [^{99m}Tc]Tc-DB14 (PEG$_4$).

In the present study, we designed two further [^{99m}Tc]Tc-DB1 mimics, with the N$_4$ coupled to the peptide chain via PEG$_x$ linkers of increasing chain-length: [^{99m}Tc]Tc-DB13 (x = 3), and [^{99m}Tc]Tc-DB14 (x = 4) (Figure 1). We were interested to investigate the effect of linker length on several biological features of resulting analogs, such as GRPR affinity, cell uptake, in vivo metabolic stability, and pharmacokinetics in mice bearing human GRPR-expressing prostate adenocarcinoma PC-3 xenografts. A further objective of this study was to assess potential improvements of the PC-3 tumor targeting and overall pharmacokinetics of [^{99m}Tc]Tc-DB7, [^{99m}Tc]Tc-DB13, and [^{99m}Tc]Tc-DB14 during transient inhibition of neprilysin (NEP) [25,26]. The latter was accomplished by coinjection of the NEP-inhibitor phosphoramidon (PA) [27] together with each radiotracer. This methodology was previously shown to enhance the metabolic stability of BBN and other peptide radioligands in peripheral blood and to improve the supply of the intact radiopeptide form to tumor sites. As a result, notably improved tumor targeting was observed in mice and recently also in patients [28–35].

2. Results

2.1. Radiolabeling and Quality Control

Radiolabeling of DB7, DB13, and DB14 with [^{99m}Tc]Tc was accomplished by 30 min incubation at room temperature in alkaline aqueous medium containing citrate anions and SnCl$_2$ as reductant. Quality control of the radiolabeled products (Figure 1) included HPLC and ITLC analysis and revealed less than 2% total radiochemical impurities ([^{99m}Tc]TcO$_4^-$, [^{99m}Tc]Tc-citrate, and [^{99m}Tc]TcO$_2 \times$ nH$_2$O). A single radiopeptide species was obtained at molecular activities of 20–40 MBq [^{99m}Tc]Tc/nmol peptide. In view of the above, [^{99m}Tc]Tc-DB7, [^{99m}Tc]Tc-DB13, and [^{99m}Tc]Tc-DB14 were used without further purification in all subsequent assays.

2.2. In Vitro Assays in PC-3 Cells

2.2.1. GRPR Affinity of Peptide Conjugates

Competition binding assays for DB7, DB13 and DB14 were performed in PC-3 cell membranes. As shown in Figure 2, all three peptides were able to displace [^{125}I][I-Tyr^4]BBN from GRPR binding sites on the membranes in a monophasic and dose-dependent manner. The binding affinities of the three analogs for the human GRPR were found comparable, DB7 (IC$_{50}$ = 0.93 ± 0.01 nM), DB13 (IC$_{50}$ = 1.03 ± 0.01 nM) and DB14 (IC$_{50}$ = 1.18 ± 0.09 nM), indicating little influence of the PEG$_x$-chain length.

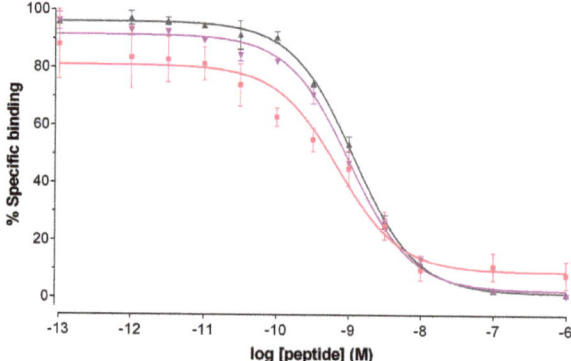

Figure 2. Displacement of [^{125}I][I-Tyr^4]BBN from GRPR binding sites in PC-3 cell membranes by increasing concentrations of DB7 (■, IC$_{50}$ = 0.93 ± 0.01 nM), DB13 (▼, IC$_{50}$ = 1.03 ± 0.01 nM), and DB14 (▲, IC$_{50}$ = 1.18 ± 0.09 nM); results represent average values ± SD of three experiments performed in triplicate.

2.2.2. Radiotracer Uptake in PC-3 Cells

The uptake of [^{99m}Tc]Tc-DB7, [^{99m}Tc]Tc-DB13, and [^{99m}Tc]Tc-DB14 in PC-3 cells is compared in Figure 3. In all cases, the bulk of radioactivity was found on the membrane of PC-3 cells with only a small portion detected within the cells, consistent with a noninternalizing radioantagonist profile [15,21]. Cell association was banned (<0.2%) in the presence of 1 µM [Tyr^4]BBN, suggesting a GRPR-mediated process (results not shown). A decline in cell uptake was observed with increasing PEG-chain length. Thus, [^{99m}Tc]Tc-DB7 (PEG$_2$, 2.6 ± 0.5%) showed superior cell uptake compared with [^{99m}Tc]Tc-DB13 (PEG$_3$, 2.0 ± 0.5%; $p < 0.05$) and [^{99m}Tc]Tc-DB14 (PEG$_4$, 1.6 ± 0.2%; $p < 0.001$).

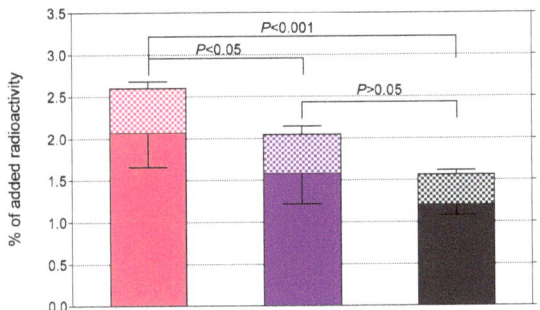

Figure 3. Specific uptake of [^{99m}Tc]Tc-DB7 (pink bars), [^{99m}Tc]Tc-DB13 (violet bars) and [^{99m}Tc]Tc-DB14 (gray bars) in PC-3 cells after 1 h incubation at 37 °C (checkered bars: internalized, solid bars: membrane bound fractions); results represent the mean ± SD of 3 experiments performed in triplicate.

2.3. In Vivo Comparison of [^{99m}Tc]Tc-DB7, [^{99m}Tc]Tc-DB13, and [^{99m}Tc]Tc-DB14

2.3.1. Metabolic Studies in Mice

The stability of [^{99m}Tc]Tc-DB7, [^{99m}Tc]Tc-DB13, and [^{99m}Tc]Tc-DB14 in peripheral mouse blood was assessed at 5 min postinjection (pi) via HPLC analysis of blood samples. Representative radiochromatograms are shown in Figure 4a, revealing a 30% radiometabolite formation, and a comparable stability across radiotracers (≈70% intact radiopeptide, $p > 0.05$). After treatment of mice with PA, radiotracer stability was significantly enhanced (Figure 4b), namely, [^{99m}Tc]Tc-DB7: 70.6 ± 1.1% to 94.5 ± 1.1% intact ($p < 0.0001$); [^{99m}Tc]Tc-DB13: 71.2 ± 3.2% to 94.2 ± 1.3% intact ($p < 0.0001$); and [^{99m}Tc]Tc-DB14: 77.9 ± 3.8% to 96.0 ± 1.0% intact ($p < 0.001$). These results implicate NEP in the partial in vivo degradation of the three radioligands [28,36,37].

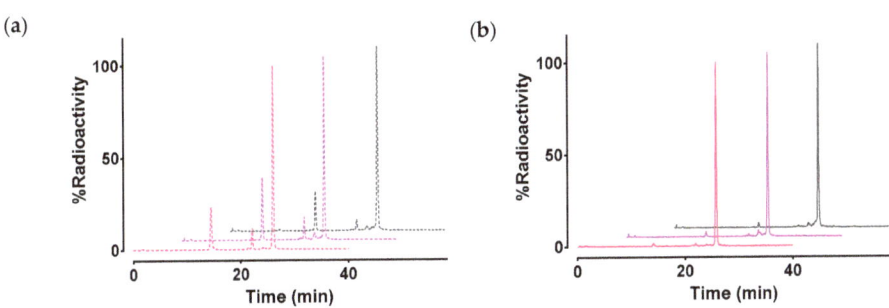

Figure 4. Representative HPLC radiochromatograms (System 2) of mouse blood collected 5 min after iv injection of [^{99m}Tc]Tc-DB7 (pink lines), [^{99m}Tc]Tc-DB13 (violet lines), and [^{99m}Tc]Tc-DB14 (gray lines) (a) without (dotted lines) or (b) with PA coinjection (solid lines); results represent average values ± SD, $n = 3$.

2.3.2. Biodistribution in PC-3 Tumor-Bearing Mice

Cumulative biodistribution data for [^{99m}Tc]Tc-DB7, [^{99m}Tc]Tc-DB13, and [^{99m}Tc]Tc-DB14 in severe combined immunodeficiency disease (SCID) mice bearing PC-3 xenografts at 4 h pi, as %injected activity per gram tissue (%IA/g) ± SD, can be found in Table 1 ([^{99m}Tc]Tc-DB7), Table 2 ([^{99m}Tc]Tc-DB13), and Table 3 ([^{99m}Tc]Tc-DB14).

Table 1. Biodistribution data for [^{99m}Tc]Tc-DB7, expressed as %injected activity per gram tissue (%IA/g) mean ± SD, n = 4, in PC-3 xenograft-bearing severe combined immunodeficiency disease (SCID) mice at 4 h pi.

Tissue	[^{99m}Tc]Tc-DB7				
	Block [1]		Controls		PA [2]
Blood	0.08 ± 0.03		0.05 ± 0.01		0.06 ± 0.02
Liver	2.11 ± 0.43		0.96 ± 0.23		0.87 ± 0.12
Heart	0.09 ± 0.03		0.04 ± 0.01		0.07 ± 0.03
Kidneys	2.38 ± 0.56		1.89 ± 0.61		1.55 ± 0.34
Stomach	0.45 ± 0.34		0.27 ± 0.17		0.22 ± 0.05
Intestines	2.85 ± 0.39		1.38 ± 0.27		1.64 ± 0.72
Spleen	1.33 ± 0.37		0.24 ± 0.07		0.28 ± 0.1
Muscle	0.02 ± 0.01		0.02 ± 0.01		0.02 ± 0.01
Lungs	0.55 ± 0.26		0.16 ± 0.12		0.14 ± 0.02
Pancreas	0.37 ± 0.09	↔ $p > 0.05$ ↔	0.56 ± 0.2	↔ $p > 0.05$ ↔	0.91 ± 0.12
Tumor	0.53 ± 0.20	↔ $p < 0.0001$ ↔	4.49 ± 1.20	↔ $p < 0.0001$ ↔	6.10 ± 1.1

All animals were injected with 180–230 kBq/10 pmol peptide; [1] animals co-injected with 50 μg [Tyr^4]BBN for in vivo GRPR-blockade; [2] animals co-injected with 300 μg PA for in situ inhibition of NEP.

Table 2. Biodistribution data for [^{99m}Tc]Tc-DB13, expressed as %IA/g mean ± SD, n = 4, in PC-3 xenograft-bearing SCID mice at 4 h pi.

Tissue	[^{99m}Tc]Tc-DB13				
	Block [1]		Controls		PA [2]
Blood	0.09 ± 0.01		0.06 ± 0.02		0.19 ± 0.19
Liver	1.8 ± 0.15		0.75 ± 0.24		0.88 ± 0.25
Heart	0.18 ± 0.07		0.07 ± 0.03		0.16 ± 0.08
Kidneys	3.5 ± 1.59		1.83 ± 0.3		2.66 ± 1.14
Stomach	1.25 ± 0.86		0.9 ± 0.18		1.09 ± 0.47
Intestines	5.37 ± 1.34		3.69 ± 0.72		4.23 ± 0.98
Spleen	1.14 ± 0.47		0.29 ± 0.07		0.36 ± 0.13
Muscle	0.04 ± 0.02		0.03 ± 0.01		0.16 ± 0.32
Lungs	0.67 ± 0.31		0.13 ± 0.01		0.24 ± 0.08
Pancreas	0.3 ± 0.09	↔ $p > 0.05$ ↔	0.75 ± 0.08	↔ $p > 0.05$ ↔	1.82 ± 1.04
Tumor	1.12 ± 0.23	↔ $p < 0.0001$ ↔	4.14 ± 0.78	↔ $p < 0.001$ ↔	5.79 ± 1.18

All animals were injected with 180–230 kBq/10 pmol peptide; [1] animals co-injected with 50 μg [Tyr^4]BBN for in vivo GRPR-blockade; [2] animals co-injected with 300 μg PA for in situ inhibition of NEP.

All three radiotracers showed a fast blood and body clearance via the kidneys into the urine, displaying low background radioactivity uptake even in the GRPR-rich pancreas [38] (<1%IA/g pancreas for all three tracers). Uptake in the experimental PC-3 tumor was comparable across compounds in control mice, [^{99m}Tc]Tc-DB7: 4.49 ± 1.20%IA/g, [^{99m}Tc]Tc-DB13: 4.14 ± 0.78%IA/g, and [^{99m}Tc]Tc-DB14 3.71 ± 1.04%IA/g ($p > 0.05$). This uptake was reduced in the block animal groups for [^{99m}Tc]Tc-DB7 ($p < 0.0001$), [^{99m}Tc]Tc-DB13 ($p < 0.0001$), and [^{99m}Tc]Tc-DB14 ($p < 0.01$), indicating GRPR-specificity.

During NEP-inhibition the uptake of the three radiotracers increased in the PC-3 tumors, [^{99m}Tc]Tc-DB7: 6.10 ± 1.20%IA/g ($p < 0.0001$); [^{99m}Tc]Tc-DB13: 5.79 ± 1.18%IA/g ($p < 0.001$); [^{99m}Tc]Tc-DB14: 4.00 ± 0.34%IA/g ($p > 0.05$). Interestingly, neither renal uptake nor pancreatic uptake showed any significant increase after treatment of mice with PA. Representative data for the three radiotracers in kidneys, pancreas and tumor are depicted in Figure 5.

Table 3. Biodistribution data for [^{99m}Tc]Tc-DB14, expressed as %IA/g mean ± SD, $n = 4$, in PC-3 xenograft-bearing SCID mice at 4 h pi.

Tissue	[^{99m}Tc]Tc-DB14					
	Block [1]		Controls		PA [2]	
Blood	0.21 ± 0.13		0.06 ± 0.03		0.07 ± 0.01	
Liver	1.69 ± 0.15		0.79 ± 0.32		0.93 ± 0.13	
Heart	0.23 ± 0.08		0.08 ± 0.05		0.15 ± 0.03	
Kidneys	7.05 ± 4.24		1.46 ± 0.25		1.79 ± 0.34	
Stomach	0.85 ± 0.10		0.75 ± 0.72		0.21 ± 0.06	
Intestines	6.25 ± 2.51		2.03 ± 0.81		1.83 ± 0.77	
Spleen	0.89 ± 0.15		0.19 ± 0.07		0.24 ± 0.05	
Muscle	0.07 ± 0.04		0.04 ± 0.03		0.03 ± 0.01	
Lungs	0.78 ± 0.25		0.15 ± 0.06		0.16 ± 0.05	
Pancreas	0.32 ± 0.09	↤ $p > 0.05$ ↦	0.36 ± 0.09	↤ $p > 0.05$ ↦	0.42 ± 0.06	
Tumor	2.16 ± 0.59	↤ $p < 0.01$ ↦	3.71 ± 1.04	↤ $p > 0.05$ ↦	4.00 ± 0.34	

All animals were injected with 180–230 kBq/10 pmol peptide; [1] animals co-injected with 50 µg [Tyr^4]BBN for in vivo GRPR-blockade; [2] animals co-injected with 300 µg PA for in situ inhibition of NEP

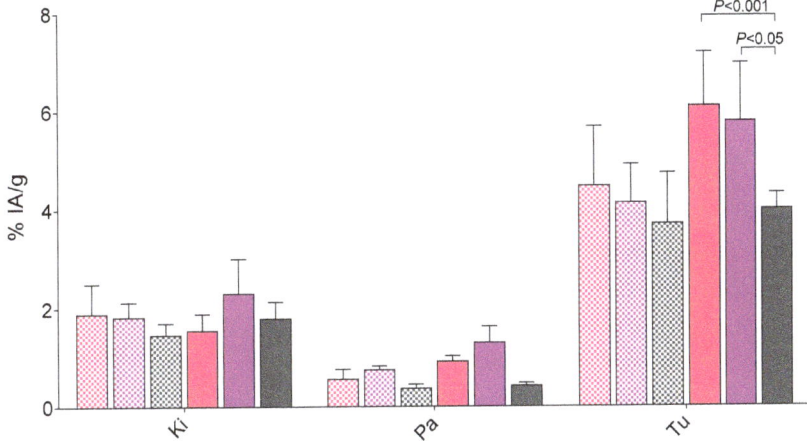

Figure 5. Selected biodistribution data for [^{99m}Tc]Tc-DB7 (pink bars), [^{99m}Tc]Tc-DB13 (violet bars), and [^{99m}Tc]Tc-DB14 (gray bars) for kidneys (Ki), pancreas (Pa), and PC-3 tumors (Tu) at 4 h pi in female SCID mice bearing subcutaneous PC-3 xenografts; results are expressed as %IA/g mean ± SD, $n = 4$, without (checkered bars) or with coinjection (solid bars) of PA. Only statistical differences in organ/tumor uptake across compounds are included in the diagram; statistical differences between PA-treated and nontreated mice are shown in Table 1, Table 2, and Table 3.

3. Discussion

Radiotracers based on GRPR antagonists have lately attracted much attention in nuclear medicine, largely because of their higher inherent safety for human use compared with agonists [20]. After intravenous injection to patients, antagonists seek and bind but do not activate the GRPR, and hence they do not elicit adverse effects. Furthermore, GRPR radioantagonists often display attractive pharmacokinetic profiles in preclinical models and in patients, as a combined result of rapid clearance from physiological tissues and good retention in tumor sites. In fact, a general concern in the application of GRPR radioligands for cancer theranostics has been the high-density expression of GRPR not only on tumors, but also in physiological tissues, especially in the pancreas [38]. Previous studies with both GRPR radioagonists and antagonists have demonstrated the significant impact of linkers, introduced between the metal chelate and the peptide part, on pharmacokinetics [23,39–42].

In line with these observations, we also noted marked differences in the pharmacokinetic profiles, and especially in the tumor vs. pancreas uptake, across a series of [^{99m}Tc]Tc-based GRPR radioantagonists, [^{99m}Tc]Tc-DB1 mimics [23]. These carry an acyclic tetraamine chelator via different types and lengths of linkers at the N-terminus of the potent GRPR antagonist [H-DPhe^6,Leu-NHEt^{13}]BBN(6-13) [17,18]. The PEG$_2$-derivatized member of this series, [^{99m}Tc]Tc-DB7 (Figure 1), displayed superior tumor over pancreas uptake, as well as considerably better metabolic stability in peripheral mouse blood.

In the present study, we were interested to find out whether the elongation of the PEG$_2$ chain would further improve the biological properties of resulting analogs. For this purpose, [^{99m}Tc]Tc-DB13 (PEG$_3$) and [^{99m}Tc]Tc-DB14 (PEG$_4$) were newly synthetized and compared to [^{99m}Tc]Tc-DB7 (Figure 1). As revealed by competitive binding assays in PC-3 cell membranes, the elongation of PEG$_2$, to PEG$_3$ and PEG$_4$ had no apparent impact on the GRPR affinity of the respective DB7, DB13, and DB14 (Figure 2). On the other hand, the resultant [^{99m}Tc]Tc radiotracers showed modest, but statistically significant, decline of uptake in PC-3 cells in vitro as the length of the PEG linker increased from PEG$_2$ to PEG$_4$ (Figure 3).

In a subsequent step, we investigated the effect of PEG-chain length on the metabolic stability of the three [^{99m}Tc]Tc radiotracers in mouse blood collected 5 min pi. This study did not reveal any statistically significant differences across radiotracers (Figure 4a). NEP has been shown to be a major protease in the rapid in vivo degradation of BBN and its analogs [28,36,37]. NEP is actually an ectoenzyme abundantly present on the epithelial cells of several tissues of the body, including vasculature walls, kidneys, lungs, and intestines, but found only in minute amounts in the blood solute [25,26]. Therefore, its action is overlooked during in vitro incubation assays of radioligands in plasma or serum. We previously demonstrated that coinjection of NEP inhibitors, such as PA [27], with BBN-like radiopeptides, both agonists and antagonists, improves their metabolic stability in circulation [28,30–34]. As a result, an appreciably higher amount of intact radiopeptide eventually reaches tumor sites. Accordingly, tumor uptake is markedly enhanced, with clear benefits to be gained both for imaging and therapy. Following this rationale, we decided to study the effects of in situ inhibition of NEP on the in vivo stability of [^{99m}Tc]Tc-DB7, [^{99m}Tc]Tc-DB13, and [^{99m}Tc]Tc-DB14 coinjected with PA in the present work. It is interesting to note that once again, significant enhancement of metabolic stability was documented for all three analogs (Figure 4b).

In order to assess how the above properties translate in terms of pharmacokinetics, biodistribution profiles of [^{99m}Tc]Tc-DB7, [^{99m}Tc]Tc-DB13, and [^{99m}Tc]Tc-DB14 were compared in mice bearing GRPR-positive experimental tumors at 4 h pi, without or with PA coinjection. Firstly, we observed declining, but not statistically significant lower, uptake in the PC-3 tumors with increasing length of the PEG chain from 2 to 4 (Figure 5), as a combined result of (i) the equivalent GRPR affinities of DB7, DB13, and DB14, (ii) the slightly declining PC-3 cell uptake capabilities of [^{99m}Tc]Tc-DB7, [^{99m}Tc]Tc-DB13, and [^{99m}Tc]Tc-DB13, and (iii) their similar metabolic stability. Secondly, overall pharmacokinetics turned out to be very comparable for all three radioantagonists, characterized by favorably low background radioactivity levels. Of particular advantage are the low radioactivity values displayed by the three radioligands in the GRPR-rich pancreas, as well as in the kidneys. Thirdly, treatment of mice with PA led to significant increase of tumor uptake compared with controls (Table 1, Table 2, and Table 3) without provoking any unfavorable rise of background activity, thereby further enhancing tumor to background contrast. Interestingly, the enhanced tumor uptake induced by PA via in situ stabilization of analogs in mouse circulation ended up being reflected in statistically significant differences between the PEG$_{2/3}$ and PEG$_4$ members of the series (Figure 5).

In conclusion, the effect of PEG$_X$ linker length (x = 2, 3 and 4) in a series of GRPR radioantagonists, ([^{99m}Tc]Tc-[N$_4$-PEGx-DPhe^6,Leu-NHEt^{13}]BBN(6-13), Figure 1) had little effect on GRPR affinity, binding in GRPR-positive PC-3 cells, metabolic stability in mouse circulation, or PC-3 tumor targeting and overall pharmacokinetics in animal models. In all cases, tumor-to-background levels were favorable, including those in the GRPR-rich pancreas and the kidneys. Similar observations have been made for other GRPR antagonists, wherein the influences of different length of PEG linkers (x = 2, 3, 4) and different chelating moieties (NOTA: 1,4,7-triazacyclononane-1,4,7-triacetic acid, NODAGA: 1,4,7-triazacyclononane-1-(glutaric acid)-4,7-diacetic acid, DOTA: 1,4,7,10-tetraazacyclododecane-1,4,7,10-tetraacetic acid, DOTAGA: 1,4,7,10-tetraazacyclododececane-1-(glutaric acid)-4,7,10-triacetic acid) were distinct [39,40,42]. The effects of the PEG spacer's length on the in vivo pharmacokinetics of resulting radioligands were found to be minor compared with those of the metal chelate or the peptide chain applied. In all cases, however, the use of PEG linkers was shown to favor in vivo metabolic stability and to boost background clearance, especially from the pancreas. Notably, the above attractive properties of [^{99m}Tc]Tc-DB7, [^{99m}Tc]Tc-DB13, and [^{99m}Tc]Tc-DB14 were further enhanced by in situ NEP inhibition, inducing tumor uptake increases without affecting the advantageously low background radioactivity levels.

4. Materials and Methods

4.1. Chemicals and Radionuclides

All chemicals were reagent-grade and were therefore used without further purification. The peptide conjugates DB7, DB13, and DB14 were synthesized on a solid support and obtained from PiChem (Graz, Austria). [Tyr^4]BBN (Pyr–Gln–Arg–Tyr–Gly–Asn–Gln–Trp–Ala–Val–Gly–His–Leu–Met–NH$_2$) was purchased from Bachem (Bubendorf, Switzerland). PA (phosphoramidon disodium dehydrate, N-(α-rhamnopyranosyloxyhydroxyphosphinyl)-L-leucyl-L-tryptophan × 2Na × 2H$_2$O) was obtained from PeptaNova GmbH (Sandhausen, Germany).

Technetium-99m in the form of [^{99m}Tc]NaTcO$_4$ was collected by elution of a [^{99}Mo]Mo/[^{99m}Tc]Tc generator (Ultra-Technekow™ V4 Generator, Curium Pharma, Petten, The Netherlands), while [^{125}I]NaI in a solution of 10^{-5} M NaOH (10 μL) was purchased from Perkin Elmer (Waltham, MA, USA).

4.1.1. Radiolabeling

The lyophilized peptide analogs were dissolved in water to a final concentration of 1 mM and 50 μL aliquots were stored at −20 °C. Labeling with [^{99m}Tc]Tc was performed in an Eppendorf vial containing 0.5 M phosphate buffer (pH 11.5, 50 μL). [^{99m}Tc]NaTcO$_4$ eluate (410 μL, 370-550 MBq) was added to the vial followed by 0.1 M sodium citrate (5 μL), the peptide stock solution (15 μL, 15 nmol), and a freshly prepared SnCl$_2$ solution in ethanol (20 μL, 20 μg). The mixture was left to react for 30 min at room temperature and the pH was neutralised with the addition of 0.1 M HCl.

Radioiodination of [Tyr^4]BBN was performed following the chloramine-T method and any sulfoxide (Met^{14}=O) that formed was reduced back to nonoxidized Met^{14} by dithiothreitol. The [^{125}I-Tyr^4]BBN was isolated in a highly pure form by HPLC and Met was added to the purified radioligand solution to prevent reoxidation of Met^{14} during storage; the resulting stock solution in 0.1% BSA-PBS was kept at −20 °C and aliquots thereof were used in competitive binding assays (74 GBq/μmol).

4.1.2. Quality Control

Quality control comprised radioanalytical high-performance liquid chromatography (HPLC) and instant thin-layer chromatography (ITLC). HPLC analyses were performed on a Waters chromatograph coupled to a 996 photodiode array UV detector (Waters, Vienna, Austria) and a Gabi gamma detector (Raytest RSM Analytische Instrumente GmbH, Straubenhardt, Germany). Data processing and chromatography were controlled with the Empower Software (Waters, Milford, MA, USA). For analyses, a Symmetry Shield RP-18 (5 μm, 4.6 mm × 150 mm) cartridge column (Waters, Eschborn, Germany) was eluted at a 1 mL/min flow rate with a linear gradient system 1 starting from 0% B and advancing to 40% B within 20 min (solvent A = 0.1% aqueous TFA and B = MeCN). ITLC analyses were performed on Whatman 3 mm chromatography paper strips (GE Healthcare, Chicago, IL, USA), developed up to 10 cm from the origin with 5 M ammonium acetate/MeOH 1/1 (v/v) for the detection of reduced hydrolyzed technetium ([^{99m}Tc]TcO$_2$ × nH$_2$O), or acetone for the detection of [^{99m}Tc]TcO$_4^-$.

All manipulations with beta- and gamma-emitting radionuclides and their solutions were performed by trained and authorized personnel behind suitable shielding in licensed laboratories, in compliance with European radiation safety guidelines and supervised by the Greek Atomic Energy Commission (license #A/435/17092/2019)

4.2. In Vitro Assays

4.2.1. Cell Lines and Culture

The human prostate adenocarcinoma PC-3 cell line spontaneously expressing GRPR [43] was obtained from LGC Standards GmbH (Wesel, Germany). All culture reagents were obtained from Gibco BRL, Life Technologies (Grand Island, NY, USA) or from Biochrom KG Seromed (Berlin, Germany). Cells were grown in Roswell Park Memorial Institute-1640 (RPMI-1640) medium with GlutaMAX-I supplemented with 10% (v/v) fetal bovine serum (FBS), 100 U/mL penicillin, and 100 µg/mL streptomycin, and kept in a controlled humidified air containing 5% CO$_2$ at 37 °C. Splitting of cells with a ratio of 1:3 to 1:5 was performed as they approached confluency, using a trypsin/EDTA (0.05%/0.02% w/v) solution.

4.2.2. Competitive Binding in PC-3 Cell Membranes

Competition binding experiments of DB7, DB14 and DB14 against [^{125}I][I-Tyr^4]BBN were conducted in PC-3 cell membranes. Increasing concentrations of tested peptide (10^{-13}–10^{-5} M) were mixed with the radioligand (40,000 cpm per assay tube, at a 50 pM concentration) and the membrane homogenate in a total volume of 300 µL binding buffer (pH 7.4, 50 mM HEPES, 1% BSA, 5.5 mM MgCl$_2$, 35 µM bacitracin). Triplicates of each concentration point were incubated for 60 min at 22 °C in an Incubator-Orbital Shaker unit (MPM Instr. SrI, Bernareggio, MI, Italy). The incubation was interrupted by adding ice-cold washing buffer (10 mM HEPES pH 7.4, 150 mM NaCl), followed by rapid filtration over glass fiber filters (Whatman GF/B, presoaked in binding buffer) on a Brandel Cell Harvester (Adi Hassel Ingenieur Büro, Munich, Germany). Filters were washed with cold washing buffer and were counted for their radioactivity content in an automated well-type gamma counter [NaI(Tl) 3" crystal, Canberra Packard Cobra Quantum series instrument]. The 50% inhibitory concentration (IC$_{50}$) values were calculated by nonlinear regression according to a one-site model applying the PRISM 6 program (Graph Pad Software, San Diego, CA, USA) and are expressed as mean ± SD of three experiments performed in triplicate.

4.2.3. Internalization of [^{99m}Tc]Tc Radiotracers in PC-3 Cells

For internalization assays with [^{99m}Tc]Tc-DB7, [^{99m}Tc]Tc-DB13, and [^{99m}Tc]Tc-DB14, PC-3 cells were seeded in six well plates (≈1 × 10^6 cells per well) 24 h before the experiment. Cells were rinsed twice with ice-cold internalization medium (RPMI-1640 GlutaMAX-I, supplemented by 1% (v/v) FBS) and then fresh medium was added (1.2 mL) at 37 °C, followed by test radiopeptide (250 fmol total peptide in 150 µL 0.5% BSA-PBS, 100,000–200,000 cpm). Nonspecific internalization was determined by a parallel triplicate series containing 1 µM [Tyr^4]BBN. After 1 h incubation at 37 °C, the plates were placed on ice, the medium was collected, and the plates were washed with 0.5% BSA-PBS (1 mL). Membrane-bound fractions were collected by incubating the cells 2 × 5 min in acid-wash solution (2 × 600 µL; 50 mM glycine buffer pH 2.8, 0.1 M NaCl) at room temperature. After rinsing the cells with 0.5% BSA-PBS (1 mL), internalized fractions were collected by lysing the cells with 1 N NaOH (2 × 600 µL). Sample radioactivity was measured on the gamma counter and the percentage, of specific internalized and membrane-bound fractions were calculated with Microsoft Excel (after subtracting nonspecific from overall internalized and membrane-bound counts). Results represent specific internalized ± SD of total added radioactivity per well from three experiments performed in triplicate.

4.3. Animal Studies

4.3.1. Metabolic Studies in Mice

A bolus containing each of [^{99m}Tc]Tc-DB7, [^{99m}Tc]Tc-DB13, and [^{99m}Tc]Tc-DB14 (100 μL, 55.5–111 MBq, 3 nmol of total peptide in vehicle: saline/EtOH 9/1 v/v) was injected into the tail vein of healthy male Swiss albino mice, together with vehicle (100 μL; control group), or PA (100 μL of vehicle containing 300 μg PA; PA group). Animals were euthanized 5 min pi and blood was collected and immediately placed in prechilled polypropylene vials containing EDTA on ice. Samples were centrifuged at 2000× g at 4 °C for 10 min, and the plasma was collected and mixed with an equal volume of MeCN and centrifuged again for 10 min at 15,000× g at 4 °C. The supernatant was collected and concentrated to a small volume under a gentle N$_2$ flux at 40 °C, diluted with physiological saline (≈400 μL), and filtered through a Millex GV filter (0.22 μm). Suitable aliquots of the filtrate were analyzed by RP-HPLC on a Symmetry Shield RP18 (5 μm, 3.9 mm × 20 mm) column (Waters, Germany), eluted at a flow rate of 1 mL/min and adopting gradient system 2: 100% A/0% B to 50% A/50% B in 50 min; A = 0.1% TFA in H$_2$O and B = MeCN (system 2). The elution time (t_R) of intact radioligand was determined by coinjection with a sample of the labeling reaction solution.

4.3.2. Biodistribution in SCID Mice Bearing PC-3 Xenografts

Suspensions of freshly harvested PC-3 cells (~150 μL, 1.4 × 10^7) in normal saline was subcutaneously inoculated into the flanks of 6 week old SCID mice (NCSR "Demokritos" Animal House, 16.01 ± 2.59 g body weight). After 3–4 weeks, palpable PC-3 tumors (108.49 ± 0.04 mg) had developed at the inoculation sites and biodistribution was performed. At the day of the experiment, a bolus of [^{99m}Tc]Tc-DB7, [^{99m}Tc]Tc-DB13, or [^{99m}Tc]Tc-DB14 (180–230 kBq, 10 pmol total peptide, in vehicle: saline/EtOH 9/1 v/v) was intravenously injected into the tail of each mouse together with either vehicle (100 μL; control group), PA (300 μg PA dissolved in 100 μL vehicle; PA group), or excess [Tyr^4]BBN (50 μg [Tyr^4]BBN dissolved in 100 μL vehicle; block group). Animals were euthanized at 4 h pi in groups of four and blood samples, organs of interest, and tumors were dissected, weighed, and counted in the gamma counter. Biodistribution data were calculated as percent of injected activity per gram of tissue (%IA/g) with the aid of suitable standards of the injected dose, using the Microsoft Excel program. Results represent average values ± SD, $n = 4$.

4.3.3. Statistical Analysis

For statistical analysis of biological results, a two-way ANOVA with multiple comparisons was used applying Tukey's post hoc analysis (GraphPad Prism Software, San Diego, CA, USA). p-values of <0.05 were considered to be statistically significant.

All animal studies were performed in compliance with European guidelines in supervised and licensed facilities (EL 25 BIO 021), and the study protocols were approved by the Department of Agriculture and Veterinary Service of the Prefecture of Athens (protocol numbers #1609 for the stability studies and #1610 for biodistribution and imaging studies)

Author Contributions: Conceptualization, B.A.N. and T.M.; methodology, B.A.N., P.K., E.L., A.K. and T.M.; validation, B.A.N., P.K., T.M., and M.d.J.; formal analysis, P.K.; investigation, P.K., A.K., E.L., B.A.N., and T.M.; writing—original draft preparation, T.M.; writing—review and editing, all authors; visualization, P.K., A.K., B.A.N., and T.M.; supervision, B.A.N., T.M., M.d.J., and E.P.K.; project administration, B.A.N., T.M. and E.P.K. All authors have read and agreed to the published version of the manuscript.

Funding: This research received no external funding.

Conflicts of Interest: The authors declare no conflict of interest.

References

1. Markwalder, R.; Reubi, J.C. Gastrin-releasing peptide receptors in the human prostate: Relation to neoplastic transformation. *Cancer Res.* **1999**, *59*, 1152–1159. [PubMed]
2. Körner, M.; Waser, B.; Rehmann, R.; Reubi, J.C. Early over-expression of GRP receptors in prostatic carcinogenesis. *Prostate* **2014**, *74*, 217–224. [CrossRef] [PubMed]
3. Beer, M.; Montani, M.; Gerhardt, J.; Wild, P.J.; Hany, T.F.; Hermanns, T.; Muntener, M.; Kristiansen, G. Profiling gastrin-releasing peptide receptor in prostate tissues: Clinical implications and molecular correlates. *Prostate* **2012**, *72*, 318–325. [CrossRef] [PubMed]
4. Gugger, M.; Reubi, J.C. Gastrin-releasing peptide receptors in non-neoplastic and neoplastic human breast. *Am. J. Pathol.* **1999**, *155*, 2067–2076. [CrossRef]
5. Halmos, G.; Wittliff, J.L.; Schally, A.V. Characterization of bombesin/gastrin-releasing peptide receptors in human breast cancer and their relationship to steroid receptor expression. *Cancer Res.* **1995**, *55*, 280–287.
6. Mattei, J.; Achcar, R.D.; Cano, C.H.; Macedo, B.R.; Meurer, L.; Batlle, B.S.; Groshong, S.D.; Kulczynski, J.M.; Roesler, R.; Dal Lago, L.; et al. Gastrin-releasing peptide receptor expression in lung cancer. *Arch. Pathol. Lab. Med.* **2014**, *138*, 98–104. [CrossRef]
7. Reubi, J.C.; Körner, M.; Waser, B.; Mazzucchelli, L.; Guillou, L. High expression of peptide receptors as a novel target in gastrointestinal stromal tumours. *Eur. J. Nucl. Med. Mol. Imaging* **2004**, *31*, 803–810. [CrossRef]
8. Maina, T.; Nock, B.A. From bench to bed: New gastrin-releasing peptide receptor-directed radioligands and their use in prostate cancer. *PET Clin.* **2017**, *12*, 205–217. [CrossRef]
9. Zhang, J.; Singh, A.; Kulkarni, H.R.; Schuchardt, C.; Müller, D.; Wester, H.J.; Maina, T.; Rosch, F.; van der Meulen, N.P.; Müller, C.; et al. From bench to bedside-the Bad Berka experience with first-in-human studies. *Semin. Nucl. Med.* **2019**, *49*, 422–437. [CrossRef]
10. Chatalic, K.L.; Kwekkeboom, D.J.; de Jong, M. Radiopeptides for imaging and therapy: A radiant future. *J. Nucl. Med.* **2015**, *56*, 1809–1812. [CrossRef]
11. Moreno, P.; Ramos-Alvarez, I.; Moody, T.W.; Jensen, R.T. Bombesin related peptides/receptors and their promising therapeutic roles in cancer imaging, targeting and treatment. *Expert. Opin. Ther. Targets* **2016**, *20*, 1055–1073. [CrossRef]
12. Delle Fave, G.; Annibale, B.; de Magistris, L.; Severi, C.; Bruzzone, R.; Puoti, M.; Melchiorri, P.; Torsoli, A.; Erspamer, V. Bombesin effects on human gi functions. *Peptides* **1985**, *6* (Suppl. 3), 113–116. [CrossRef]
13. Bruzzone, R.; Tamburrano, G.; Lala, A.; Mauceri, M.; Annibale, B.; Severi, C.; de Magistris, L.; Leonetti, F.; Delle Fave, G. Effect of bombesin on plasma insulin, pancreatic glucagon, and gut glucagon in man. *J. Clin. Endocrinol. Metab.* **1983**, *56*, 643–647. [CrossRef] [PubMed]
14. Bitar, K.N.; Zhu, X.X. Expression of bombesin-receptor subtypes and their differential regulation of colonic smooth muscle contraction. *Gastroenterology* **1993**, *105*, 1672–1680. [CrossRef]
15. Maina, T.; Nock, B.A.; Kulkarni, H.; Singh, A.; Baum, R.P. Theranostic prospects of gastrin-releasing peptide receptor-radioantagonists in oncology. *PET Clin.* **2017**, *12*, 297–309. [CrossRef] [PubMed]
16. Fani, M.; Nicolas, G.P.; Wild, D. Somatostatin receptor antagonists for imaging and therapy. *J. Nucl. Med.* **2017**, *58* (Suppl. 2), 61S–66S. [CrossRef] [PubMed]
17. Wang, L.H.; Coy, D.H.; Taylor, J.E.; Jiang, N.Y.; Moreau, J.P.; Huang, S.C.; Frucht, H.; Haffar, B.M.; Jensen, R.T. Des-Met carboxyl-terminally modified analogues of bombesin function as potent bombesin receptor antagonists, partial agonists, or agonists. *J. Biol. Chem.* **1990**, *265*, 15695–15703. [PubMed]
18. Wang, L.H.; Coy, D.H.; Taylor, J.E.; Jiang, N.Y.; Kim, S.H.; Moreau, J.P.; Huang, S.C.; Mantey, S.A.; Frucht, H.; Jensen, R.T. Desmethionine alkylamide bombesin analogues: A new class of bombesin receptor antagonists with potent antisecretory activity in pancreatic acini and antimitotic activity in Swiss 3T3 cells. *Biochemistry* **1990**, *29*, 616–622. [CrossRef] [PubMed]
19. Nock, B.; Nikolopoulou, A.; Chiotellis, E.; Loudos, G.; Maintas, D.; Reubi, J.C.; Maina, T. [^{99m}Tc]demobesin 1, a novel potent bombesin analogue for grp receptor-targeted tumour imaging. *Eur. J. Nucl. Med. Mol. Imaging* **2003**, *30*, 247–258. [CrossRef]
20. Maina, T.; Bergsma, H.; Kulkarni, H.R.; Mueller, D.; Charalambidis, D.; Krenning, E.P.; Nock, B.A.; de Jong, M.; Baum, R.P. Preclinical and first clinical experience with the gastrin-releasing peptide receptor-antagonist [^{68}Ga]SB3 and PET/CT. *Eur. J. Nucl. Med. Mol. Imaging* **2016**, *43*, 964–973. [CrossRef]

21. Cescato, R.; Maina, T.; Nock, B.; Nikolopoulou, A.; Charalambidis, D.; Piccand, V.; Reubi, J.C. Bombesin receptor antagonists may be preferable to agonists for tumor targeting. *J. Nucl. Med.* **2008**, *49*, 318–326. [CrossRef] [PubMed]
22. Bakker, I.L.; van Tiel, S.T.; Haeck, J.; Doeswijk, G.N.; de Blois, E.; Segbers, M.; Maina, T.; Nock, B.A.; de Jong, M.; Dalm, S.U. In vivo stabilized SB3, an attractive GRPR antagonist, for pre- and intra-operative imaging for prostate cancer. *Mol. Imaging Biol.* **2018**, *20*, 973–983. [CrossRef] [PubMed]
23. Nock, B.A.; Charalambidis, D.; Sallegger, W.; Waser, B.; Mansi, R.; Nicolas, G.P.; Ketani, E.; Nikolopoulou, A.; Fani, M.; Reubi, J.C.; et al. New gastrin releasing peptide receptor-directed [^{99m}Tc]demobesin 1 mimics: Synthesis and comparative evaluation. *J. Med. Chem.* **2018**, *61*, 3138–3150. [CrossRef] [PubMed]
24. Nock, B.; Maina, T. Tetraamine-coupled peptides and resulting ^{99m}Tc-radioligands: An effective route for receptor-targeted diagnostic imaging of human tumors. *Curr. Top. Med. Chem.* **2012**, *12*, 2655–2667. [CrossRef]
25. Roques, B.P.; Noble, F.; Dauge, V.; Fournie-Zaluski, M.C.; Beaumont, A. Neutral endopeptidase 24.11: Structure, inhibition, and experimental and clinical pharmacology. *Pharmacol. Rev.* **1993**, *45*, 87–146.
26. Roques, B.P. Zinc metallopeptidases: Active site structure and design of selective and mixed inhibitors: New approaches in the search for analgesics and anti-hypertensives. *Biochem. Soc. Trans.* **1993**, *21 Pt 3*, 678–685. [CrossRef]
27. Suda, H.; Aoyagi, T.; Takeuchi, T.; Umezawa, H. Letter: A thermolysin inhibitor produced by actinomycetes: Phosphoramidon. *J. Antibiot. (Tokyo)* **1973**, *26*, 621–623. [CrossRef]
28. Nock, B.A.; Maina, T.; Krenning, E.P.; de Jong, M. "To serve and protect": Enzyme inhibitors as radiopeptide escorts promote tumor targeting. *J. Nucl. Med.* **2014**, *55*, 121–127. [CrossRef]
29. Kaloudi, A.; Lymperis, E.; Kanellopoulos, P.; Waser, B.; de Jong, M.; Krenning, E.P.; Reubi, J.C.; Nock, B.A.; Maina, T. Localization of ^{99m}Tc-GRP analogs in GRPR-expressing tumors: Effects of peptide length and neprilysin inhibition on biological responses. *Pharmaceuticals* **2019**, *12*, 42. [CrossRef]
30. Lymperis, E.; Kaloudi, A.; Sallegger, W.; Bakker, I.L.; Krenning, E.P.; de Jong, M.; Maina, T.; Nock, B.A. Radiometal-dependent biological profile of the radiolabeled gastrin-releasing peptide receptor antagonist SB3 in cancer theranostics: Metabolic and biodistribution patterns defined by neprilysin. *Bioconjug. Chem.* **2018**, *29*, 1774–1784. [CrossRef]
31. Lymperis, E.; Kaloudi, A.; Kanellopoulos, P.; Krenning, E.P.; de Jong, M.; Maina, T.; Nock, B.A. Comparative evaluation of the new GRPR-antagonist ^{111}In-SB9 and ^{111}In-AMBA in prostate cancer models: Implications of in vivo stability. *J. Labelled Comp. Radiopharm.* **2019**, *62*, 646–655. [CrossRef] [PubMed]
32. Lymperis, E.; Kaloudi, A.; Kanellopoulos, P.; de Jong, M.; Krenning, E.P.; Nock, B.A.; Maina, T. Comparing Gly^{11}/dAla^{11}-replacement vs. the in-situ neprilysin-inhibition approach on the tumor-targeting efficacy of the ^{111}In-SB3/^{111}In-SB4 radiotracer pair. *Molecules* **2019**, *24*, 1015. [CrossRef] [PubMed]
33. Chatalic, K.L.; Konijnenberg, M.; Nonnekens, J.; de Blois, E.; Hoeben, S.; de Ridder, C.; Brunel, L.; Fehrentz, J.A.; Martinez, J.; van Gent, D.C.; et al. In vivo stabilization of a gastrin-releasing peptide receptor antagonist enhances PET imaging and radionuclide therapy of prostate cancer in preclinical studies. *Theranostics* **2016**, *6*, 104–117. [CrossRef] [PubMed]
34. Mitran, B.; Rinne, S.S.; Konijnenberg, M.W.; Maina, T.; Nock, B.A.; Altai, M.; Vorobyeva, A.; Larhed, M.; Tolmachev, V.; de Jong, M.; et al. Trastuzumab cotreatment improves survival of mice with PC-3 prostate cancer xenografts treated with the GRPR antagonist ^{177}Lu-DOTAGA-PEG2-RM26. *Int. J. Cancer* **2019**, *145*, 3347–3358. [CrossRef] [PubMed]
35. Valkema, R.; Froberg, A.; Maina, T.; Nock, B.A.; de Blois, E.; Melis, M.L.; Konijnenberg, M.W.; Koolen, S.L.W.; Peeters, R.P.; de Herder, W.W.; et al. Clinical translation of the pepprotect: A novel method to improve the detection of cancer and metastases by peptide scanning under the protection of enzyme inhibitors. *Eur. J. Nucl. Med. Mol. Imaging* **2019**, *46* (Suppl. 1), S701–S702. [CrossRef]
36. Shipp, M.A.; Tarr, G.E.; Chen, C.Y.; Switzer, S.N.; Hersh, L.B.; Stein, H.; Sunday, M.E.; Reinherz, E.L. CD10/neutral endopeptidase 24.11 hydrolyzes bombesin-like peptides and regulates the growth of small cell carcinomas of the lung. *Proc. Natl. Acad. Sci. USA* **1991**, *88*, 10662–10666. [CrossRef]
37. Linder, K.E.; Metcalfe, E.; Arunachalam, T.; Chen, J.; Eaton, S.M.; Feng, W.; Fan, H.; Raju, N.; Cagnolini, A.; Lantry, L.E.; et al. In vitro and in vivo metabolism of Lu-AMBA, a GRP-receptor binding compound, and the synthesis and characterization of its metabolites. *Bioconjug. Chem.* **2009**, *20*, 1171–1178. [CrossRef]
38. Fleischmann, A.; Laderach, U.; Friess, H.; Buechler, M.W.; Reubi, J.C. Bombesin receptors in distinct tissue compartments of human pancreatic diseases. *Lab. Investig.* **2000**, *80*, 1807–1817. [CrossRef]

39. Varasteh, Z.; Rosenstrom, U.; Velikyan, I.; Mitran, B.; Altai, M.; Honarvar, H.; Rosestedt, M.; Lindeberg, G.; Sörensen, J.; Larhed, M.; et al. The effect of mini-PEG-based spacer length on binding and pharmacokinetic properties of a ^{68}Ga-labeled NOTA-conjugated antagonistic analog of bombesin. *Molecules* **2014**, *19*, 10455–10472. [CrossRef]
40. Jamous, M.; Tamma, M.L.; Gourni, E.; Waser, B.; Reubi, J.C.; Maecke, H.R.; Mansi, R. PEG spacers of different length influence the biological profile of bombesin-based radiolabeled antagonists. *Nucl. Med. Biol.* **2014**, *41*, 464–470. [CrossRef]
41. Gourni, E.; Mansi, R.; Jamous, M.; Waser, B.; Smerling, C.; Burian, A.; Buchegger, F.; Reubi, J.C.; Maecke, H.R. N-terminal modifications improve the receptor affinity and pharmacokinetics of radiolabeled peptidic gastrin-releasing peptide receptor antagonists: Examples of ^{68}Ga- and ^{64}Cu-labeled peptides for PET imaging. *J. Nucl. Med.* **2014**, *55*, 1719–1725. [CrossRef] [PubMed]
42. Abiraj, K.; Mansi, R.; Tamma, M.L.; Fani, M.; Forrer, F.; Nicolas, G.; Cescato, R.; Reubi, J.C.; Maecke, H.R. Bombesin antagonist-based radioligands for translational nuclear imaging of gastrin-releasing peptide receptor-positive tumors. *J. Nucl. Med.* **2011**, *52*, 1970–1978. [CrossRef] [PubMed]
43. Reile, H.; Armatis, P.E.; Schally, A.V. Characterization of high-affinity receptors for bombesin/gastrin releasing peptide on the human prostate cancer cell lines PC-3 and DU-145: Internalization of receptor bound ^{125}I-(Tyr^4)bombesin by tumor cells. *Prostate* **1994**, *25*, 29–38. [CrossRef] [PubMed]

Sample Availability: Samples of the compounds are not available from the authors.

© 2020 by the authors. Licensee MDPI, Basel, Switzerland. This article is an open access article distributed under the terms and conditions of the Creative Commons Attribution (CC BY) license (http://creativecommons.org/licenses/by/4.0/).

Article

Design, Synthesis, Computational, and Preclinical Evaluation of natTi/45Ti-Labeled Urea-Based Glutamate PSMA Ligand

Kristina Søborg Pedersen [1], Christina Baun [2,3], Karin Michaelsen Nielsen [1], Helge Thisgaard [2,3], Andreas Ingemann Jensen [1] and Fedor Zhuravlev [1,*]

1. Department of Health Technology, Technical University of Denmark, Frederiksborgvej 399, Building 202, 4000 Roskilde, Denmark; krped@dtu.dk (K.S.P.); kamnie@dtu.dk (K.M.N.); atije@dtu.dk (A.I.J.)
2. Department of Clinical Research, University of Southern Denmark, Sønder Boulevard 29, DK-5000 Odense, Denmark; Christina.Baun@rsyd.dk (C.B.); Helge.Thisgaard@rsyd.dk (H.T.)
3. Department of Nuclear Medicine, Odense University Hospital, DK-5000 Odense, Denmark
* Correspondence: fezh@dtu.dk; Tel.: +45-4677-5337

Received: 31 January 2020; Accepted: 28 February 2020; Published: 2 March 2020

Abstract: Despite promising anti-cancer properties in vitro, all titanium-based pharmaceuticals have failed in vivo. Likewise, no target-specific positron emission tomography (PET) tracer based on the radionuclide ^{45}Ti has been developed, notwithstanding its excellent PET imaging properties. In this contribution, we present liquid–liquid extraction (LLE) in flow-based recovery and the purification of ^{45}Ti, computer-aided design, and the synthesis of a salan-^{nat}Ti/^{45}Ti-chelidamic acid (CA)-prostate-specific membrane antigen (PSMA) ligand containing the Glu-urea-Lys pharmacophore. The compound showed compromised serum stability, however, no visible PET signal from the PC3+ tumor was seen, while the ex vivo biodistribution measured the tumor accumulation at 1.1% ID/g. The in vivo instability was rationalized in terms of competitive citrate binding followed by Fe(III) transchelation. The strategy to improve the in vivo stability by implementing a unimolecular ligand design is presented.

Keywords: titanium-45; PSMA; PET

---

1. Introduction

Since its inception in 1975 [1], PET has become one of the most useful and rapidly developing diagnostic modalities in the field of oncology, cardiology, and neurology [2]. Following its clinical success and rapid development, the global market for PET radiopharmaceuticals grew from $1.5 billion in 2014 to an estimated $2.3 billion in 2019 and is expected to reach $2.8 billion by 2022 [3]. PET radiopharmaceuticals (tracers) provide functional imaging of disease by precise molecular targeting of the affected tissue. To match the time-scale of the biological process under investigation, PET tracers utilize a broad spectrum of positron-emitting radionuclides. Among these, radiometals are particularly attractive as they combine widely varying half-lives with an ease of radiolabeling via kit formulations [4]. While the radionuclides ^{68}Ga and ^{82}Rb still dominate the traditional clinical setting, the use of ^{64}Cu [5] and ^{89}Zr [6] is on the rise in university clinics and clinical trials, and a plethora of PET tracers based on more unconventional PET radiometals such as ^{44}Sc, ^{45}Ti, ^{55}Co, and ^{86}Y are in development [7]. The radionuclide ^{45}Ti occupies a special place among unconventional PET radiometals, featuring 85% β$^+$ decay, negligible secondary radiation, and low β$^+$ endpoint energy (1.04 MeV), which translates into high spatial resolution as evidenced by sharp Derenzo phantom images [8–10]. With its 3.1 h half-life, ^{45}Ti is well-suited for the radiolabeling of small molecules, peptides, and antibody fragments. The three hours half-life also allows for regional distribution and longer transport distances compared to ^{68}Ga, and makes it possible to perform delayed PET imaging. The case in point is PSMA-based PET diagnostics, where imaging 3 h post-injection (p.i.) is considered to produce the highest contrast allowing the discovery of additional lesions [11]. The production of ^{45}Ti via the ^{nat}Sc(p,n)^{45}Ti reaction is convenient and high-yielding, allowing for 4.6 GBq of activity after one hour of irradiation of Sc foil using medical cyclotron (11.8 MeV protons, 20 µA) [10]. The naturally monoisotopic Sc is inexpensive and readily available.

The imaging and production advantages of ^{45}Ti are offset by the chemical challenges of recovering the radionuclide from the irradiated Sc matrix. The high oxophilicity of Ti^{4+} (Θ = 1.0) [12] and propensity for hydrolysis require strong acidities of the digestion solution ([H$^+$] > 8 M) to avoid the formation of titanyl species [13]. The separation of ^{45}Ti from Sc has been previously performed using various resins. The cation exchange resin AG 50W-X8 allowed for as much as 92.3% recovery of the ^{45}Ti activity from the 6 M HCl. However, the IR analysis indicated the predominance of the titanyl [^{45}Ti]TiOCl$_2$ species in the dried-down sample [8]. A hydroxamic acid resin with oxalic acid elution was also used, but the yields were modest (42%–56%) [10,14]. Higher recovery yields (93%) could be obtained on a HypoGel 1,3-diol resin in organic solvents, but the procedure required on-resin chelation chemistry. Recently, we reported the efficient LLE of ^{45}Ti from Sc-containing 12 M HCl using a guaiacol/anisole mixture [15]. The extraction is very fast and can be performed manually at low radioactivity levels. By using a membrane-based separator with integrated pressure control, we achieved a reliable and robust extraction with clean phase separation at high flow rates and GBq levels of activity. However, the purification of the product from the guaiacol/anisole mixture could be tedious and require preparative HPLC.

The efficient and stable chelation of titanium also presents a challenge. The cyclen family of popular chelators, such as 1,4,7,10-Tetraazacyclododecane-1,4,7,10-tetraacetic acid (DOTA) and its derivatives, has not demonstrated any utility for chelating Ti^{4+}. Although the tetradentate diamino bis-phenolato family of ligands (salan) do allow for the fast and quantitative chelation of Ti^{4+}, the resulting cytotoxic Ti(salan) complexes are prone to hydrolysis [16]. Supplementing salan with dipicolinic acid (dipic) leads to a heptacoordinate geometry around titanium and to a significant boost in the hydrolytic stability of the resulting salan-Ti-dipic, which retains its cytotoxicity [17]. Disappointingly, the PET radio-isotopomer [^{45}Ti]salan-Ti-dipic showed negligible accumulation in the tumor tissue and fast hepatobiliary excretion [18]. The difficulties in recovery and purification and the absence of an in-vivo stable bifunctional chelator for ^{45}Ti highlights the challenges of bringing this otherwise promising PET radionuclide into clinics and explains why no target-specific ^{45}Ti PET tracers have been reported so far.

In this study, we aimed at overcoming the ^{45}Ti purification challenge by the recovery of radiotitanium using LLE. We attempted to develop a hydrolytically stable, target-specific ^{45}Ti-based tracer by modifying the previously reported salan-dipic scaffold into a bifunctional chelating moiety facilitating the labeling of a targeting vector. Given the growing importance of PSMA as a molecular targeting vector, we sought to design the first ^{45}Ti-containing PSMA PET tracer by accessing its molecular compatibility with the glutamate carboxypeptidase II (GCPII, also known as PSMA) active site in silico, synthesize the compound, and test its performance in vivo using ^{45}Ti PET and ex vivo organ biodistribution.

2. Results

2.1. Flexible Docking of Salan-Ti-CA-PSMA into the Active Site of GCPII

The active site of the human recombinant GCPII was modeled using the crystal structure from the Protein Data Bank (PDB ID: 5O5T, 1.43Å resolution) containing PSMA-1007 (**1**) as a cognate ligand. Since the OPLS3 force field lacked the necessary parameters for Ti, the structure (**2**), containing the Glu-urea-Lys pharmacophore, the linker (Scheme 1, blue), and the chelidamic acid (CA) residue (Scheme 1, red) were used for molecular docking. Then, the docked structure was further functionalized by attaching the salan-Ti-CA moiety at the benzylic ester linkage forming compound (3), which is henceforth referred to as salan-Ti-CA-PSMA (3).

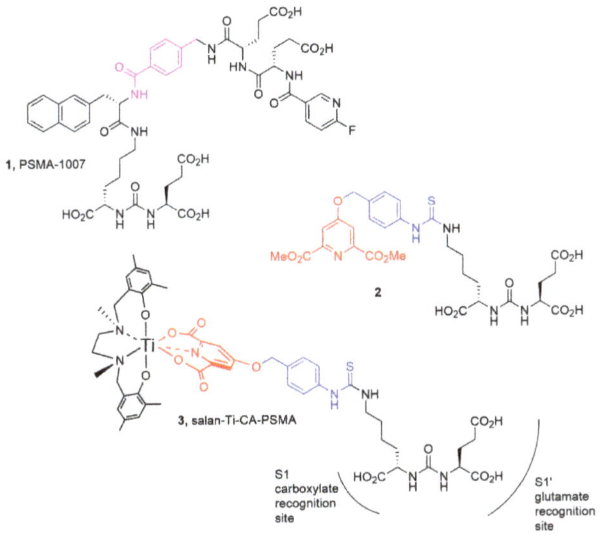

Scheme 1. The structures used in the flexible docking studies (protein: PDB ID: 5O5T). The linker fragments are colored blue and pink; the chelidamic acid (CA) residue is red. PDB: Protein Data Bank.

The docking studies revealed striking similarities between the mode of active site binding in (3) and in the co-crystallized (1), PSMA-1007 (Figure 1). The carbonyl group of the urea is anchored to Zn^{2+} (815), in both structures, while the carboxylate groups of the P1' formed hydrogen bonds to Arg 210, Asn 257, Tyr 552, Tyr 700, and Lys 699 of the S1'glutamate recognition site of the protein. The interaction with Arg 463 locks in place the thiourea carbonyl in (3) and the corresponding amide carbonyl in (1) forcing the benzylic groups to become coplanar and occupy almost the same space in the active site. The significant difference in binding between (1) and (3) is in the positioning of the terminal residues. While the fluoropyridine ring of (1) is forced into a small pocket formed by Trp 541, Arg 511, and Arg 463, the bulky salan-Ti-CA chelate is protruding into a much larger opening shaped by Arg 511, Tyr 709, Arg 181, and Lys 545. The structure (3) also lacks the stabilizing interaction of the naphthalene residue with the hydrophobic accessory pocket of the protein, which can affect the overall binding.

2.2. Synthesis of CA-PSMA and Nonradioactive Salan-Ti-CA-PSMA

The synthesis of the chelating pharmacophore CA-PSMA (10) is presented in Scheme 2. The room temperature O-alkylation of the chelidamic acid diethyl ether (4) [19] with benzylic bromide (5) [20] yielded the corresponding isothiocyanate (6), which was coupled with (7) and synthesized as described by Maresca et al. [21] to yield (8). The ethyl esters were saponified with LiOH in tetrahydrofuran (THF)/water, giving (9) followed by removal of the *tert*-butyl groups with trifluoroacetic acid (TFA) in dichloromethane (DCM) to give CA-PSMA (10). Overall, the synthesis of CA-PSMA (10) was performed in eight steps with the longest linear sequence of six steps and the combined yield of 28%. The synthesis duration and relatively low overall yield challenge the potential translation into routine production, and further optimizations of the synthesis are necessary to facilitate an efficient upscaling.

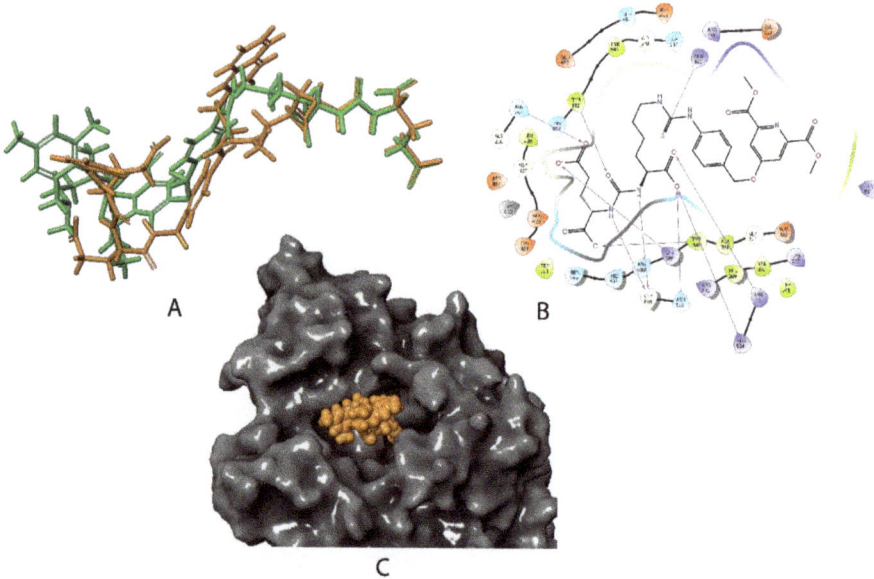

Figure 1. (**A**): Overlay of the best poses for **2** (green) with the cognate ligand **1** (PSMA-1007) (orange) as it is co-crystallized with glutamate carboxypeptidase II (GCPII) in the active site of PSMA (PDB ID: 5O5T); (**B**): the corresponding binding interaction of **2** with the active site schematically depicted in 2D; (**C**): the molecular surface of the protein (gray) and the best pose for **3** represented as CPK (Corey-Pauling-Koltun) model (orange).

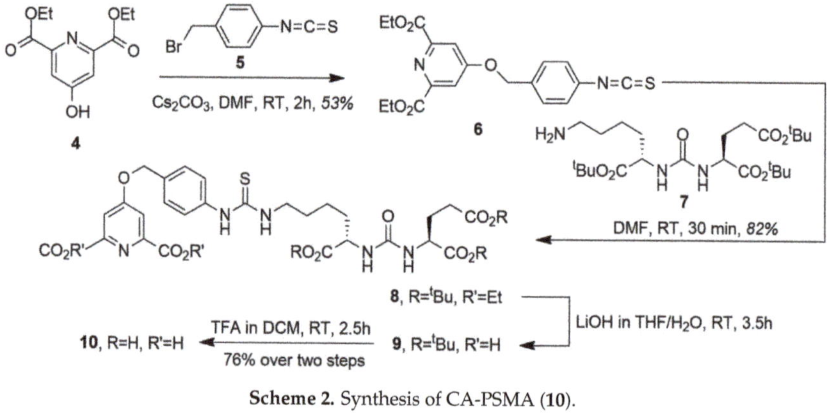

Scheme 2. Synthesis of CA-PSMA (**10**).

The (salan)Ti(O^{i}Pr)$_2$ complex (**11**) synthesized as described by Chmura et al. [22] was chosen as a precursor for the synthesis of the target compound (**3**) (Scheme 3). In the crowded steric environment imposed by salan, the reaction of (**11**) with (**10**) required elevated temperature producing after 30 min the orange-red salan-Ti-CA-PSMA (**3**) in 66% yield. The compound was further used as reference material for the radio-TLC and radio-HPLC identification of [^{45}Ti]salan-Ti-CA-PSMA ([^{45}Ti]-**3**).

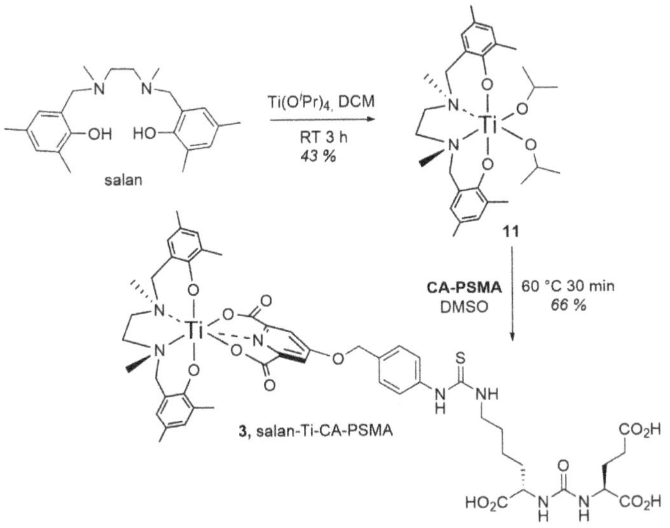

Scheme 3. Synthesis of salan-Ti-CA-PSMA (3).

2.3. Production of ^{45}Ti and Synthesis of [^{45}Ti]salan-Ti-CA-PSMA

^{45}Ti was produced by the (p,n) nuclear reaction from scandium foil on a 16.5 MeV GE PETtrace cyclotron. The proton energy was degraded with an aluminum foil to approximately 13 MeV [23] in order to decrease the co-production of titanium-44 ($T_{\frac{1}{2}}$ = 60 y) by the (p, 2n) nuclear reaction [24]. An experimental saturation yield of 731 ± 234 MBq/µA for thin scandium foil (127 µm) and 1429 ± 397 MBq/µA for thick scandium foil (250 µm) was obtained from irradiation for 10–30 min at 15–20 µA. This corresponded to 66 ± 21% and 64 ± 18%, respectively, of the theoretical saturation yield, which was estimated from cross-sections from the EXFOR database and calculated stopping power using SRIM (Stopping and Range of Ions in Matter) software. It was attempted to achieve a larger ratio of produced ^{45}Ti to the mass of scandium to facilitate the purification by LLE. A large amount of scandium challenges the purification of ^{45}Ti, since the HCl concentration decreases when dissolving scandium due to the formation of $ScCl_3$ and H_2, and the extraction of ^{45}Ti is less efficient at HCl concentrations below 11 M. Furthermore, the amount of nonradioactive titanium introduced by the scandium foil could be reduced and thereby increase the molar activity. Hence, the thin scandium foil was preferred and the size of the foil was reduced from 1–0.7 cm^2 to 0.4 cm^2, which decreased the yield of the ^{45}Ti production due to the increased risk of not hitting the foil with the entire proton beam profile.

^{45}Ti was separated from scandium by LLE with guaiacol/anisole 9/1 (v/v) either in batch or by using a Zaiput membrane separator, as described previously [15]. The extraction efficiency with different ratios between the aqueous and organic phases is shown in Table 1. The extraction efficiency was improved with the increased organic phase volume. However, for the further purification of the radiotracer, a large volume of organic solvent was challenging. Hence, a ratio of 1:1.33 (aqueous/organic phase) (v/v) was preferred. Another caveat is the addition of a small amount (16 % v/v) of pyridine to the ^{45}Ti-containing guaiacol/anisole mixture. This neutralizes the traces of HCl in the organic phase, which otherwise might interfere with subsequent radiotitanation.

The synthesis of [^{45}Ti]salan-Ti-CA-PSMA ([^{45}Ti]-3) generally followed a previously developed procedure [18], where the radionuclide was consecutively treated with salan and then with CA-PSMA (10) dissolved in DMSO. Different conditions were tested including reaction temperature (60 and 80 °C) and time (5–60 min), concentration of salan and CA-PSMA (2–15 mM), and the addition of pyridine.

Table 1. The average extraction efficiency of ^{45}Ti at different ratios between the aqueous and organic phases. The mass of scandium foil, the HCl concentration after dissolving the scandium foil, the mass of scandium per volume, and the number of experiments are listed.

Sc Foil (mg)	Concentration of HCl after Dissolving Foil (M)	Sc/HCl (mg/mL)	Aq:org Ratio	Average Extraction %	Number of Experiments
42–80	10.9–12.0	2.6–18.7	1:3	77.7 ± 10.8	7
22–36	11.4–11.7	7.3–12.0	1:1.33	68.0 ± 6.5	5
31–57	11.2–12.0	2.3–14.3	1:1	59.0 ± 5.8	8

A radiolabeling yield of 47% was obtained after 60 min at 60 °C with 15 mM of salan and CA-PSMA. No radiolabeling was observed when the concentration of salan and CA-PSMA was decreased to 7.6 mM. When increasing the temperature to 80 °C, a radiolabeling yield of 80%–85% was observed after 15–60 min with 15 mM of salan and CA-PSMA, while no radiolabeling was observed with 2 mM of salan and CA-PSMA. A higher radiolabeling yield could be obtained with 4 mM of salan and CA-PSMA at 80 °C by adding pyridine 1/5 (v/v) to the guaiacol/anisole phase, and we found that a reaction time of 15 min was suitable for the formation of [^{45}Ti]salan-Ti-CA-PSMA ([^{45}Ti]-3) (Scheme 4). A total of 86 ± 4 % of ^{45}Ti was present as [^{45}Ti]salan-Ti-CA-PSMA, which was quantified and identified by radio-HPLC/HPLC with the nonradioactive salan-Ti-CA-PSMA (3) as reference material (Figure 2).

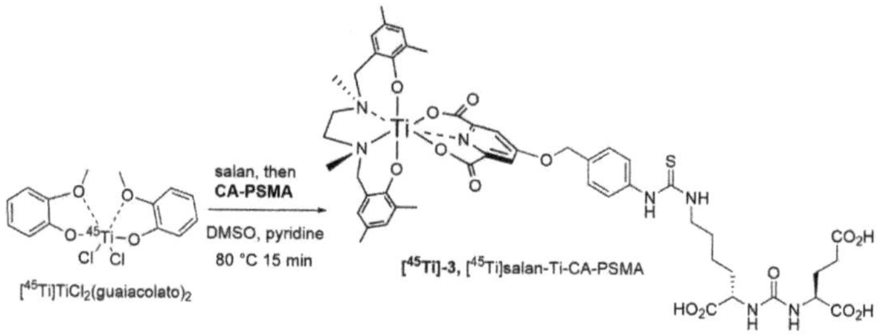

Scheme 4. Radiosynthesis of [^{45}Ti]salan-Ti-CA-PSMA ([^{45}Ti]-3).

Since the radiolabeling was performed in a relatively large volume (6 mL) of high boiling point solvents, an efficient solvent change method was sought after. To that end, the absorption on the silica, alumina N, C18, and QMA (CO_3^{2-}) cartridges was tested first. However, the [^{45}Ti]salan-Ti-CA-PSMA ([^{45}Ti]-3) had either low affinity for the cartridge (C18 and QMA (CO_3^{2-})) or was difficult to elute afterwards (silica and alumina N). The preparative HPLC (C18 column, eluents: 0.1% formic acid in Milli-Q water (A) and ACN (B)) proved to be the method of choice for the purification of [^{45}Ti]salan-Ti-CA-PSMA. The fractions collected from the preparative HPLC were analyzed by analytical radio-HPLC/HPLC in order to identify the fractions containing the product by comparing the retention time to the one for the nonradioactive salan-Ti-CA-PSMA (3) reference material. The fractions containing the product were collected and concentrated on a C18 cartridge and eluted in EtOH/H_2O. The final formulation of [^{45}Ti]salan-Ti-CA-PSMA was prepared in phosphate-buffered saline (PBS)/EtOH 9/1 (v/v) at pH 7.5. A radiochemical yield (RCY) of 13.0 ± 5.6% decay corrected (d.c.) and 5.1 ± 2.3% non-decay corrected (n.d.c.) was obtained after purification and formulation. The relatively low RCY was partially due to the LLE step, where 25%–40% of the total ^{45}Ti was not extracted into the organic phase, and it was partially due to the preparative HPLC, where only 20%–50% of the purified [^{45}Ti]salan-Ti-CA-PSMA had a radiochemical purity (RCP) above 95%.

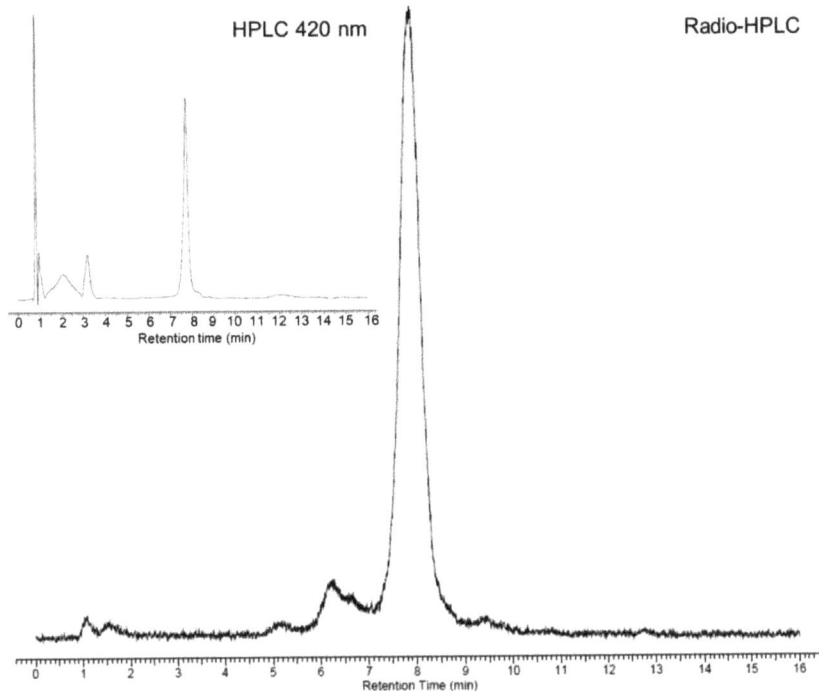

Figure 2. The radio-HPLC chromatogram of the radiolabeling solution after 15 min at 80 °C. The retention time of the peak corresponding to [^{45}Ti]salan-Ti-CA-PSMA ([^{45}Ti]-3) was 7.9 min. Insert: The UV signal at 420 nm of the HPLC chromatogram for the nonradioactive Salan-Ti-CA-PSMA (3). The retention time was 7.7 min. The time difference is due to the radioactivity detector being placed after the UV detector.

2.4. Analyses and Stability Study of [^{45}Ti]salan-Ti-CA-PSMA

The RCP of the final product in PBS buffer was 96 ± 3% (n = 7). The radionuclidic purity of the product was determined by gamma spectroscopy using a germanium detector and was found to be 100% as only background radiation and the three most abundant gamma lines of ^{45}Ti (511.0, 719.6, and 1408.1 keV) [25] were detected. The molar activity of the product was 110 ± 59 GBq/µmol (n = 7) according to the amount of nonradioactive salan-Ti-CA-PSMA (3) in the final formulation of [^{45}Ti]salan-Ti-CA-PSMA ([^{45}Ti]-3) calculated by HPLC. The molar activity was also calculated from the amount of titanium in the product found by inductively coupled plasma optical emission spectrometry (ICP-OES) and was calculated to be 107 GBq/µmol. The nonradioactive titanium content was expected to arise mainly from impurities from the scandium foil; however, the molar activity was anticipated to be sufficient for the animal studies. The amount of other selected metals was also measured by ICP-OES. The amount of scandium was <8.3 µg, the amount of zinc was <5.0 µg, and the amount of iron was 2.0 ± 2.1 µg as found by ICP-OES.

The octanol/water partition coefficient (log P) of [^{45}Ti]salan-Ti-CA-PSMA ([^{45}Ti]-3) was measured at 1.7 in PBS buffer at pH = 7.5.

The stability of [^{45}Ti]salan-Ti-CA-PSMA in PBS buffer and in mouse serum/PBS buffer 1/1 (v/v) at 37 °C and in PBS buffer at room temperature was studied over 4 h by radio-TLC (Figure 3). The percentage of [^{45}Ti]salan-Ti-CA-PSMA compared to other [^{45}Ti]Ti-compounds was gradually decreasing over time both in PBS buffer and in mouse serum to 33% in PBS buffer and to 80% in mouse serum/PBS buffer 1/1 (v/v) after 4 h at 37 °C. The decomposition rate of [^{45}Ti]salan-Ti-CA-PSMA in PBS buffer was appreciably slower at room temperature with 69% [^{45}Ti]salan-Ti-CA-PSMA remaining intact after 4 h. The radio-TLC showed that the ^{45}Ti-containing decomposition products remained at the baseline, suggesting titanium de-chelation and hydrolysis.

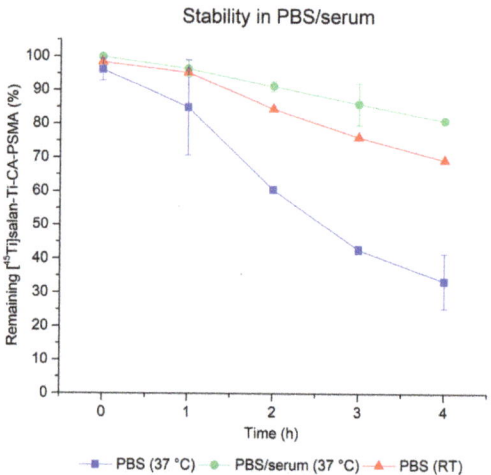

Figure 3. The stability of [^{45}Ti]salan-Ti-CA-PSMA ([^{45}Ti]-**3**) in PBS buffer at 37 °C (**blue squares**) and at room temperature (**red triangles**) and in PBS/mouse serum 1/1 (*v/v*) at 37 °C (**green circles**). The [^{45}Ti]salan-Ti-CA-PSMA was quantified by radio-TLC using nonradioactive salan-Ti-CA-PSMA (**3**) as reference material.

2.5. In Vivo Study in Mice

The in vivo distribution of [^{45}Ti]salan-Ti-CA-PSMA ([^{45}Ti]-**3**) in PSMA-positive tumor-bearing mice was conducted with four animals, which all received the same batch of [^{45}Ti]salan-Ti-CA-PSMA. The radiotracer was administered at 4, 5.5, 6, and 7.5 h after end of synthesis (EOS), respectively, with a maximal radiochemical purity of 69% (4 h after EOS). The molar activity was 14–30 GBq/µmol depending on the injection time. The microPET scans (Figure 4) showed a high radioactivity accumulation in the gallbladder and intestine. However, no visible uptake of the radiotracer in the tumor, located at the left shoulder of each mouse, was observed. The ex vivo biodistribution (post-mortem) for all four mice was analyzed by measuring the injected dose per gram tissue (% ID/g) of the different organs analyzed (Figure 4). A low amount of radioactivity (1.1 ± 0.1% ID/g) was observed in the tumor tissue; however, this amount was not notably higher than that for the other organs, and it may arise from the blood content. A significant amount of radioactivity per weight of organ was found in the blood (4.0 ± 0.6% ID/g) and in the liver (3.4 ± 0.3% ID/g), while the highest accumulation per weight of organ was observed in the gallbladder (92.7 ± 40.5% ID/g). The gallbladder of the mouse no. 1 was not analyzed since the high accumulation of radioactivity in the gallbladder was noticed after the analysis of the microPET scan of this mouse. A relative large variation between the three other mice was observed for this organ. This variation was expected to be due to the low mass of the gallbladder (6–7 mg), which could cause uncertainty on the mass of the organ. The PET images and the ex vivo biodistribution of the four mice indicated that [^{45}Ti]salan-Ti-CA-PSMA ([^{45}Ti]-**3**) was metabolized and/or excreted through the hepatobiliary system.

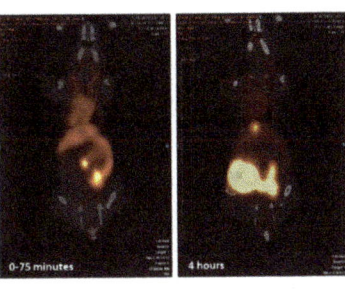
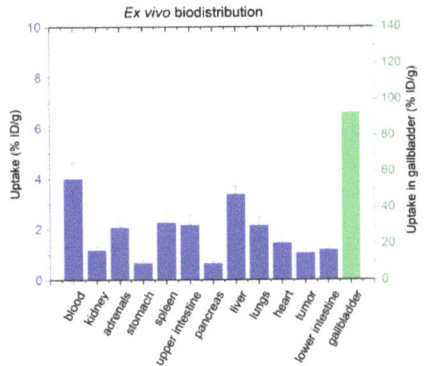

Figure 4. (**left**) PET/CT images (coronal plane) of one of the tumor-bearing mice (mouse no. 1) receiving [^{45}Ti]Salan-Ti-CA-PSMA ([^{45}Ti]-3). Mouse no. 1 was dynamically scanned 0–75 min (left) and four hours (right) post-injection (p.i.); (**right**) The ex vivo biodistribution 4 h p.i. shown as the uptake of ^{45}Ti in the percentage of injected dose per weight of tissue. The values were an average for the four mice, and the error bars represent the standard deviation. The uptake in the gallbladder (green) can be read off the right y-axis.

3. Discussion

The design of the presented PSMA ligand was guided by two major considerations. Given the hydrolytic stability observed in previous studies [18], we based our chelator design on the salan-Ti-dipic molecular scaffold. The functionalization with the Glu-urea-Lys residue required an exchange of dipic for a structurally related chelidamic acid ester (**4**), which underwent smooth O-alkylation with the benzylic isothiocyanate (**5**) and consecutive room temperature coupling with the pharmacophore residue Glu-urea-Lys (**7**). The order of deprotection was important requiring the saponification to (**9**) in order to prevent significant loss of yield, followed by the removal of three ^{t}Bu groups to give CA-PSMA (**10**). The complexation with salan of both nonradioactive and ^{45}Ti was almost immediate at room temperature, but the attachment of CA-PSMA required forcing conditions, which was likely due to the crowded steric environment imposed by the octahedrally coordinated salan. The LLE of ^{45}Ti proved to be a reliable way to recover and purify the radionuclide from 12 M HCl containing a significant amount of Sc. The extraction was found to be scalable and could be performed manually at low levels of activity, or via the syringe pump-driven membrane separator [15]. The latter could be remotely controlled and implemented as a part of an automated radiosynthesis module.

For tracers relying on tight active site binding such as PSMA, the geometrical requirements of the chelator and the length of the linker may play the deciding role. The molecular docking studies we performed on (**2**) clearly indicated that the P1' and the P1 residues of (**2**) bind in the same active conformation, as does the co-crystallized ligand PSMA-1007 (Figure 1A). The signature interactions with the Zn^{2+}; Arg 210, Asn 257, Tyr 552, and Tyr 700; and Lys 699 of the S1'glutamate recognitions site, as well as with Asn 519 and Arg 534 of the S1 carboxylate recognition site of the protein were well reproduced (Figure 1B). Furthermore, the aromatic linker of (**2**) (Scheme 1, blue) was found to be co-planar to the benzylic linker in the co-crystallized PSMA-1007 (Scheme 1, pink), making the backbones of (**1**) and (**2**) well-aligned. This alignment fidelity was also reflected in the comparable DockScores of those compounds, (−8.9 and −12.7, respectively), predicting (**2**) to be a strong binder (comparing those values with the DockScore −0.7 for a typical non-binding decoy molecule). However, the salan-Ti-CA-PSMA (**3**) structurally differs from (**2**) in the attachment of the chelate salan-Ti-CA. The obvious concern was that the chelate might be too sterically big to fit inside the active site of the enzyme. Using quantum chemistry calculations we estimated that salan-Ti-CA will occupy 780 Å^3 of space. This volume is significantly larger than that Ga-DOTA occupies in the strong binder Ga-PSMA-617 (560 Å^3). However, the molecular modeling indicated that the linker is long enough for the chelate to extend beyond the narrow binding channel of the active site and occupy a large funnel-shaped entrance of the enzyme (Figure 1C). The closest neighboring side-chains of the enzyme are situated as far as 5 Å from the chelate, relieving it from any steric clashes and the need for conformational changes in the protein.

One feature shared by many clinically tested PSMA PET tracers is their high hydrophilicity: ^{68}Ga-PSMA-11 (logP = −4.3) and ^{68}Ga-PSMA-617 (logP = −3.4). This predisposition to aqueous media leads to increased accumulation in the kidney and the bladder and can interfere with the detection of prostate lesions. Therefore, the second design consideration was an attempt to moderate the high hydrophilicity imposed by the Glu-urea-Lys pharmacophore. The previous efforts to increase hydrophobicity to improve tumor uptake and internalization focused on placing the hydrophobic residues within the backbone of the inhibitor [26]. This affected binding often in an unpredictable way. To avoid this unpredictability of modifying the linker moiety, we limited ourselves to the non-binding chelate functionality and retained the four methyl groups the salan ligand provided. The measured logP = 1.7 puts salan-Ti-CA-PSMA (3) in the category of highly hydrophobic PSMA agents, yet still within the general drug-like Lipinski logP range of values (−1 to +5). Overall, the salan-Ti-CA-PSMA structure was found to be sterically and electronically compatible with the active site of GCII and judged to be an appropriate candidate for in-vivo testing.

The decrease in the concentration of [^{45}Ti]salan-Ti-CA-PSMA ([^{45}Ti]-3) in PBS/serum at 37 °C (approximately 20% loss in 4 h) was a sign of the limited stability of the synthesized compound in vitro (Figure 3). This was later confirmed by the PET and the ex vivo studies. The disappointing uptake of [^{45}Ti]salan-Ti-CA-PSMA in the tumor parallels that observed for salan-Ti-dipic we reported earlier [18]. In both cases, the uptake of the radiolabeled material monitored by PET was heavily dominated by the liver and gallbladder with no visible accumulation in the tumor (Figure 4, left). The comparison of the *ex-vivo* biodistribution between the two ^{45}Ti-labeled compounds showed that tumor uptake rose from 0.18% ID/g for [^{45}Ti]salan-Ti-dipic to 1.1% ID/g for [^{45}Ti]salan-Ti-CA-PSMA (Figure 4, right). The similar PET and ex vivo biodistribution profile for the two compounds, which share essentially the same chelate functionality, strongly suggests that the chelate is the culprit of the in vivo instability and hepatobiliary excretion we observed. The competitive binding hypothesis can be used to rationalize the observed behavior. Citrate is known to be a potent molecular binder for Ti(IV) [27] present in blood in approximately 100 μM concentrations [28]. The competitive substitution of the bidentate dipic/CA by the citrate would be close to thermoneutral and driven by the μM concentration of citrate versus sub-nM amounts of ^{45}Ti-labeled material. The Tinoco group has recently reported that for highly hydrolytically stable Ti(deferasirox)$_2$, the citrate binding facilitates the transmetallation of Ti(IV) by labile Fe(III), which is present inside the cytosol [29]. A similar process is likely to occur here. After the citrate substitution, [^{45}Ti]salan-Ti-CA-PSMA ([^{45}Ti]-3) loses its CA-PSMA (10) moiety converting into hydrophobic [^{45}Ti]salan-Ti(citrate), which is swept from the bloodstream by hepatocytes. A small portion of [^{45}Ti]salan-Ti(citrate) reacts, further releasing [^{45}Ti]Ti(citrate)$_2$$^{2-}$, which is picked up by transferrin [28] and remains in the blood. In the hepatocytes, [^{45}Ti]salan-Ti(citrate) loses its radiotitanium by Fe(III) transmetallation in the cytosol. The hypothesis presented above suggests a strategy for improving the in vivo stability of the titanium chelate. Finding a stronger enthalpic binder for titanium would be an obvious chemical modification. A more practical and perhaps more productive solution is to design a unimolecular chelator by tethering the salan and the CA functionality. This would make the intermolecular citrate substitution thermodynamically non-competitive, effectively blocking the transmetallation of the chelated titanium.

4. Materials and Methods

4.1. Synthesis of CA-PSMA (10) and salan-Ti-CA-PSMA (3)

4.1.1. General

All commercially available materials were used as received without further purification. The materials were purchased from Sigma Aldrich (Schnelldorf, Germany), except for glutamic acid di-*t*-butyl ester hydrochloride (98%), O^{t}Bu-Lys-Cbz · HCl (97%), *p*-Tolyl isothiocyanate (97%), carbon tetrachloride (99%), chelidamic acid monohydrate (95%), and lithium hydroxide (anhydrous, 95%), which were purchased from abcr GmbH (Karlsruhe, Germany). Anhydrous ethanol from Fluka, methanol (LiChroSolv), DMSO (99%), and TLC plates (silica gel and silica gel 60 RP-18, F$_{254}$) were purchased from Merck (Darmstadt, Germany), and heptane (99.7%), *n*-hexane (97%), chloroform (99.2%), dichloromethane (100%), ethyl acetate (99.9%), acetonitrile (99.9% gradient grade for liquid chromatography), and toluene (HiPer Chromanorm) from VWR were used. Various cartridges (alumina N, C18 plus, QMA, silica plus) were purchased from Waters (Milford, MA, USA) and from ABX (Radeberg, Germany) (QMA preconditioned with CO_3^{2-}).

NMR spectra were recorded with an Agilent 400 MR spectrometer (Santa Clara, CA, USA) operating at 400.445 MHz (^{1}H). For analytical HPLC, a Hitachi Chromaster (Tokyo, Japan) equipped with a Carroll and Ramsey 105-S radio-detector and a Hitachi 5430 double diode array detector with a Phenomenex Luna 3 μ C18 (2) column was applied. The UV-signal at 270 and 420 nm was recorded. The eluents used were 0.1% TFA in Milli-Q water (A) and 0.1% TFA in ACN (B). The following gradient was used with a flow rate of 0.5 mL/min: 0–0.1 min: 0% B; 0.1–1 min: 0%–45% B; 1–7 min: 45%–55% B; 7–9 min: 55%–60% B; 9–10 min: 60%–100% B, 10–12 min: 100% B; 12–14: 100%–0% B; 14–16 min: 0% B.

A Shimadzu LC-MS system (Tokyo, Japan) consisting of a CBM-20A communications bus module, a SPD-20A UV-vis detector, a FRC-10A fraction collector, a LC-20AP preparative liquid chromatograph, a FCV-200AL Prep Quaternary Valve, and an LC-MS-2020 was applied for preparative HPLC and Mass spectrometry (MS). A Shim-pack GIST 5 μm C18, 20 × 250mm column was used. The eluents were 0.1% formic acid in Milli-Q water (A) and ACN (B). A flow rate of 15 mL/min and the following gradient were applied: 0–5 min: 0% B; 5–25 min: 0%–60 % B; 25–34 min: 60%–70% B, 34–35 min: 70%–90% B; 35–39 min: 90% B; 39–42 min: 90%–60% B. Fractions of 3 mL were collected automatically with the fraction collector.

4.1.2. Synthesis of diethyl-4-hydroxypyridine-2,6-dicarboxylate (4)

Compound **4** was synthesized as described previously [19]. First, 12.6 mL (173 mmol) thionyl chloride was added to 50 mL of anhydrous ethanol in a round-bottom flask under argon flow cooled in an ice bath. Then, 5.03 g (27 mmol) chelidamic acid was added in small portions, and the mixture was stirred for 5 min at 0 °C. The mixture was stirred for 20 h at room temperature, followed by 2 h at reflux. The solvent was removed under reduced pressure. The flask with the crude product was placed in an ice bath, and 50 mL water was added. The mixture was neutralized with 5 mL of 10% aqueous Na$_2$CO$_3$ and 5 mL of 50% aqueous ethanol. The precipitate was filtered and dried under reduced pressure. The product was obtained as a white solid (5.49 g (85%)). ^{1}H NMR (400 MHz, ACN-d$_3$) δ 7.65 (s, 2H), 4.39 (q, *J* = 7.1 Hz, 4H), 1.37 (t, *J* = 7.1 Hz, 6H).

4.1.3. Synthesis of 1-(bromomethyl)-4-isothiocyanatobenzene (5)

Compound **5** was synthesized as described previously [20]. First, 2.6 g (17.5 mmol) of *p*-Tolyl isothiocyanate was dissolved in 40 mL of CCl$_4$ in a round-bottom flask. Then, 3.2 g (18 mmol) *N*-bromosuccinimide was added, followed by a small amount of benzoyl peroxide. The mixture was stirred at reflux for 8 h and then filtered when cooled down to room temperature. The solvent was evaporated, and the crude product was recrystallized in methanol. The product was obtained as pale yellow needle-like crystals (1.9 g (48%)). ^{1}H NMR (400 MHz, DMSO-d$_6$) δ 7.52 (d, J = 8.6 Hz, 2H), 7.42 (d, J = 8.6 Hz, 2H), 4.72 (s, 2H). ^{13}C NMR (100 MHz, acetone-d$_6$) δ 141.1, 141.1, 139.0, 131.6, 126.9, 33.2.

4.1.4. Synthesis of diethyl 4-((4-isothiocyanatobenzyl)oxy)pyridine-2,6-dicarboxylate (6)

All starting materials and reagents were dried under vacuum. First, 1.0 g (4.2 mmol, 1 equiv.) of compound **4** was dissolved in 70 mL of dry N,N-dimethylformamide (DMF) in a round-bottom flask. Then, 2.7 g (8.3 mmol, 2 equiv.) of Cs$_2$CO$_3$ was added. The mixture was stirred for 30 min at room temperature. Then, 1.42 g (6.2 mmol, 1.5 equiv.) of compound **5** was dissolved in 6 mL of dry DMF and added dropwise. The mixture was stirred for 2 h at room temperature. The reaction mixture was filtered and DMF was removed. The crude product was dissolved in DCM, filtered, and dried. The product was isolated as a white solid (0.85 g (53%)) by column chromatography with chloroform/ethyl acetate 9/1 (v/v). ^{1}H NMR (400 MHz, DMSO-d6) δ 7.81 (s, 2H), 7.57 (d, J = 8.6 Hz, 2H), 7.49 (d, J = 8.6 Hz, 2H), 5.39 (s, 2H), 4.37 (q, J = 7.1 Hz, 4H), 1.33 (t, J = 7.1 Hz, 6H). ^{13}C NMR (100 MHz, DMSO-d6) δ 166.0, 164.1, 149.7, 135.5, 133.6, 129.8, 129.3, 126.2, 114.4, 69.3, 61.7, 14.1.

4.1.5. Synthesis of (S)-di-*tert*-butyl 2-(3-((S)-6-amino-1-*tert*-butoxy-1-oxohexane-2-yl)ureido) pen-tanedioate (7)

The method described by Maresca et al. was applied for the synthesis [21]. First, 1.24 g (4.2 mmol) of triphosgene was dissolved in 20 mL of dry DCM at 0 °C. Then, 4.22 g (11 mmol) of O^{t}Bu-Lys-Cbz · HCl and 4.3 mL (24 mmol) of N,N-diisopropylethylamine (DIPEA) in 20 mL dry DCM was added dropwise over 1 h. A solution of 3.35 g (11 mmol) OtBu$_2$-Glu · HCl and 4.3 mL (24 mmol) DIPEA in 20 mL of dry DCM was added in one portion, and the mixture was stirred for 1 h. The solvent was evaporated, and the crude product was dissolved in 60 mL of ethyl acetate and washed twice with 80 mL of 2 M NaHSO$_4$. The organic phase was washed with brine and dried over Na$_2$SO$_4$. Na$_2$SO$_4$ was removed by filtration, and the crude product was dried. The product (2.47 g (36%)) was obtained by column chromatography with hexane/ethyl acetate 2/1 (v/v) as eluent. (The product was also synthesized by first adding OtBu$_2$-Glu · HCl dropwise to the triphosgene solution and then adding O^{t}Bu-Lys-Cbz · HCl in one portion afterwards.) ^{1}H NMR (400 MHz, CDCl$_3$) δ 7.32 (m, 5H), 5.24–5.12 (m, 3H), 5.09 (d, J = 5.8 Hz, 2H), 4.37–4.28 (m, 2H), 3.22–3.12 (m, 2H), 2.36–2.20 (m, 2H), 2.10–2.00 (m, 1H), 1.88–1.71 (m, 2H), 1.66–1.56 (m, 1H), 1.54–1.47 (m, 2H), 1.46–1.41 (m, 27H), 1.40–1.30 (m, 2H). ^{13}C NMR (100 MHz, CDCl$_3$) δ 172.6, 172.5, 172.4, 157.0, 156.7, 136.8, 128.6, 128.2, 128.2, 82.3, 81.9, 80.7, 66.7, 53.4, 53.1, 40.8, 32.7, 31.7, 29.5, 28.5, 28.2, 28.1, 22.4.

For Cbz deprotection, 0.63 g (1.0 mmol) of ammonium formate was suspended in 10 mL of EtOH. Then, 0.62 g (1 mmol) of the Cbz-protected compound from the previous step was dissolved in 10 mL of ethanol and added to the suspension. Then, 67 mg of 10% Pd-C was added, and the mixture was stirred at room temperature overnight. The mixture was filtered through a celite pad, and the solvent was removed under reduced pressure. The remaining ammonium formate was removed by dissolving the crude product in 50 mL of DCM and washing with 50 mL of 1 M Na$_2$CO$_3$ aqueous solution. The organic phase was washed with brine and dried over Na$_2$SO$_4$. The solvent was removed under reduced pressure, and the product was obtained (0.41 g (84%)). ^{1}H NMR (400 MHz, CDCl$_3$) δ 6.41–6.20 (m, 2H), 4.36–4.26 (m, 2H), 3.07–2.92 (m, 2H), 2.34–2.89 (m, 2H), 2.12–2.01 (m, 1H), 1.89–1.61 (m, 5H), 1.55–1.38 (m, 28H), 1.28–1.23 (m, 1H). ^{13}C NMR (100 MHz, DMSO-d$_6$) δ 172.3, 172.0, 171.5, 157.3, 80.5, 80.3, 79.8, 53.1, 52.2, 38.6, 31.3, 30.9, 27.8, 27.7, 27.6, 22.1.

4.1.6. Synthesis of diethyl 4-((4-(3-((S)-6-(*tert*-butoxy)-5-(3-((S)-1,5-di-*tert*-butoxy-1,5-dioxopentan-2-yl) ureido)-6-oxohexyl)thioureido)benzyl)oxy)pyridine-2,6-dicarboxylate (8)

First, 0.75 g (19 mmol) of compound **6** was dissolved in 20 mL of dry DMF in a round-bottom flask. Then, 0.95 g (19 mmol) of compound **7** was dissolved in 20 mL of dry DMF and added dropwise to the flask. The mixture was stirred at room temperature for 30 min. The solvent was evaporated, and the product was dried under reduced pressure. The product was purified by column chromatography with hexane/ethyl acetate 1/2 (v/v) and 1.36 g (82%) of compound **8** was obtained. ^{1}H NMR (400 MHz, DMSO-d$_6$) δ 7.81 (s, 2H), 7.44 (m, 4H), 6.28 (m, 2H), 5.31 (s, 2H), 4.38 (q, J = 7.1 Hz, 4H), 4.07–3.96 (m, 2H), 3.46 (m, 2H), 2.29–2.15 (m, 2H), 1.91–1.82 (m, 1H), 1.71–1.50 (m, 5H), 1.40–1.37 (m, 28H), 1.34 (t, J = 7.1Hz, 6H), 1.36–1.29 (m, 1H). ^{13}C NMR (100 MHz, DMSO-d$_6$) δ 180.3, 172.2, 171.9, 171.4, 166.1, 164.1, 157.1, 149.7, 139.5, 130.8, 128.6, 122.7, 114.3, 80.6, 80.3, 79.7, 70.0, 61.6, 53.1, 52.1, 43.6, 31.8, 28.1, 27.7, 27.6, 22.5, 14.1.

4.1.7. Synthesis of 4-((4-(3-((S)-6-(*tert*-butoxy)-5-(3-((S)-1,5-di-*tert*-butoxy-1,5-dioxopentan-2-yl)ureido)-6-oxohexyl)thioureido)benzyl)oxy)pyridine-2,6-dicarboxylic acid (9)

First, 0.25 g (0.29 mmol) of compound **8** was dissolved in 7 mL of THF and 14 mg (0.58 mmol, 2.2 equiv.) LiOH in 3 mL water was added. The mixture was stirred for 3.5 h at room temperature. The solvent was removed under reduced pressure, and the product was obtained as the lithium salt (0.22 g (91%)) without further purification. ^{1}H NMR (400 MHz, DMSO-d$_6$) δ 7.60 (d, J = 8.0 Hz, 2H), 7.55 (s, 2H), 7.36 (d, J = 8.4 Hz 2H), 6.40 (m, 2H), 5.20 (s, 2H), 4.00 (m, 2H), 3.46 (m, 2H), 2.30–2.14 (m, 2H), 1.90–1.81 (m, 1H), 1.71–1.50 (m, 5H), 1.40–1.36 (m, 27H), 1.36–1.27 (m, 2H). ^{13}C NMR (100 MHz, DMSO-d$_6$) δ 180.4, 172.3, 171.9, 171.5, 167.6, 166.2, 157.2, 156.4, 140.0, 128.0, 122.5, 110.0, 80.6, 80.3, 79.7, 69.2, 53.2, 52.2, 43.5, 31.8, 30.9, 30.4, 28.2, 27.7, 27.7, 22.7.

4.1.8. Synthesis of 4-((4-(3-((S)-5-carboxy-5-(3-((S)-1,3-dicarboxypropyl)ureido)pentyl)thioureido)-benzyl)oxy)pyridine-2,6-dicarboxylic acid (CA-PSMA) (10)

First, 0.20 g (0.24 mmol) of compound **9** was mixed with 6 mL of TFA/DCM 1/1 (v/v) and stirred at room temperature for 2.5 h. The product was precipitated in 30 mL of cold diethyl ether and washed three times with cold diethyl ether. The product was dried and the product was obtained as a white solid (0.13 g (84%)). ^{1}H NMR (400 MHz, DMSO-d$_6$) δ 7.74 (s, 2H), 7.48 (d, J = 8.7 Hz, 2H), 7.41 (d, J = 8.5 Hz, 2H), 6.36 (m, 2H), 5.32 (s, 2H), 4.08 (m, 2H), 3.44 (m, 2H), 2.29–2.19 (m, 2H), 1.96–1.87 (m, 1H), 1.76–1.64 (m, 2H), 1.60–1.50 (m, 3H), 1.38–1.28 (m, 2H). ^{13}C NMR (100 MHz, DMSO-d$_6$) δ 180.3, 174.5, 174.2, 173.7, 170.1, 167.4, 164.5, 157.3, 150.3, 139.4, 128.6, 122.6, 113.2, 70.2, 52.2, 51.6, 43.7, 31.9, 29.9, 28.2, 27.5, 22.8.

4.1.9. Synthesis of (6,6′-((ethane-1,2-diylbis(methylazanediyl))bis(methylene))bis(2,4-dimethyl-phenol))-diisopropoxide titanium(IV) (salan)Ti(O^{i}Pr)$_2$ (11)

(salan)Ti(O^{i}Pr)$_2$ (**11**) was synthesized as described by Chmura et al. [22]. First, 0.44 g (1.5 mmol) of titanium isopropoxide was transferred to a round-bottom flask under argon flow and dissolved in 12 mL of dry DCM. Then, 0.54 g (1.5 mmol) of salan was dissolved in 3 mL of dry DCM and added to the round-bottom flask. The mixture was stirred for 3 h at room temperature. The solvent was evaporated, and the crude product was recrystallized in heptane. The product was obtained as yellow crystals (0.33 g (43 %)). ^{1}H NMR (400 MHz, CDCl$_3$) δ 6.89 (d, J = 1.5 Hz, 2H), 6.61 (d, J = 1.5 Hz, 2 H), 5.17 (sept., J = 6.1 Hz, 2H), 4.69 (d, J = 13.3 Hz, 2H), 3.05 (d, J = 13.3 Hz, 2H), 3.04 (d, J = 9.4 Hz, 2H), 2.41 (s, 6H), 2.24 (s, 6H), 2.22 (s, 6H), 1.73 (d, J = 9.5 Hz, 2H), 1.28 (d, J = 6.1 Hz, 6H), 1.20 (d, J = 6.1 Hz, 6H).

4.1.10. Formation of salan-Ti-CA-PSMA (3)

First, 20 mg (0.04 mmol) (salan)Ti(^{i}OPr)$_2$ (**11**) and 25 mg (0.04 mmol) of CA-PSMA (**10**) were dissolved in 1.5 mL of DMSO. The reaction mixture was stirred for 30 min at 60 °C. The product was precipitated in diethyl ether and washed twice with diethyl ether. Then, 20 mg of the crude product was purified by preparative TLC (C18 plates) using ACN/H$_2$O 1/1 (v/v) as eluent. The product was collected from the plate (R$_f$: 0.5-0.6), eluted from the C18 silica with acetone, filtered, and dried (13 mg (66 %)). ^{1}H NMR (400 MHz, DMSO-d$_6$) δ 7.69 (s, 2H), 7.45 (m, 4H), 6.78 (s, 2H), 6.76 (s, 2H), 6.33

(m, 2H), 5.47 (d, J = 11.7 Hz, 1H), 5.42 (d, J = 11.7 Hz, 1H), 4.92 (d, J = 13.7 Hz, 2H), 4.08 (m, 2H), 3.44 (m, 2H), 3.32 (d, J = 13.8 Hz, 2H), 3.12 (d, J = 8.8 Hz, 2H), 2.6 (s, 6H), 2.3 (d, J = 8.9 Hz, 2H), 2.27–2.20 (m, 2H), 2.15 (s, 6H), 1.95–1.87 (m, 1H), 1.83 (s, 6H), 1.75–1.64 (m, 2H), 1.58–1.49 (m, 3H), 1.35–1.28 (m, 2H). ^{13}C NMR (100 MHz, DMSO-d_6) δ 180.2, 174.6, 174.3, 173.8, 170.6, 167.9, 157.3, 155.5, 150.9, 139.7, 135.5, 130.2, 129.6, 128.9, 127.9, 127.2, 123.8, 122.8, 111.4, 71.4, 53.1, 52.3, 51.7, 45.9, 31.9, 30.0, 28.2, 27.6, 22.8, 20.3, 15.4. MS (m/z): [M + H]$^+$ calcd for $C_{49}H_{59}N_7O_{14}S$, 1050.96; found, 1050.

The radiolabeling conditions were tested, where 2 mL of 0.01 M (0.02 mmol) TiCl$_4$ in 12 M HCl was mixed with 2 mL of guaiacol/anisole 9/1 (v/v) for 10 min. The mixture was centrifuged, and the two phases were separated with a pipette. Then, 0.4 mL pyridine was added to 1 mL of the organic phase. Then, 4 mg (0.01 mmol) salan in 0.4 mL of DMSO was added, and the mixture was stirred at 80 °C for 5 min. Then, 7.5 mg (0.01 mmol) of CA-PSMA (**10**) in 0.4 mL of DMSO was added, and the reaction mixture was stirred for 30 min. The product was purified using preparative HPLC. The fractions were analyzed by analytical HPLC, the fractions containing the product were collected, and the solvent was evaporated. The product was dried and analyzed by NMR, which gave identical spectra to Salan-Ti-CA-PSMA (**3**) synthesized by the previous method.

4.2. Production of ^{45}Ti and Synthesis of [^{45}Ti]salan-Ti-CA-PSMA ([^{45}Ti]-3)

4.2.1. General

The scandium foil (250 μm and 127 μm, 99.9% pure, rare earth analysis) was purchased from Alfa Aesar, and the aluminum foil (500 μm, 99.9%) was purchased from VWR. Hydrochloric acid (37%, trace metal basis) was purchased from Honeywell, and guaiacol and anisole (99%) from Sigma Aldrich were used for the LLE of ^{45}Ti. In addition, 15 mL plastic centrifuge tubes with screw caps (SuperClear) were purchased from VWR, and the membrane separator was purchased from Zaiput Flow Technologies. Perfluoroalkoxy alkane (PFA) diaphragms from McMaster Carr., Pall polytetrafluoroethylene (PTFE) membranes (0.2 μm pore size, 139 μm thickness, PTFE, polypropylene (PP) support), PFA tubing (1/16" OD, 0.03" ID and 1/8" OD, 1/16" ID) from Idex Health and Science, and static mixers (PTFE, 10 element, 1.7 cm length, placed inside a piece of 1/8" OD, 1/16" ID PFA tubing) from Stamixco were used. Pyridine (anhydrous, 99.8%) and ICP standards (multi-element standard solution 1 for ICP (TraceCert in 10% HNO$_3$), titanium standard for ICP (1000 mg/L, TraceCert in 2% HNO$_3$), and scandium standard for ICP (1000 mg/L, TraceCert in 5% HNO$_3$)) from Sigma Aldrich were purchased.

^{45}Ti was produced with a GE 16.5 MeV PETtrace cyclotron. A CRC-55tR, CII Capintec, Inc. dose calibrator and a Princeton Gammatech LGC 5 germanium detector were used to measure the radioactivity. For the batch LLE, an IKA ROCKER 3D digital shaker for phase mixing and an Eppendorf 5702 centrifuge for phase separation were applied. For the LLE in flow, KDS 100 Legacy Syringe Pumps were used. A Perkin Elmer Cyclone Plus imager and a Wolf Trimline LED illuminator were applied for radio-TLC. A Thermo Scientific iCAP 6000 Series ICP Optical Emission Spectrometer was used for ICP-OES measurements. For microPET/CT scans, a Siemens INVEON multimodality scanner in a docked mode (Siemens pre-clinical solutions, Knoxville, TN, US) was used.

4.2.2. Production and Purification of ^{45}Ti

First, 0.3–2.0 GBq ^{45}Ti was produced from 250 μm or 127-μm thick scandium foil by the ^{45}Sc(p,n)^{45}Ti nuclear reaction using a 16.5 MeV GE PETtrace cyclotron. A 500-μm thick aluminum foil was used to degrade the beam energy to approximately 13 MeV. The mass of the scandium foil used in the target was between 22 and 80 mg and the foil was cut into squares with scissors. The scandium foil was irradiated for 20–30 min at 15–20 μA and was dissolved in 3–6 mL 12 M HCl within 5-10 min. The solution was filtered and centrifuged. ^{45}Ti was separated from scandium by LLE with guaiacol/anisole 9/1 (v/v) either in flow by the membrane-based separation method or in batch mode. A Zaiput membrane separator with a 2 mil thick PFA diaphragm for pressure control was used for the LLE in flow. The organic and the aqueous phase were loaded in respective syringes and pumped through PFA tubing using two syringe pumps. An organic flow rate of 0.75 mL/min and aqueous flow rate of 0.25 mL/min were applied. The two phases were mixed when entering a polyether ether ketone (PEEK) tee and passed two 10-element static mixers placed in a piece of tubing. Hereafter, steady slug flow was developed throughout 108 cm of PFA tubing, where the two phases were mixed further. Then, the mixed phases entered the separator, where the organic phase permeated a hydrophobic PTFE/PP) membrane with a pore size of 0.2 µm, while the aqueous phase was retained. The organic phase was collected from the permeate outlet and the aqueous phase was collected from the retentate outlet.

For the LLE in batch mode, the two phases were mixed for 10 min using a digital shaker, centrifuged, and separated with a pipette. The ratio between the organic and aqueous phase was 1:1, 1:1.33, or 1:3 (aq:org) with the dissolved, irradiated scandium foil in 12 M HCl as the aqueous phase and guaiacol/anisole 9/1 (v/v) as the organic phase.

4.2.3. Formation and Purification of [^{45}Ti]salan-Ti-CA-PSMA ([^{45}Ti]-3)

The ^{45}Ti-labeling was performed by mixing 0.5 mL or 2 mL of the organic phase (guaiacol/anisole) containing ^{45}Ti (60–130 MBq/mL, molar activity: 0.5–1.1 GBq/µmol) from the LLE with 8 mg (0.02 mmol) of salan dissolved in 0.5 mL of DMSO at 60 °C for 5 min. Then, 15 mg (0.02 mmol) of CA-PSMA (**10**) dissolved in 0.5 mL of DMSO was added to the reaction mixture and stirred at 60 °C for 60 min.

The ^{45}Ti-labeling was also performed at 80 °C, where the same amount of organic phase and of salan and CA-PSMA (**10**) were applied. Salan and CA-PSMA (**10**) were each dissolved in 0.5 mL or 2 mL of DMSO. Finally, the labeling was performed with 4 mL of organic phase containing ^{45}Ti (190–225 MBq/mL, molar activity: 43–220 GBq/µmol), where 0.8 mL of pyridine and 8 mg (0.02 mmol) of salan dissolved in 0.8 mL DMSO were added and stirred at 80 °C for 5 min. Then, 15 mg (0.02 mmol) of CA-PSMA (**10**) dissolved in 0.8 mL of DMSO was added to the reaction mixture and stirred for 15 min at 80 °C.

The formation of [^{45}Ti]Salan-Ti-CA-PSMA ([^{45}Ti]-3) was followed by radio-HPLC, where the retention time of the radio peak was compared to the UV signal of salan-Ti-CA-PSMA (**3**). The radiolabeling solution was diluted with toluene 1/1 (v/v) and loaded on a silica plus or an alumina N cartridge. The cartridge was washed with toluene, and the product was eluted with 20%–50% H$_2$O in MeOH or BuOH/H$_2$O/AcOH 4/1/1. It was also attempted to trap the product on a C18 plus cartridge (conditioned with 5–10 mL EtOH followed by 5–10 mL water) and elute the product with ACN/H$_2$O 6/4 (v/v). The radiolabeling solution was also loaded on a QMA cartridge preconditioned with carbonate solution and washed with water. The cartridge was washed with water and DMSO.

The radiolabeling solution was also injected onto a preparative HPLC with a C18 column, and fractions were collected automatically. The fractions were analyzed by measuring the radioactivity with a dose calibrator and evaluating the radiochemical identity by analytical radio-HPLC. The fractions containing the product were collected and diluted with H$_2$O 1/1 (v/v) before being passed through a C18 plus cartridge (conditioned with 5–10 mL ethanol followed by 5–10 mL H$_2$O). The cartridge was washed with H$_2$O and [^{45}Ti]Salan-Ti-CA-PSMA ([^{45}Ti]-3) was eluted in 1.5 mL ethanol/H$_2$O 9/1 (v/v). This solution was dried down to 50–100 µL under argon flow and was diluted with PBS buffer (pH 7.5).

4.3. Analyses and Stability Studies of [^{45}Ti]salan-Ti-CA-PSMA ([^{45}Ti]-3)

The RCP of the final product in PBS buffer after the end of synthesis and purification was analyzed by radio-HPLC. The HPLC-peak expected to arise from the [^{45}Ti]Salan-Ti-CA-PSMA ([^{45}Ti]-3) was identified by comparing the retention time to the one of the nonradioactive Salan-Ti-CA-PSMA (**3**). The RCP was calculated by dividing the area of the [^{45}Ti]Salan-Ti-CA-PSMA peak with the total peak area.

The radionuclidic purity was determined by gamma spectroscopy. A Germanium detector, which was calibrated with barium-133 and europium-152 sources, was applied. The samples containing approximately 3 MBq of ^{45}Ti were measured in a distance of 1 m for 0.3–3 h. ^{45}Ti was identified from the three gamma lines with highest intensity (511.0 keV (169.6%), 719.6 keV (0.154%), and 1408.1 keV

(0.085%) [25]). The radionuclidic purity was calculated by dividing the activity of ^{45}Ti with the total activity of the sample.

The amount of metals (Ti (323.4 nm), Sc (357.2 nm), Fe (239.5, 240.4, 259.8 nm), and Zn (202.5, 206.2, 213.8 nm)) in the final product was analyzed by ICP-OES at the listed wavelengths. Then, 50 µL of the [^{45}Ti]Salan-Ti-CA-PSMA ([^{45}Ti]-3) in the final eluate solution (EtOH/H$_2$O 9/1 (v/v)) was diluted to 10 mL with 1% HCl. The concentrations of metals were calculated from a standard curve of samples with known concentrations. The molar activity was calculated from the amount of titanium found by ICP-OES or from the concentration of Salan-Ti-CA-PSMA (3) calculated from the UV-signal of the product seen on the analytical HPLC from a standard curve.

The octanol/water partition coefficient of [^{45}Ti]Salan-Ti-CA-PSMA ([^{45}Ti]-3) was measured by taking 0.5 mL of the eluate from the C18 cartridge, which was evaporated to dryness. Then, 1 mL of PBS buffer and 1 mL octanol were added to the dried product. The two phases were mixed for 2 min and subsequently separated. The radioactivity of 0.5 mL of both phases was measured with a dose calibrator. The partition coefficient was calculated by dividing the activity in the octanol phase ($A_{45Ti,octanol}$) with the activity in the aqueous phase ($A_{45Ti,water}$), and the log P value was calculated by Equation (1).

$$\log P = \log \frac{A_{45Ti,octanol}}{A_{45Ti,water}} \tag{1}$$

The stability of [^{45}Ti]Salan-Ti-CA-PSMA ([^{45}Ti]-3) was analyzed in PBS buffer at room temperature and 37 °C and in PBS buffer/mouse serum 1/1 (v/v) and PBS/H$_2$O 1/1 (v/v) at 37 °C. The samples were analyzed by radio-TLC after 0, 1, 2, 3, and 4 h using C18 TLC plates and ACN/H$_2$O 7/3 (v/v) as eluent. The nonradioactive Salan-Ti-CA-PSMA (3) was used as a TLC reference, and the plates were analyzed using a Perkin Elmer Cyclone Plus imager.

4.4. Computational Studies

The geometry optimization for the chelate salan-Ti-CA and Ga-DOTA were performed using Turbomole 7.4.1 at the BP/TZVP level, and the molecular volume was calculated using CosmothermX 19.0.1. For molecular docking studies performed using the Schrodinger 2018 suite of programs, the crystal structure from the Protein Data Bank (PDB ID: 5O5T, 1.43Å resolution) was used. All waters of crystallization were removed except for water 1233, which was complexed to the two zinc atoms of the active site. The protein and the ligand were pre-processed using the default settings of the protein and ligand preparation routines. The flexible docking was carried out at standard precision (SP) using Glide and then refined using extra precision (XP) settings.

4.5. In Vivo Study in Mice

The in vivo studies in mice were performed at Odense University Hospital. The animal experiment license number was 2016-15-0201-01027.

Four male Balb/c nude mice (Janvier) received an injection with 5 million PC3+ cells at the left shoulder. The tumors were allowed to grow for 24 days. The tumor sizes were between 7 × 6 and 9 × 9.5 mm, and the mice weighed 25–27 g two days before the in vivo study. A Siemens INVEON multimodality scanner (Siemens pre-clinical solutions, Knoxville, TN, US) was applied for the PET/CT scans. A CT scan of 15 min was performed prior to each PET scan. The mice were anesthetized with a mixture of 1.5%–2% isoflurane and 100% oxygen during injection of the radiotracer and during the scans, where the mice were placed feet first in a prone position on a heated PET/CT animal bed. Each mouse received 0.5–1.4 MBq of [^{45}Ti]Salan-Ti-CA-PSMA ([^{45}Ti]-3) in PBS buffer containing 10% EtOH administered intravenously through the tail vein. The first mouse received the [^{45}Ti]Salan-Ti-CA-PSMA tracer 4 h EOS, while the fourth mouse received the tracer 7.5 h EOS. The two first mice were scanned by dynamic acquisition from 0–75 min followed by 20 min static scan four hours p.i. CT and PET images were co-registered using a transformation matrix and CT-based attenuation correction was applied to the PET data. The PET data was reconstructed using an OSEM3D/MAP algorithm (matrix 128 × 128, 2 OSEM3D and 18 MAP iterations). The two last mice received a 15 min static scan one hour p.i., followed by a 20 min static scan four hours p.i. After the last scanning, the mice were euthanized by cervical dislocation and dissected. The organs were collected and weighed separately. The amount of ^{45}Ti in the organs was measured with a gamma well counter.

5. Conclusions

The development of titanium pharmaceuticals and radiopharmaceuticals so far has been unsuccessful with all drug candidates consistently failing in vivo. Bringing together new methods of ^{45}Ti purification, computer-aided design, and synthesis of the new ^{45}Ti-containing PSMA ligand, this work takes another step toward creating the first target-specific ^{45}Ti PET tracer. The molecular docking of salan-Ti-CA-PSMA (3) into the active site of GCPII indicated the possibility of the signature binding of the Glu-urea-Lys pharmacophore without creating prohibitive steric interaction of the protein with the bulky salan-Ti-CA chelator. Although the new compound yielded no visible PET signature in the tumor and poor ex vivo tumor accumulation, the analysis of the data and the literature may suggest the involvement of competitive citrate substitution and Ti transchelation as the pathway for in vivo de-titanation. A chemical elaboration of the structure to create a unimolecular chelator could be a practical way to increase the in vivo stability of future ^{45}Ti-containing radiopharmaceuticals.

Author Contributions: Conceptualization: F.Z.; chemical synthesis: K.S.P.; animal studies, PET, and ex vivo analysis: C.B.; H.T., and K.S.P, isotope production: K.M.N.; molecular modeling: F.Z., supervision: F.Z., A.I.J., and H.T.; interfacing: A.I.J; manuscript drafting: F.Z. and K.S.P. All authors have read and agreed to the published version of the manuscript.

Funding: K.M.N. was supported by the Independent Research Fund Denmark, Grant 8022-00111B.

Conflicts of Interest: The authors declare no conflict of interest

References

1. Wagner, H. A brief history of positron emission tomography (PET). *Semin. Nucl. Med.* **1998**, *28*, 213–220. [CrossRef]
2. Jones, T.; Townsend, D.W. History and future technical innovation in positron emission tomography. *J. Med. Imaging* **2017**, *4*, 011013. [CrossRef] [PubMed]
3. Singh, V. *World Radiopharmaceuticals Market: Opportunities and Forecasts, 2014–2022*; Allied Market Research: London, UK, 2016.
4. Brandt, M.; Cardinale, J.; Aulsebrook, M.L.; Gasser, G.; Mindt, T.L. An overview of PET radiochemistry, part 2: Radiometals. *J. Nucl. Med.* **2018**, *59*, 1500–1506. [CrossRef] [PubMed]
5. Boschi, A.; Martini, P.; Janevik-Ivanovska, E.; Duatti, A. The emerging role of copper-64 radiopharmaceuticals as cancer theranostics. *Drug Discov. Today* **2018**, *23*, 1489–1501. [CrossRef] [PubMed]
6. La, M.T.; Tran, V.H.; Kim, H.K. Progress of Coordination and Utilization of Zirconium-89 for Positron Emission Tomography (PET) Studies. *Nucl. Med. Mol. Imaging* **2019**, *53*, 115–124. [CrossRef] [PubMed]
7. Costa, P.; Metello, L.; Alves, F.; Duarte Naia, M. Cyclotron Production of Unconventional Radionuclides for PET Imaging: The Example of Titanium-45 and Its Applications. *Instruments* **2018**, *2*, 8. [CrossRef]
8. Vāvere, A.L.; Laforest, R.; Welch, M.J. Production, processing and small animal PET imaging of titanium-45. *Nucl. Med. Biol.* **2005**, *32*, 117–122. [CrossRef]
9. Chaple, I.F.; Lapi, S.E. Production and Use of the First-Row Transition Metal PET Radionuclides 43,44Sc, 52Mn, and 45Ti. *J. Nucl. Med.* **2018**, *59*, 1655–1659. [CrossRef]
10. Chen, F.; Valdovinos, H.F.; Hernandez, R.; Goel, S.; Barnhart, T.E.; Cai, W. Intrinsic radiolabeling of Titanium-45 using mesoporous silica nanoparticles. *Acta Pharmacol. Sin.* **2017**, *38*, 907–913. [CrossRef]
11. Umbricht, C.A.; Benešová, M.; Schmid, R.M.; Türler, A.; Schibli, R.; van der Meulen, N.P.; Müller, C. 44Sc-PSMA-617 for radiotheragnostics in tandem with 177Lu-PSMA-617—Preclinical investigations in comparison with 68Ga-PSMA-11 and 68Ga-PSMA-617. *Ejnmmi Res.* **2017**, *7*, 1–10. [CrossRef]

12. Kepp, K.P. A Quantitative Scale of Oxophilicity and Thiophilicity. *Inorg. Chem.* **2016**, *55*, 9461–9470. [CrossRef] [PubMed]
13. Kislik, V. Competetive Complexation/Solvation Theory of solvent extraction. II. Solvent extraction of metals by acidic extractants. *Sep. Sci. Technol.* **2002**, *37*, 2623–2657. [CrossRef]
14. Gagnon, K.; Severin, G.W.; Barnhart, T.E.; Engle, J.W.; Valdovinos, H.F.; Nickles, R.J. 45 Ti extraction using hydroxamate resin. In Proceedings of the AIP Conference, AIP, Playa del Carmen, Mexico, 26–29 August 2012; Volume 1509, pp. 211–214.
15. Pedersen, K.S.; Imbrogno, J.; Fonslet, J.; Lusardi, M.; Jensen, K.F.; Zhuravlev, F. Liquid–liquid extraction in flow of the radioisotope titanium-45 for positron emission tomography applications. *React. Chem. Eng.* **2018**, *3*, 898–904. [CrossRef]
16. Tshuva, E.Y.; Ashenhurst, J.A. Cytotoxic Titanium(IV) Complexes: Renaissance. *Eur. J. Inorg. Chem.* **2009**, *2009*, 2203–2218. [CrossRef]
17. Immel, T.A.; Grutzke, M.; Spate, A.-K.; Groth, U.; Ohlschlager, P.; Huhn, T. Synthesis and X-ray structure analysis of a heptacoordinate titanium(iv)-bis-chelate with enhanced in vivo antitumor efficacy. *Chem. Commun.* **2012**, *48*, 5790–5792. [CrossRef]
18. Severin, G.W.; Nielsen, C.H.; Jensen, A.I.; Fonslet, J.; Kjær, A.; Zhuravlev, F. Bringing Radiotracing to Titanium-Based Antineoplastics: Solid Phase Radiosynthesis, PET and ex Vivo Evaluation of Antitumor Agent [45Ti](salan)Ti(dipic). *J. Med. Chem.* **2015**, *58*, 7591–7595. [CrossRef]
19. Rizeq, N.; Georgiades, S.N. Linear and Branched Pyridyl—Oxazole Oligomers: Synthesis and Circular Dichroism Detectable Effect on c-Myc G-Quadruplex Helicity. *Eur. J. Org. Chem.* **2016**, *2016*, 122–131. [CrossRef]
20. Wang, N.; Kähkönen, A.; Ääritalo, T.; Damlin, P.; Kankare, J.; Kvarnström, C. Polyviologen synthesis by self-assembly assisted grafting. *RSC Adv.* **2015**, *5*, 101232–101240. [CrossRef]
21. Maresca, K.P.; Hillier, S.M.; Femia, F.J.; Keith, D.; Barone, C.; Joyal, J.L.; Zimmerman, C.N.; Kozikowski, A.P.; Barrett, J.A.; Eckelman, W.C.; et al. A Series of Halogenated Heterodimeric Inhibitors of Prostate Specific Membrane Antigen (PSMA) as Radiolabeled Probes for Targeting Prostate Cancer. *J. Med. Chem.* **2009**, *52*, 347–357. [CrossRef]
22. Chmura, A.J.; Davidson, M.G.; Jones, M.D.; Lunn, M.D.; Mahon, M.F.; Johnson, A.F.; Khunkamchoo, P.; Roberts, S.L.; Wong, S.S.F. Group 4 Complexes with Aminebisphenolate Ligands and Their Application for the Ring Opening Polymerization of Cyclic Esters. *Macromolecules* **2006**, *39*, 7250–7257. [CrossRef]
23. Severin, G.W.; Fonslet, J.; Zhuravlev, F. Hydrolytically Stable Titanium-45. In Proceedings of the 15th International Workshop on Targetry and Target Chemistry, Prague, Czech Republic, 18–21 August 2014.
24. EXFOR: Experimental Nuclear Reaction Data. Available online: https://www-nds.iaea.org/exfor/ (accessed on 31 October 2019).
25. NuDat 2.7, National Laboratory National Nuclear Data Center Brookhaven. Available online: https://www.nndc.bnl.gov/nudat2/ (accessed on 3 July 2019).
26. Wirtz, M.; Schmidt, A.; Schottelius, M.; Robu, S.; Günther, T.; Schwaiger, M.; Wester, H.J. Synthesis and in vitro and in vivo evaluation of urea-based PSMA inhibitors with increased lipophilicity. *Ejnmmi Res.* **2018**, *8*, 84. [CrossRef] [PubMed]
27. Panagiotidis, P.; Kefalas, E.T.; Raptopoulou, C.P.; Terzis, A.; Mavromoustakos, T.; Salifoglou, A. Delving into the complex picture of Ti(IV)-citrate speciation in aqueous media: Synthetic, structural, and electrochemical considerations in mononuclear Ti(IV) complexes containing variably deprotonated citrate ligands. *Inorg. Chim. Acta* **2008**, *361*, 2210–2224. [CrossRef]

28. Saxena, M.; Loza-Rosas, S.A.; Gaur, K.; Sharma, S.; Pérez Otero, S.C.; Tinoco, A.D. Exploring titanium(IV) chemical proximity to iron(III) to elucidate a function for Ti(IV) in the human body. *Coord. Chem. Rev.* **2018**, *363*, 109–125. [CrossRef] [PubMed]
29. Loza-Rosas, S.A.; Vázquez-Salgado, A.M.; Rivero, K.I.; Negrón, L.J.; Delgado, Y.; Benjamín-Rivera, J.A.; Vázquez-Maldonado, A.L.; Parks, T.B.; Munet-Colón, C.; Tinoco, A.D. Expanding the Therapeutic Potential of the Iron Chelator Deferasirox in the Development of Aqueous Stable Ti(IV) Anticancer Complexes. *Inorg. Chem.* **2017**, *56*, 7788–7802. [CrossRef]

Sample Availability: Samples of the compounds are not available from the authors.

© 2020 by the authors. Licensee MDPI, Basel, Switzerland. This article is an open access article distributed under the terms and conditions of the Creative Commons Attribution (CC BY) license (http://creativecommons.org/licenses/by/4.0/).

Communication

Revisiting the Radiosynthesis of [^{18}F]FPEB and Preliminary PET Imaging in a Mouse Model of Alzheimer's Disease

Cassis Varlow [1,2], Emily Murrell [1], Jason P. Holland [3,4], Alina Kassenbrock [3], Whitney Shannon [1,5], Steven H. Liang [3], Neil Vasdev [1,3,6,*] and Nickeisha A. Stephenson [1,3,7,*]

1. Azrieli Centre for Neuro-Radiochemistry, Brain Health Imaging Centre, Centre for Addiction and Mental Health, Toronto, ON M5T 1R8, Canada; cassis.varlow@mail.utoronto.ca (C.V.); emily.murrell@camhpet.ca (E.M.); wes728@mail.usask.ca (W.S.)
2. Institute of Medical Science, University of Toronto, Toronto, ON M5S1A8, Canada
3. Division of Nuclear Medicine and Molecular Imaging, Massachusetts General Hospital & Department of Radiology, Harvard Medical School, Boston, MA 02114, USA; jason.holland@chem.uzh.ch (J.P.H.); alinakassenbrock@gmail.com (A.K.); Liang.Steven@mgh.harvard.edu (S.H.L.)
4. Department of Chemistry, University of Zurich, 8057 Zurich, Switzerland
5. Department of Chemistry, University of Saskatchewan, Saskatoon, SK S7N OX2, Canada
6. Department of Psychiatry, University of Toronto, Toronto, ON M5T-1R8, Canada
7. Department of Chemistry, The University of West Indies at Mona, Kingston 7, Jamaica
* Correspondence: neil.vasdev@utoronto.ca (N.V.); nickeisha.stephenson@uwimona.edu.jm (N.A.S.); Tel.: +416-535-8501 (ext. 30988) (N.V.); +1-876-927-1910 (N.A.S.)

Academic Editor: Svend Borup Jensen
Received: 30 January 2020; Accepted: 20 February 2020; Published: 22 February 2020

Abstract: [^{18}F]FPEB is a positron emission tomography (PET) radiopharmaceutical used for imaging the abundance and distribution of mGluR5 in the central nervous system (CNS). Efficient radiolabeling of the aromatic ring of [^{18}F]FPEB has been an ongoing challenge. Herein, five metal-free precursors for the radiofluorination of [^{18}F]FPEB were compared, namely, a chloro-, nitro-, sulfonium salt, and two spirocyclic iodonium ylide (SCIDY) precursors bearing a cyclopentyl (SPI5) and a new adamantyl (SPIAd) auxiliary. The chloro- and nitro-precursors resulted in a low radiochemical yield (<10% RCY), whereas both SCIDY precursors and the sulfonium salt precursor produced [^{18}F]FPEB in the highest RCYs of 25% and 36%, respectively. Preliminary PET/CT imaging studies with [^{18}F]FPEB were conducted in a transgenic model of Alzheimer's Disease (AD) using B6C3-Tg(APPswe,PSEN1dE9)85Dbo/J (APP/PS1) mice, and data were compared with age-matched wild-type (WT) B6C3F1/J control mice. In APP/PS1 mice, whole brain distribution at 5 min post-injection showed a slightly higher uptake (SUV = 4.8 ± 0.4) than in age-matched controls (SUV = 4.0 ± 0.2). Further studies to explore mGluR5 as an early biomarker for AD are underway.

Keywords: [^{18}F]FPEB; mGluR5; positron emission tomography (PET); iodonium-ylide; Alzheimer's Disease (AD)

---

1. Introduction

L-Glutamate is the primary excitatory neurotransmitter at the majority of excitatory synapses in the mammalian central nervous system (CNS). Signaling processes involving metabotropic glutamate occur via membrane-bound G-protein coupled receptors (GPCRs) known as metabotropic glutamate receptors (mGluRs). There are eight mGluR subtypes, of which the mGluR5 subtype is involved in the excitatory signaling cascade of intracellular calcium release, playing a vital role in brain development, learning, memory, and in maintaining synaptic plasticity [1]. Changes in mGluR5 expression have been identified in several neuropathological diseases and disorders including addiction, Parkinson's disease, post-traumatic stress disorder (PTSD), epilepsy, Huntington's disease, and Alzheimer's disease (AD), making mGluR5 an attractive target for the development of new therapeutics and for monitoring the progression of several CNS disease states [2–4]. Positron emission tomography (PET) serves as a highly sensitive and non-invasive means of monitoring the changes in mGluR5 distribution and regulation associated with pathophysiological conditions in the CNS [5,6]. [^{18}F]3-Fluoro-5-[(pyridin-3-yl)ethynyl] benzonitrile ([^{18}F]FPEB) was developed as a mGluR5 PET ligand and is widely used in clinical research [7,8].

The routine radiosynthesis of [^{18}F]FPEB has traditionally been low yielding because nucleophilic aromatic substitution by [^{18}F]fluoride is not favored as the nitrile group is positioned meta to the leaving group on the aromatic ring of the precursor. Furthermore, high temperature reactions are generally required, leading to decomposition products and radiochemical impurities. [^{18}F]FPEB is commonly synthesized from a chloro- or nitro-precursor in low radiochemical yields (RCYs, <10%) as shown in Table 1 (1 and 2, respectively). When initially reported by Merck Research Laboratories, [^{18}F]FPEB was prepared via a S_NAr reaction of an aryl-chloro precursor (**1**) with labeled potassium cryptand fluoride, [K$_{222}$][^{18}F], and K$_2$CO$_3$ as the base [9]. This reaction resulted in 5% RCY and required microwave conditions. Like the majority of others, our development of [^{18}F]FPEB for human use was initiated with the commercially available nitro-precursor and a low RCY (4%) was achieved [10]. Compound **3** is a boronic acid precursor and **4** is an arylstannane precursor that were used in copper-mediated radiofluorinations to synthesize [^{18}F]FPEB in decay-corrected automated radiochemical conversions (RCC; not isolated) of 5% and 11% respectively [11,12]. Translating these precursors for the radiopharmaceutical production of [^{18}F]FPEB will be challenging as lower RCY can be expected upon purification and formulation, while additional testing for residual metals will be required.

In our efforts to optimize the efficiency of ^{18}F-C$_{sp2}$ bond formation, we discovered that the use of harsh reaction conditions such as high temperatures and base concentrations resulted in the increased formation of radiolabeled impurities over time. This included the hydrolyzed product 3-fluoro-5-(pyridin-2-ylethynyl)benzamide, which contributed to the low yields previously obtained for both the chloro- and nitro-precursors [13]. Our development of spirocyclic iodonium ylide (SCIDY) precursors for radiofluorination of non-activated aromatic rings [14] was successfully applied to the synthesis of (**5**) as a novel precursor for [^{18}F]FPEB [13,15]. The ylide precursor showed a ten-fold increase in the radiochemical yield and a five-fold increase in molar activity (A$_m$) of [^{18}F]FPEB, compared with our traditional S_NAr reaction with the nitro-precursor [10], and was validated and translated for human use [12,13].

The aim of this study was to compare the chloro- and nitro-precursors for [^{18}F]FPEB, with our first-generation SCIDY precursor (cyclopentyl auxiliary (**5**)) and new second-generation SCIDY precursor (adamantyl auxiliary (**7**)) and a newly reported sulfonium salt precursor (**6**) for the routine production of [^{18}F]FPEB [16,17]. Only metal-free precursors were considered for this work to avoid the need for additional quality control testing of residual metals in routine radiopharmaceutical production. In light of our pilot PET imaging studies that showed increased [^{18}F]FPEB binding in a patient with early mild cognitive impairment [10], we also explored the use of [^{18}F]FPEB to detect early changes in mGluR5 expression in a transgenic murine model of AD.

Table 1. Reported Radiosyntheses of [18F]FPEB.

Compound	Precursor (X)	Reported [18F]FPEB Radiochemical Yields (RCYs)	Radiolabeling Method	Validated for Human Use	Reference
(1)	-Cl	5%	Manual	✓	[9]
(2)	-NO_2	4–10%	Automated	✓	[8,10,18,19]
(3)	-$B(OH)_2$	5% *	Automated		[11]
(4)	-$SnBu_3$	11% *	Manual		[12]
(5)	(spirocyclic iodonium ylide)	29%	Automated	✓	[13,15]
(6)	(thiophene sulfonium triflate)	55%	Manual		[16]

* not isolated.

2. Results and Discussion

2.1. Radiosyntheses of [18F]FPEB with Five Different Precursors

The chloro-precursor (1) was synthesized following a literature procedure, with the modification that conventional heating was used to drive the radiofluorination instead of the previously reported microwave conditions [9]. In our hands, a maximum 3% RCC was observed by manual synthesis at 200 °C from a 10 min reaction. In light of the low radiofluorination conversion, this precursor was not considered for automated radiosynthesis (Table 2, Entry I). While manual radiosynthesis of [18F]FPEB via the commercially available nitro-precursor (2) showed an RCC based on radio-HPLC analysis of 33% ($n = 3$) at 150 °C for a reaction time of 5 min, upon automation and isolation, the highest RCY was only 4% (Table 2, Entry II). This low RCY from the nitro-precursor is consistent with other reported production yields [8,10,18,19]. Losses experienced when translated to an automated synthesis unit may be attributed to transfer to tubing and reaction vessels. The low yielding reaction, together with the presence of radiochemical and UV active impurities, observed by HPLC, led us to abandon the optimization of [18F]FPEB radiosynthesis using the nitro-precursor and explore alternative precursors.

We recently showed that using the SCIDY chemistry with a cyclopentyl auxiliary (SPI5) (5) as the precursor for [18F]FPEB production led to a five-fold increase in RCY and a three-fold increase in molar activity (A_m) compared to the nitro-precursor. The SCIDY precursor enabled the displacement reaction with [18F]fluoride to be conducted at milder conditions (lower temperatures and shorter times), thereby minimizing the formation of radiochemical impurities. [18F]FPEB synthesis via the SCIDY SPI5 auxiliary precursor was conducted on a GE TRACERlab™ FX2 N automated synthesis module. Varying temperatures and reaction times resulted in RCYs of [18F]FPEB ranging from 19% to 23% after HPLC purification and formulation (Table 2, Entry III). [18F]FPEB production was repeated via the SPI5 precursor using the optimal conditions (100 °C, Et_4NHCO_3 as the base and a 5 min reaction time) which resulted in a RCY of 25% ± 2%, with a molar activity (Am) of = 37 ± 13 GBq/µmol (n = 3), consistent with our previous reports [13,15].

Table 2. Summary of automated radiosyntheses of [^{18}F]FPEB (this work).

Entry	Precursor (X=)	Base	Solvent	Temperature (°C)	Reaction Time (min)	RCY (%)	Am (GBq/µmol)
I	(1)	K_2CO_3/K_{222}	DMSO	200	10	N/A	N/A
II	(2)	K_2CO_3/K_{222}	DMSO	150	5	4	N/A
III	(5)	Et_4NHCO_3	DMF	80	5	23	22
		Et_4NHCO_3	DMF	80	10	19	54
		Et_4NHCO_3	DMF	100	5	25 ± 2 *	37 ± 13 *
IV	(7)	Et_4NHCO_3	DMF	100	5	24	21
V	(6)	Et_4NHCO_3	CH_3CN	80	5	15	70
		Et_4NHCO_3	CH_3CN	100	5	26	168
		$KHCO_3/K_{222}$	CH_3CN	80	5	36 ± 6 *	77 ± 35 *

* n = 3.

These automated reaction conditions were then applied to a novel SCIDY precursor for [^{18}F]FPEB radiosynthesis, bearing an adamantyl-based auxiliary (SPIAd). Experimental and mechanistic studies have shown that radiofluorination with the bulkier SPIAd auxiliary in SCIDY precursors can improve the RCY compared to the SPI5 precursors in conventional nucleophilic aromatic substitution with [^{18}F]fluoride [17,20]. The SPIAd precursor (compound 7) also resulted in a 24% RCY for [^{18}F]FPEB, and is similar to that of the SPI5 precursor (Table 2, Entry IV). Our initial semi-preparative HPLC purification conditions did not adequately separate the adamantyl precursor from [^{18}F]FPEB, and given the equivalent RCY between the two different SCIDY auxiliaries, optimization of the SPIAd precursor reaction and/or HPLC conditions were not pursued (Supplementary Materials).

A recently reported precursor (6) based on a sulfonium salt was also applied to the manual synthesis of [^{18}F]FPEB and gave a reported 55% RCY [16]. Inspired by this report, we adapted the radiochemistry precursor and methodology for automated radiofluorination with the GE TRACERlab™ FX2 N synthesis module. The reaction was conducted at varying temperatures (80 or 100 °C) for 5 min and with varying bases (Et_4NHCO_3 or $KHCO_3/K_{222}$) to establish optimal temperature and base conditions, as summarized in Table 2, with the literature synthesis providing the highest yield [16]. Using [K_{222}][^{18}F] and the milder base ($KHCO_3$) at 80 °C for 5 min (Table 2, Entry V), the sulfonium salt precursor produced [^{18}F]FPEB with a RCY of 36% ± 6% and Am = 77 ± 35 GBq/µmol (n = 3).

When considering which chemical route should be used to produce [^{18}F]FPEB for clinical research, many factors will impact this choice, including precursor availability, radiochemical yield, molar activity, ease of automation and purification, etc. Although the nitro-precursor is currently the only commercially available compound for [^{18}F]FPEB production, the resulting yields are much lower than the SCIDY and sulfonium salt precursors. The SPI5 auxiliary and the sulfonium salt precursors appear to the be the best suited for routine radiopharmaceutical production of [^{18}F]FPEB.

2.2. Small Animal PET/CT Imaging

mGluR5 has emerged as an imaging target in AD pathogenesis. It has been demonstrated that soluble oligomeric amyloid-β (Aβo) induces an accumulation and over-stabilization of mGluR5, and that Aβo up-regulates mGluR5, leading to an abnormal increase in the release of intracellular Ca^{2+} [6,21–23]. Preliminary PET imaging data with [^{18}F]FPEB showed increased brain uptake in the transgenic model of AD versus the age-matched controls (10 month data shown in Figure 1). Dynamic PET/CT imaging was carried out to investigate the difference in [^{18}F]FPEB binding between an established preclinical model of AD using 10 month old transgenic APP/PS1 mice and their age-matched wild-type (WT) controls. Axial, coronal, and sagittal images of the murine brains acquired at 20 min post-injection of [^{18}F]FPEB are presented in Figure 1A. A marked increased uptake of radiotracer binding was observed in the brains of the transgenic mice compared with WT controls. The time–activity curves revealed a similar initial peak uptake and brain penetration for both groups of mice (Figure 1B). One minute after the time of injection (TOI), maximum uptake was observed in both genotypes, (SUV > 6; whole brain VOI). By 10 min post-injection, a modest higher [^{18}F]FPEB retention was observed in the brain of transgenic mice (SUV = 4.8) versus WT controls (SUV = 4.0), and as the tracer cleared from normal tissues, a significant difference ($p < 0.05$) was apparent at <5 min post-injection. A comparison of the area under the curve (AUC) analysis (Figure 1C) over time for the respected genotypes underscores the trends observed in the TACs.

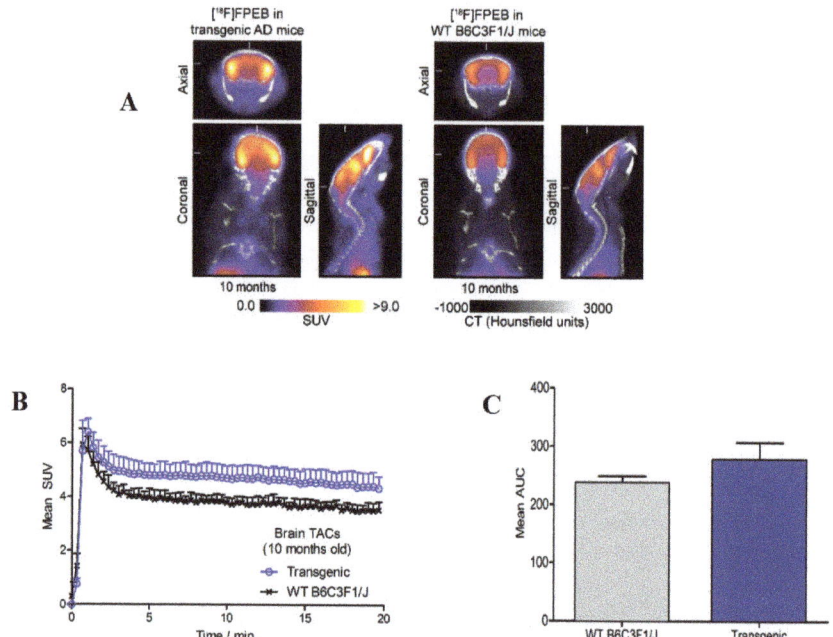

Figure 1. (**A**) PET/CT images of [^{18}F]FPEB 20 min post-injection in 10 month old APP/PS1 (transgenic) mice and aged-matched wild-type (WT) B6C3F1/J (control) mice, $n = 3$/group. (**B**) Time–activity curve for whole-brain. (**C**) Mean area under the curve (AUC).

In a pilot study, we found that patients with early mild cognitive impairment had an increased brain uptake of [^{18}F]FPEB, and that radiotracer uptake in the brain was reflective of increased mGluR5 density [10]. This observation supports the hypothesis that mGluR5 may be implicated in the early stages of AD pathogenesis [24]. Consistent with this hypothesis and clinical research indication, PET/CT imaging studies of [^{18}F]FPEB uptake in a transgenic mouse model of AD also showed an increased radiotracer uptake and retention in the brain of the APP/PS1 mice, compared with wild-type controls. This preliminary work provides support that mGluR5 levels measured by [^{18}F]FPEB are potentially useful as an early biomarker of AD. Further [^{18}F]FPEB imaging studies and biological evaluations are underway including regional analysis of imaging data as well as ex vivo biodistribution and autoradiography studies to evaluate this functional link between mGluR5 and AD.

3. Materials and Methods

3.1. Materials and General Methods

Unless otherwise stated, all reagents were obtained from commercially available sources and used without further purification. The identification of all radiochemical products was determined by HPLC co-elution with an authentic non-radioactive standard. All RCC and RCY values are reported as decay corrected, relative to starting [^{18}F]fluoride (ca. 100 mCi). [^{18}F]Fluoride was produced using a Scanditronix MC17 cyclotron from enriched [^{18}O]H$_2$O through the ^{18}O(p,n)^{18}F reaction. Reverse-phase high performance liquid chromatography (HPLC) was used to isolate and purify [^{18}F]FPEB. Dynamic PET/CT imaging experiments were conducted on a dedicated small-animal PET/CT scanner (eXplore Vista-CT, *Sedecal*, Algete, Spain) equipped with VISTA-CT version 4.11 software.

3.1.1. General Chemistry and Radiochemistry Methods

Preparation of the chloro-precursor was performed as previously reported [9]. The nitro-precursor (Lot No.: 20190401) and FPEB standard (Lot No.: 20130101) were purchased from ABX. Synthesis of the sulfonium salt precursor and radiolabeling were performed as previously reported [16]. For manual labeling, azeotropically dried potassium cryptand [^{18}F]fluoride was dissolved in DMSO (3.0 mL), while 400 µL aliquots (typically 3–5 mCi) were used per reaction to dissolve the precursor in a 1 dram vial. The reaction was heated at 150 °C for 5 min, then quenched with water and cooled for 3 min. Product identity and radiochemical conversion were determined as the ratio of free [^{18}F]fluoride to [^{18}F]FPEB as integrated by radio-HPLC. The manual radiolabeling of the chloro- (**1**) and nitro- (**2**) precursors was performed as previously reported, with slight modification to the concentration and reaction time. One milligram of **2** was dissolved in 0.4 mL [^{18}F]fluoride in DMSO for 5 min instead of 1.5 mL for 15 min [8,18], and the chloro-reaction was conducted at high temperatures instead of under microwave conditions [9]. For automated radiosyntheses, a GE TRACERlab™ FX2 N module with **2** and the SPI5 auxiliary SCIDY precursor (**5**) were performed as previously reported [13,15]. Flash chromatography was performed on a Biotage Isolera One automated flash purification system. Biotage SNAP KP-Sil 50 g cartridges (45–60 micron) were used with a flow rate of 50 mL/min for gradient solvent systems. Fractions were monitored and collected by UV absorbance using the internal UV detector set at 254 and 280 nm.

3.1.2. SCIDY-SPIAd Auxilary Synthesis and Characterization

The titled compound was prepared using a modified literature procedure [15]. Trifluoroacetic acid (0.9 mL) was added to a solution of IPEB (120 mg, 0.36 mmol) in chloroform (0.12 mL). Oxone (179 mg, 0.58 mmol) was added and the reaction mixture was stirred for 3 h, until full conversion of starting materials was determined by TLC (SiO$_2$ coated on polyethylene, 250 µm, with 100% EtOAc). Volatile contents were then removed by rotary evaporation. The round bottom flask was covered in foil and further dried under high vacuum for 5 h. The dried residue was suspended in ethanol (1.5 mL) and (1*r*,3*r*,5*r*,7*r*)-spiro[adamantane-2,2'-[1,3]dioxane]-4',6'-dione (67 mg, 0.54 mmol). SPIAd was added followed by 10% Na$_2$CO$_3$(aq) (*w/v*, 1.5 mL, 0.33 M solution) in ~0.2 mL aliquots until the pH of the reaction mixture was equal to pH 10. The reaction mixture was stirred for 5 h until full conversion to the iodonium ylide was determined by TLC (SiO$_2$ coated on polyethylene, 250 µm, with 10% EtOH in EtOAc, (1:9 mL *v/v*)). The reaction mixture was then diluted with water and extracted with chloroform. The chloroform extracts were combined and washed with water (4 × 10 mL) and brine (1 × 10 mL). The organic layer was dried with anhydrous MgSO$_4$, filtered, and concentrated. The final compound was purified by flash chromatography using a gradient 60% EtOAc in hexanes to 100% EtOAc to 5% methanol in EtOAc. Compound **7** (56 mg, 0.11 mmol) was isolated as an off-white powder with a 41% yield.

^{1}H NMR (500 MHz, dmso-d6) δ (ppm): 8.63 (dt, *J* = 4.7, 1.3 Hz, 1H), 8.31 (t, *J* = 1.5 Hz, 1H), 8.25 (t, *J* = 1.6 Hz, 1H), 8.17 (t, *J* = 1.6 Hz, 1H), 7.89 (td, *J* = 7.7, 1.8 Hz, 1H), 7.73–7.67 (m, 1H), 7.47 (ddd, *J* = 7.7,

4.8, 1.2 Hz, 1H), 2.35 (s, 2H), 1.93 (d, J = 12.3 Hz, 5H), 1.80–1.75 (m, 2H), 1.68–1.62 (m, 7H). ^{13}C NMR (126 MHz, dmso-d6) δ (ppm):163.10, 150.88, 141.60, 139.17, 137.54, 137.49, 136.20, 128.34, 125.05, 124.86, 117.02, 116.90, 114.37, 105.94, 92.58, 84.86, 59.00, 36.92, 35.31, 33.64, 26.38. HRMS (m/z): [M + Na]$^+$ calcd. for $C_{22}H_{15}IN_2O_4Na$, 520.9974; found 520.9967.

3.1.3. General GE TRACERlab™ FX2 N Automated Synthesis Method

Cyclotron produced [^{18}F]fluoride in enriched [^{18}O]H$_2$O was delivered to a GE TRACERlab™ FX2 N automated synthesis module. The [^{18}F]fluoride was trapped on a HCO$_3$$^-$ anion exchange cartridge (Chromafix® [^{18}F] SPE) without additional conditioning and eluted with one of the following base mixtures: Et$_4$NHCO$_3$ in CH$_3$CN/H$_2$O, 4,7,13,16,21,24-hexaoxa-1,-10,diazabicyclo [8.8.8]hexacosane (Kryptofix 222) and KHCO$_3$ in CH$_3$CN/H$_2$O or Kryptofix 222 and K$_2$CO$_3$ in MeOH/H$_2$O. The [^{18}F]fluoride was released from the cartridge and dried under nitrogen gas at 90 °C, followed by azeotropic drying with acetonitrile under nitrogen gas at 110 °C, producing either dry [^{18}F]Et$_4$NF or [^{18}F]KF/K$_{222}$ for radiofluorination. Automated radiosyntheses were carried out with precursors **2, 5, 6,** and **7**. Each precursor was dissolved in either DMSO, DMF, or CH$_3$CN and eluted into the reaction vial with the dry [^{18}F]fluoride, where they were heated for the specified reaction time at varying temperatures. The reactions were then cooled to room temperature and quenched with HPLC buffer. The mixture was diluted with water and the crude reaction mixture was injected onto a semi-preparative HPLC column. The eluent was monitored by UV, at a wavelength of 254 nm, and radiochemical detectors in series. The desired radiochemical product was collected and formulated by dilution with sterile water and loaded onto a C18 Sep-Pak cartridge (pre-activated with 10 mL EtOH, followed by 10 mL H$_2$O). The cartridge was washed with water to remove impurities and then eluted with dehydrated EtOH, and finally, diluted with 0.9% sodium chloride. The final radiochemical yield of this formulated product was determined (50–60 min; see ESI). Molar activity measurements were carried out as described in [12].

3.1.4. Preliminary Small Animal PET/CT Imaging Studies

All animal experiments were conducted in compliance with Institutional Animal Care and Use Committee (IACUC) guidelines and the Guide for the Care and Use of Laboratory Animals. Female wild-type B6C3F1/J mice and female transgenic B6C3-Tg(APPswe,PSEN1dE9)85Dbo/J (APP/PS1; Stock No.: 034829-JAX or MMRRC No. 034) mice were obtained from Jackson Laboratory (Bar Harbor, ME). Mice were provided with food and water ad libitum. Mice were subjected to PET/CT imaging studies after being aged to 10 months. Mice were administered formulations of [^{18}F]FPEB (~3.1–14.0 MBq [~84–378 μCi], specific activity of 5 Ci/μmol, in 200 μL sterile PBS, pH7.4, ≤5% v/v EtOH) via intravenous (i.v.) tail-vein injection using a catheter. Approximately 5 min prior to recording PET images, the mice were anesthetized by inhalation of 3–4% isoflurane (Baxter Healthcare, Deerfield, IL)/oxygen gas mixture, and a catheter was inserted into the tail-vein, with the mice then transferred to the scanner bed and placed in the prone position. Anesthesia was maintained with 1–2% isoflurane/oxygen gas mixture (flow rate ~5 L/min). Co-registered dynamic PET/CT images were recorded for a total of 20 min post-injection radiotracer injection (n = 3 per group). List-mode data were acquired for 20 min per scan using a γ-ray energy window of 250–700 keV. To ensure that the activity bolus was measured, PET/CT data acquisition was initiated 20 s prior to injecting the radioactivity. Data were processed by 3-dimensional Fourier re-binning (3D-FORE), and images were reconstructed using the 2-dimensional ordered-subset expectation maximum (2D-OSEM) algorithm. Image data were normalized to correct for the non-uniformity of response of the PET, dead-time count losses, positron branching ratio, and physical decay to the time of injection, but no attenuation, scatter, or partial-volume averaging correction were applied. An empirically determined system calibration factor (in units of Bq/cps) combined with the decay corrected administered activity and the animals' weights were used to parameterize image activity in terms of the standardized uptake value (SUV). Manually drawn 2-dimensional regions-of-interest (ROIs) or 3-dimensional volumes-of-interest (VOIs) were used to determine the maximum and mean SUV radiotracer uptake in various tissues. Time–activity curves (TACs) were generated from the ROI analysis on dynamic PET/CT data using 20 s frames. CT images were recorded using an X-ray current of 300 µA, 360 projections, and an image size of 63.8 mm × 63.8 mm × 46.0 mm. Data were acquired using the Vista CT 4.11 Build 701 software, and reconstructed images were analyzed by using ASIPro VM^{TM} software (Concorde Microsystems, Siemens Preclinical Solutions, LLC, Knoxville, TN, USA) and VivoQuant® 1.23 (InviCRO, LLC, Boston, MA, USA).

3.2. Data Analysis and Statistics

Data and statistical analyses were performed using GraphPad Prism 5.01 (GraphPad Software, Inc., La Jolla, CA, USA) and Microsoft Excel spreadsheets. Differences at the 95% confidence level ($p < 0.05$) were considered to be statistically significant.

Supplementary Materials: The following are available online, Figures S1–S3 Semi-preparative HPLC traces of a typical radiosynthesis of [^{18}F]FPEB using precursor (**5–7**).

Author Contributions: C.V., E.M. and W.S. and N.A.S. conducted the chemistry radiochemical experiments. A.K. and J.P.H. conducted the PET imaging experiments and analyzed the data. C.V., N.A.S and N.V. wrote the manuscript. N.A.S., J.P.H., S.H.L. and N.V. conceived and supervised the project. All authors have read and agreed to the published version of the manuscript.

Funding: N.V. thanks the National Institute on Ageing of the NIH (R01AG054473 and R01AG052414), the Azrieli Foundation, the Canada Foundation for Innovation, the Ontario Research Fund, and the Canada Research Chairs Program for support. J.P.H. thanks the Swiss National Science Foundation (SNSF Professorship PP00P2_163683 and PP00P2_190093), the European Union's Horizon 2020 research and innovation programme/from the European Research Council under the Grant Agreement No 676904, ERC-StG-2015, NanoSCAN, and the University of Zurich (UZH) for financial support.

Conflicts of Interest: N.A.S., S.H.L. and N.V. have an issued patent entitled, "Iodine(III)-Mediated Radiofluorination; US 9.434,699 B2, that is licensed to a third party; S.H.L. and N.V. have a patent entitled, "Method of Fluorination Using Iodonium Ylides, US 9.434,699 B2. The authors declare no other conflicts of interest.

References

1. Pillai, R.; Tipre, D. Metabotropic glutamate receptor 5—A promising target in drug development and neuroimaging. *Eur. J. Nucl. Med. Mol. Imaging* **2016**, *43*, 1151–1170. [CrossRef] [PubMed]
2. Terbeck, S.; Akkus, F.; Chesterman, L.P.; Hasler, G. The role of metabotropic glutamate receptor 5 in the pathogenesis of mood disorders and addiction: Combining preclinical evidence with human Positron Emission Tomography (PET) studies. *Front. Mol. Neurosci.* **2015**. [CrossRef] [PubMed]
3. Marino, M.J.; Valenti, O.; Conn, P.J.; Conn, P.J. Glutamate Receptors and Parkinson's Disease. *Drugs Aging* **2003**, *20*, 377–397. [CrossRef] [PubMed]
4. Alagille, D.; Dacosta, H.; Chen, Y.; Hemstapat, K.; Rodriguez, A.; Baldwin, R.M.; Conn, J.P.; Tamagnan, G.D. Potent mGluR5 antagonists: Pyridyl and thiazolyl-ethynyl-3,5-disubstituted-phenyl series. *Bioorg. Med. Chem. Lett.* **2011**, *21*, 3243–3247. [CrossRef]
5. Rook, J.M.; Tantawy, M.N.; Ansari, M.S.; Felts, A.S.; Stauffer, S.R.; Emmitte, K.A.; Kessler, R.M.; Niswender, C.M.; Daniels, J.S.; Jones, C.K.; et al. Relationship between In Vivo Receptor Occupancy and Efficacy of Metabotropic Glutamate Receptor Subtype 5 Allosteric Modulators with Different In Vitro Binding Profiles. *Neuropsychopharmacology* **2014**, *40*, 755–765. [CrossRef]
6. Majo, V.J.; Prabhakaran, J.; Mann, J.J.; Kumar, J.S.D. PET and SPECT tracers for glutamate receptors. *Drug Discov. Today* **2013**, *18*, 173–184. [CrossRef]
7. Wong, D.F.; Waterhouse, R.; Kuwabara, H.; Kim, J.; Chamroonrat, W.; Stabins, M.; Holt, D.P.; Dannals, R.F.; Hamill, T.G.; Mozley, P.D.; et al. 18F-FPEB, a PET Radiopharmaceutical for Quantifying Metabotropic Glutamate 5 Receptors: A First-in-Human Study of Radiochemical Safety, Biokinetics, and Radiation Dosimetry. *J. Nucl. Med.* **2013**, *54*, 388–396. [CrossRef]
8. Sullivan, J.M.; Lim, K.; Labaree, D.; Lin, S.-F.; McCarthy, T.J.; Seibyl, J.P.; Tamagnan, G.; Huang, Y.; Carson, E.R.; Ding, Y.-S.; et al. Kinetic analysis of the metabotropic glutamate subtype 5 tracer [18F]FPEB in bolus and bolus-plus-constant-infusion studies in humans. *Br. J. Pharmacol.* **2012**, *33*, 532–541. [CrossRef]

9. Hamill, T.G.; Krause, S.; Ryan, C.; Bonnefous, C.; Govek, S.; Seiders, T.J.; Cosford, N.; Roppe, J.; Kamenecka, T.; Patel, S.; et al. Synthesis, characterization, and first successful monkey imaging studies of metabotropic glutamate receptor subtype 5 (mGluR5) PET radiotracers. *Synapse* **2005**, *56*, 205–216. [CrossRef]
10. Liang, S.H.; Yokell, D.L.; Jackson, R.N.; Rice, P.A.; Callahan, R.; Johnson, K.A.; Alagille, D.; Tamagnan, G.; Collier, L.; Vasdev, N. Microfluidic continuous-flow radiosynthesis of [18F]FPEB suitable for human PET imaging. *MedChemComm* **2013**, *5*, 432–435. [CrossRef]
11. Mossine, A.V.; Brooks, A.F.; Makaravage, K.J.; Miller, J.M.; Ichiishi, N.; Sanford, M.S.; Scott, P.J.H. Synthesis of [18F]Arenes via the Copper-Mediated [18F]Fluorination of Boronic Acids. *Org. Lett.* **2015**, *17*, 5780–5783. [CrossRef] [PubMed]
12. Makaravage, K.J.; Brooks, A.F.; Mossine, A.V.; Sanford, M.S.; Scott, P.J.H. Copper-Mediated Radiofluorination of Arylstannanes with [18F]KF. *Org. Lett.* **2016**, *18*, 5440–5443. [CrossRef] [PubMed]
13. Stephenson, N.A.; Holland, J.P.; Kassenbrock, A.; Yokell, D.L.; Livni, E.; Liang, S.H.; Vasdev, N. Iodonium ylide-mediated radiofluorination of 18F-FPEB and validation for human use. *J. Nucl. Med.* **2015**, *56*, 489–492. [CrossRef] [PubMed]
14. Rotstein, B.; Stephenson, N.A.; Vasdev, N.; Liang, S.H. Spirocyclic hypervalent iodine(III)-mediated radiofluorination of non-activated and hindered aromatics. *Nat. Commun.* **2014**, *5*. [CrossRef]
15. Liang, S.H.; Wang, L.; Stephenson, N.; Rotstein, B.; Vasdev, N. Facile 18F labeling of non-activated arenes via a spirocyclic iodonium(III) ylide method and its application in the synthesis of the mGluR5 PET radiopharmaceutical [18F]FPEB. *Nat. Protoc.* **2019**, *14*, 1530–1545. [CrossRef]
16. Gendron, T.; Sander, K.; Cybulska, K.; Benhamou, L.; Sin, P.K.B.; Khan, A.; Wood, M.; Porter, M.; Årstad, E. Ring-Closing Synthesis of Dibenzothiophene Sulfonium Salts and Their Use as Leaving Groups for Aromatic 18F-Fluorination. *J. Am. Chem. Soc.* **2018**, *140*, 11125–11132. [CrossRef]
17. Rotstein, B.; Wang, L.; Liu, R.Y.; Patteson, J.; Kwan, E.E.; Vasdev, N.; Liang, S.H. Mechanistic studies and radiofluorination of structurally diverse pharmaceuticals with spirocyclic iodonium(iii) ylides† †Electronic supplementary information (ESI) available: Detailed experimental procedures, characterization of compounds, NMR spectra and computational studies. *Chem. Sci.* **2016**, *7*, 4407–4417. [CrossRef]
18. Wang, J.Q.; Tueckmantel, W.; Zhu, A.; Pellegrino, D.; Brownell, A.L. Synthesis and preliminary biological evaluation of 3-[18F]fluoro-5-(2-pyridinylethynyl)benzonitrile as a PET radiotracer for imaging metabotropic glutamate receptor subtype 5. *Synapse* **2007**, *61*, 951–961. [CrossRef]
19. Lim, K.; Labaree, D.; Li, S.; Huang, Y. Preparation of the Metabotropic Glutamate Receptor 5 (mGluR5) PET Tracer [18F]FPEB for Human Use: An Automated Radiosynthesis and a Novel One-Pot Synthesis of its Radiolabeling Precursor. *Appl. Radiat. Isot.* **2014**, *94*, 349–354. [CrossRef]
20. Hill, D.E.; Holland, J.P. Computational studies on hypervalent iodonium(III) compounds as activated precursors for 18F radiofluorination of electron-rich arenes. *Comput. Theor. Chem.* **2015**, *1066*, 34–46. [CrossRef]
21. Casley, C.S.; Lakics, V.; Lee, H.-G.; Broad, L.M.; Day, T.A.; Cluett, T.; Smith, M.A.; O'Neill, M.J.; Kingston, A.E. Up-regulation of astrocyte metabotropic glutamate receptor 5 by amyloid-β peptide. *Brain Res.* **2009**, *1260*, 65–75. [CrossRef] [PubMed]
22. Um, J.W.; Kaufman, A.C.; Kostylev, M.; Heiss, J.K.; Stagi, M.; Takahashi, H.; Kerrisk, M.E.; Vortmeyer, A.; Wisniewski, T.; Koleske, A.J.; et al. Metabotropic Glutamate Receptor 5 Is a Coreceptor for Alzheimer Aβ Oligomer Bound to Cellular Prion Protein. *Neuron* **2013**, *79*, 887–902. [CrossRef] [PubMed]
23. Niswender, C.M.; Conn, P.J. Metabotropic glutamate receptors: Physiology, pharmacology, and disease. *Annu. Rev. Pharmacol. Toxicol.* **2010**, *50*, 295–322. [CrossRef] [PubMed]
24. Kumar, A.; Dhull, D.K.; Mishra, P.S. Therapeutic potential of mGluR5 targeting in Alzheimer's disease. *Front. Neurosci.* **2015**, *9*, 215. [CrossRef] [PubMed]

Sample Availability: Samples of the compounds are not available from the authors.

© 2020 by the authors. Licensee MDPI, Basel, Switzerland. This article is an open access article distributed under the terms and conditions of the Creative Commons Attribution (CC BY) license (http://creativecommons.org/licenses/by/4.0/).

Article

Radiosynthesis of [^{18}F]-Labelled Pro-Nucleotides (ProTides)

Alessandra Cavaliere [1,2], Katrin C. Probst [2], Stephen J. Paisey [2], Christopher Marshall [2], Abdul K. H. Dheere [3], Franklin Aigbirhio [3], Christopher McGuigan [1] and Andrew D. Westwell [1,*]

1. School of Pharmacy & Pharmaceutical Sciences, Cardiff University, Redwood Building, King Edward VII Avenue, Cardiff CF10 3NB, Wales, UK; alessandra.cavaliere@yale.edu (A.C.); McGuigan@cardiff.ac.uk (C.M.)
2. Wales Research & Diagnostic Positron Emission Tomography Imaging Centre (PETIC), School of Medicine, Cardiff University, University Hospital of Wales, Heath Park, Cardiff CF14 4XN, Wales, UK; katrin.probst@gmx.de (K.C.P.); PaiseySJ@cardiff.ac.uk (S.J.P.); MarshallC3@cardiff.ac.uk (C.M.)
3. Wolfson Brain Imaging Centre and Department of Clinical Neurosciences, University of Cambridge, Cambridge CB2 0QQ, UK; akhd@kcl.ac.uk (A.K.H.D.); fia20@wbic.cam.ac.uk (F.A.)
* Correspondence: WestwellA@cf.ac.uk; Tel.: +44-(0)29-2087-5800

Received: 20 December 2019; Accepted: 1 February 2020; Published: 6 February 2020

Abstract: Phosphoramidate pro-nucleotides (ProTides) have revolutionized the field of anti-viral and anti-cancer nucleoside therapy, overcoming the major limitations of nucleoside therapies and achieving clinical and commercial success. Despite the translation of ProTide technology into the clinic, there remain unresolved in vivo pharmacokinetic and pharmacodynamic questions. Positron Emission Tomography (PET) imaging using [^{18}F]-labelled model ProTides could directly address key mechanistic questions and predict response to ProTide therapy. Here we report the first radiochemical synthesis of [^{18}F]ProTides as novel probes for PET imaging. As a proof of concept, two chemically distinct radiolabelled ProTides have been synthesized as models of 3′- and 2′-fluorinated ProTides following different radiosynthetic approaches. The 3′-[^{18}F]FLT ProTide was obtained via a late stage [^{18}F]fluorination in radiochemical yields (RCY) of 15–30% ($n = 5$, decay-corrected from end of bombardment (EoB)), with high radiochemical purities (97%) and molar activities of 56 GBq/µmol (total synthesis time of 130 min.). The 2′-[^{18}F]FIAU ProTide was obtained via an early stage [^{18}F]fluorination approach with an RCY of 1–5% ($n = 7$, decay-corrected from EoB), with high radiochemical purities (98%) and molar activities of 53 GBq/µmol (total synthesis time of 240 min).

Keywords: fluorination; ProTides; fluorine-18; radiolabelling; PET imaging

---

1. Introduction

Clinically approved nucleoside analogues occupy a unique place in drug therapy due to their ability to interfere in biosynthetic and metabolic pathways fundamental to aberrant cellular replication and growth. This is particularly apparent in conditions such as cancer [1] and viral infections [2], where nucleoside analogues are able to inhibit essential human or viral enzymes such as thymidylate synthase or ribonucleotide reductase.

Despite their well-established value in drug therapy, nucleosides suffer from a number of drawbacks as therapeutic agents. Cellular entry of nucleosides through the outer cell membrane requires the active participation of concentrative and equilibrative nucleoside transporters; down-regulation of transporters in cancer cells for example constitutes a known drug resistance mechanism [3]. Following cellular uptake, nucleosides require activation via (normally) three successive enzyme-mediated phosphorylation steps (Figure S1) [4]. The first kinase-mediated phosphorylation is most frequently the rate-limiting step prior to incorporation of the active therapeutic nucleotide tri-phosphate within targets such as DNA/RNA. The requirements for transporter-mediated cellular entry and subsequent phosphorylation to the active form limit the therapeutic efficacy of nucleoside analogues. For example, the anticancer drug gemcitabine, widely used in pancreatic, non-small cell lung, ovarian and breast cancer therapy, is associated with levels of patient response as low as 20% [5].

Given the well-established drawbacks associated with nucleoside analogues, a great deal of research effort has been devoted towards the development of pro-nucleotides [6]. These nucleotide derivatives are defined as analogues able to by-pass the requirement for nucleoside membrane transporters and deliver a masked monophosphate that subsequently breaks down to a nucleotide monophosphate, or nucleotide derivative, within the cell. The most successful approach to this problem has been the phosphoramidate pro-nucleotides (ProTides) pioneered by McGuigan and colleagues at Cardiff University, U.K. [6]. The ProTide approach has revolutionized the field, delivering greatly enhanced concentrations of active nucleotide triphosphate within the diseased cell and improving clinical efficacy in important areas of anticancer and antiviral therapy. Examples of success in the antiviral field include the hepatitis C drug sofosbuvir (Gilead Sciences, Inc., Foster City, CA, USA), launched in 2014 to radically address the high level of unmet medical need in this common disease and achieving the status of the world's top-selling drug worldwide following launch [7]. Within the anticancer field, several agents have progressed to clinical evaluation with very promising early clinical data. These include the gemcitabine ProTide NUC-1031 (Acelarin), discovered and developed through a collaboration between Cardiff University and NuCana plc. Acelarin is currently in Phase III clinical evaluation in pancreatic cancer [8].

Despite the emerging clear clinical advantages of the ProTide approach, evidence at the molecular level for the accumulation of ProTides at the in vivo site of action is currently lacking. A potential solution to this problem would be to label ProTides such that they were amenable to non-invasive molecular imaging in vivo. In this regard, Positron Emission Tomography (PET) represents an attractive solution. PET imaging is a sensitive and rapidly emerging molecular imaging technology, widely used in prognostic and diagnostic clinical applications [9]. The basis of PET imaging is the incorporation of a radioactive PET-emitting nuclide into the biomarker or drug molecule of interest. The positron released then annihilates close to the site of emission by electron collision, generating two γ-rays (511 Kev) that are detected by coincidence measurement followed by 3D image reconstruction. This annihilation event locates the site of origin of the target molecule with a high level of sensitivity [10]. The choice of PET isotope is crucial; amongst the many options available ^{18}F has emerged as the non-metal nuclide of choice due to its intermediate half-life (110 min), relatively low positron energy and exclusive positron mode of decay [11].

The incorporation of ^{18}F into small molecules provides significant challenges for the radiochemistry community. The routinely utilized ^{18}F-fluoride (dried and purified from aqueous fluoride solution from the cyclotron nuclear reactor source) is poorly nucleophilic, and electrophilic options derived from fluorine gas are disfavoured [12]. In addition, the half-life of ^{18}F (110 min) means that ^{18}F-incorporation normally has to occur at a late stage of multi-step syntheses, where purification, QC and formulation for patient administration is necessary within a few hours to generate a significant PET signal. These challenges are not insurmountable, and a range of ^{18}F-labelled biomarkers and drugs have been produced and applied in clinical medicine. This is best exemplified by the routine use of the gold standard PET biomarker [^{18}F]FDG (fluorodeoxyglucose) in cancer diagnosis, staging and monitoring of response to therapy [13].

Building on our previous experience of ^{18}F-radiolabelling of nucleosides for PET imaging [14,15], we set out to apply our technology to the ProTide field. Here we report the first radiosynthetic routes to both 3'- and 2'-fluorinated model ProTides. Application of this technology could help to improve both diagnostic and prognostic applications of ProTide technology in the clinic, with profound implications for this exciting area of anticancer and antiviral therapy.

2. Results and Discussion

As a proof of concept, two [^{18}F]-radiolabelled ProTides have been synthesised. The [^{18}F]**Fluoro Thymidine (FLT) ProTide (1)** and the [^{18}F]Fluoro-Iodo-ArabinofuranosylUracil (FIAU) ProTide (**2**) (Figure 1) were our chosen model standards for 3′-fluorinated and 2′-fluorinated ProTides, respectively.

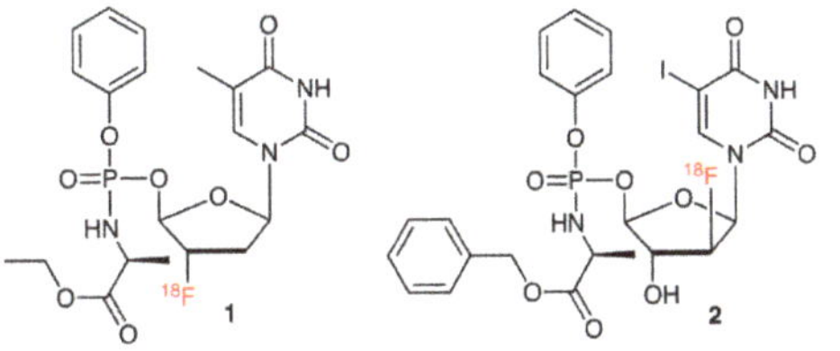

Figure 1. Structures of the [^{18}F]FLuoroThymidine (FLT) ProTide (**1**) and [^{18}F]Fluoro-Iodo-ArabinofuranosylUracil (FIAU) (ProTide (**2**).

2.1. [^{18}F]FLT- a Prototypical 3′-fluorinated ProTide

[^{18}F]FLT, a 3′-fluorinated nucleoside, is an established PET imaging agent used as a tumour proliferation biomarker [16]. Synthetic approaches involving a late stage [^{18}F]-fluorination of different precursor molecules of [^{18}F]FLT have been extensively studied [15]. Moreover, our group has previously reported the synthesis of a series of (non-radiolabelled) FLT ProTides that showed a relatively safe toxicological profile compared to the parent nucleoside as well as moderate anti-HIV activity [17]. For these reasons, [^{18}F]FLT ProTide has been selected as a target compound in this study to represent the class of 3′-fluorinated ProTides. The choice of the phenol group as aromatic moiety and the L-alanine ethyl ester as the amino acid ester on the phosphoramidate moiety were dictated by the accessibility of the starting materials as well as the favourable yields associated with the coupling reactions involved in the synthesis [17]. The radiochemical synthesis of [^{18}F]FLT ProTide (**1**) was planned accordingly taking into account the short half-life of fluorine-18 (110 min), with the [^{18}F]fluorination occurring at a late stage in the synthesis (Scheme 1). The challenge for this radiosynthetic plan was therefore to identify a precursor molecule with a balanced reactivity towards the weakly nucleophilic [^{18}F]fluoride with stability in the harsh thermal conditions used for the radiolabelling step. A series of good leaving groups {methanesulfonyl (mesyl); p-toluenesulfonyl (tosyl); p-nitrophenylsulfonyl (nosyl)} for the key nucleophilic fluoride displacement reaction (intermediates **4–7**) were selected for reaction optimization.

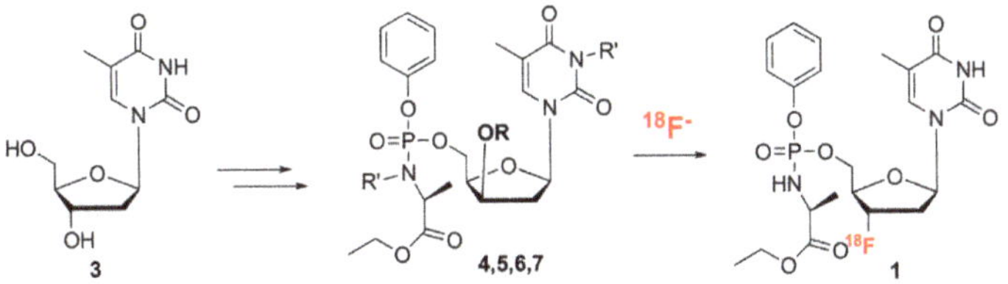

4. R: SO$_2$CH$_3$ (Mesyl); R′: H
5. R: SO$_2$PhCH$_3$ (Tosyl); R′: H
6. R: SO$_2$PhNO2 (Nosyl); R′: H
7. R: SO$_2$PhNO2 (Nosyl); R′: BOC

Scheme 1. Late stage fluorination approach for the synthesis of the [^{18}F]FLT ProTide (**1**).

2.1.1. Synthesis of the Cold FLT ProTide Standard

A cold non-radioactive standard of the [^{18}F]FLT ProTide was synthesized according to established ProTide chemistry protocols (Scheme 2) [18]. The phosphorochloridate intermediate (**10**) was first obtained from the L-alanine ethyl ester hydrochloride salt and the commercially available dichlorophosphate (**9**) using triethylamine as base. Compound **10** was obtained as a mixture of diastereoisomers because of the formation of a new stereocenter at the phosphorus atom in a 1:1 Rp: Sp ratio. Commercially available FLT (**8**; Carbosynth) was then reacted with the phosphorylating reagent using *tert*-butyl magnesium chloride (*t*-BuMgCl) as a hindered base. The desired product (**11**) was obtained as a mixture of diastereoisomers (1:1 Rp:Sp) with a yield of 24% and was used as standard analytical control for studies of the radiochemical synthesis of [^{18}F]FLT ProTide.

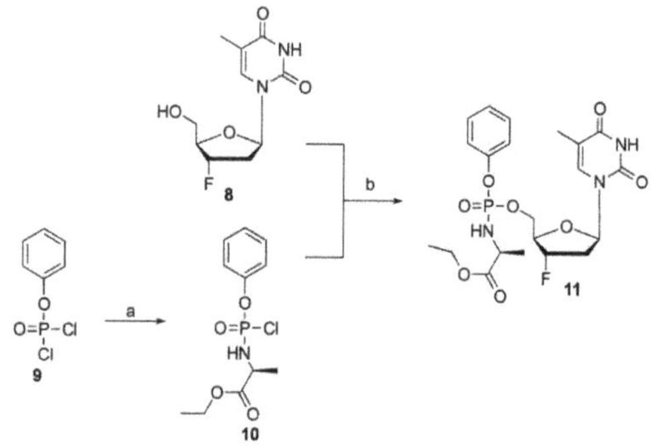

Scheme 2. Synthesis of the cold standard FLT ProTide. Reagents and conditions: (a) L-alanine ethyl ester hydrochloride salt, Et$_3$N, −78 °C to rt, anh. CH$_2$Cl$_2$, 3 h, 92%; (b) *t*-BuMgCl, anh. THF, 18 h, 24%.

Stability studies were performed on the non-radioactive standard to test the susceptibility of the ProTide phosphoramidate to the high temperatures used during the [^{18}F]fluorination. The dynamic behaviour of the phosphoramide moiety was therefore monitored at temperatures ranging from 50 °C to 120 °C. ^{31}P NMR spectroscopy is particularly well suited for this purpose considering the characteristic chemical shifts at around δ 4 ppm of the phosphoramidate backbone of the FLT ProTide [17]. The two ^{31}P NMR peaks of the diasteroisomeric mixture were observed to be stable when the compound was heated up to 120 °C, confirming its stability to the high temperatures that were used during the radiolabelling step (see Supplementary Materials, Figure S2).

2.1.2. Synthesis of a Series of Organophosphates as Precursor Molecules of the [^{18}F]FLT ProTide

To design a late stage fluorination for the class of 3′-substituted ProTides, a multi-step synthesis was performed to obtain a thymidine based ProTide with an anhydroxylic group in the 3′-β position of the ribose ring. The first step consisted of the formation of the 3′-β hydroxy intermediate (**13**) via a Mitzunobu reaction [18] followed by hydrolysis of the intermediate compound **12** to obtain inversion of the stereochemistry of the hydroxylic group at the 3′ position of the thymidine. The intermediate **14** was again synthesised following the standard procedure previously described. N-methylimidazole (NMI) was used at this time as the coupling reagent because of the presence of the free hydroxylic group in the 3′-position that could compete with the 5′-OH group for the phosphorylation [19].

The hydroxyl group at the 3′-β position of **14** is a poor leaving group for the nucleophilic substitution reaction with anhydrous [^{18}F]fluoride. Therefore, the 3′-hydroxyl group was selectively activated with a series of sulfonic esters to produce good leaving groups (**4–6**) for reaction with the weakly nucleophilic [^{18}F]fluoride, in accordance with literature precedent [11,20]. The intermediate **14** was reacted with mesyl chloride, tosyl chloride and nosyl chloride respectively in presence of a weak base such as pyridine or Et$_3$N with or without AgOTf as a catalyst (Scheme 3). To improve the stability of precursor **6** and avoid competitive cyclization/elimination reaction upon reaction with the fluorine-18 [21], protection of the NH group of the pyrimidine ring was performed with the *tert*-butoxycarbonyl group (Boc). Surprisingly, the major product of the reaction observed was the di-protected ProTide (**7**) bearing Boc groups at both the NH of the pyrimidine ring and the phosphoramidate moiety. The abundance of the di-Boc protected compound compared to the mono-protected product as result of the Boc-protection reaction, together with the need for a stable fluorination precursor, led us to choose the di-Boc protected product (7) for the following radio-fluorination step.

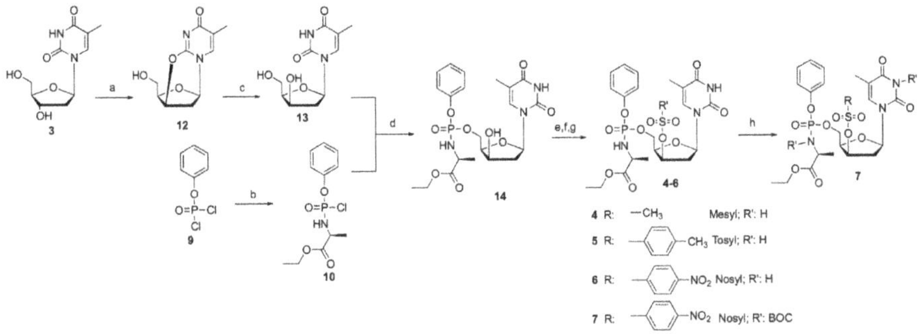

Scheme 3. Synthetic procedure for the mesyl, tosyl and nosyl precursors. Reagents and conditions: (a) PPh₃, DIAD, anh. CH₃CN, −20 °C to 0 °C, 5 h, 52%; (b) L-alanine ethyl ester hydrochloride salt, Et₃N, −78 °C to rt, anh. CH₂Cl₂, 3 h, 92%; (c) NaOH (1.5 M), CH₃OH, 90 °C, 3 h, 64%; (d) Phosphorochloridate, NMI, anh. THF, 25 °C, 18 h, nitrogen atm, 18.4%; (e) Mesyl chloride, Et₃N, anh. CH₂Cl₂, nitrogen atm., 0 °C to 25 °C, 1.5 h, 29.5%; (f) Tosyl chloride, pyridine, AgOTf, 0 °C to rt, 2 h, 35%; (g) Nosyl chloride, pyridine, AgOTf, 0 °C to rt, 2 h, 60%; (h) Di-*tert*-butyldicarbonate, pyridine, rt, 16 h, 56%.

2.1.3. Radiochemical Synthesis of the [^{18}F]FLT ProTide

The Eckert & Ziegler modular lab was used for the [^{18}F]-fluorination following the schematic described in the Supplementary Material (Figure S3). K [^{18}F]F/K$_{222}$/K$_2$CO$_3$ was used as the fluorinating agent and a series of solvents and temperatures were tested to establish the best conditions for the radio-fluorination. The methanesulfonyl (mesyl) precursor (4) and the *p*-toluenesulfonyl (tosyl) precursor (5) did not give the expected [^{18}F]-radiolabelled compound as observed from the radio HPLC chromatograms (Figures S4 and S5 and Tables S1 and S2). The *p*-nitrobenzenesulfonate (nosyl) precursor 6 was, as expected, the most reactive among the three organosulfonate leaving groups for the S$_N$2 reaction with the weak nucleophile [^{18}F]fluoride [11], but lacked stability with the formation of multiple radiolabeled polar compounds and a radiochemical yield < 1% as determined by analytical HPLC (Figure S6 and Table S3).

To increase the stability of the precursor, two Boc protecting groups were added, as previously described, leading to the formation of compound 7. This precursor proved to be the best substrate for [^{18}F]-fluorination. Radio-HPLC showed a major product (15) with a retention time at around 15 min (Figure S7). This suggested that the desired Boc radiolabelled product was formed therefore supporting the hypothesis that the Boc double protection provides improved stability for the nosyl precursor. The deprotection step was then carried out by adding 2 N HCl for 10 min at 95 °C [22] and the final compound was then neutralised with a 2 M NaOH solution. Gratifyingly the major product of this reaction was the [^{18}F]FLT ProTide (1) with few other minor by-products (Scheme 4, Figure S8).

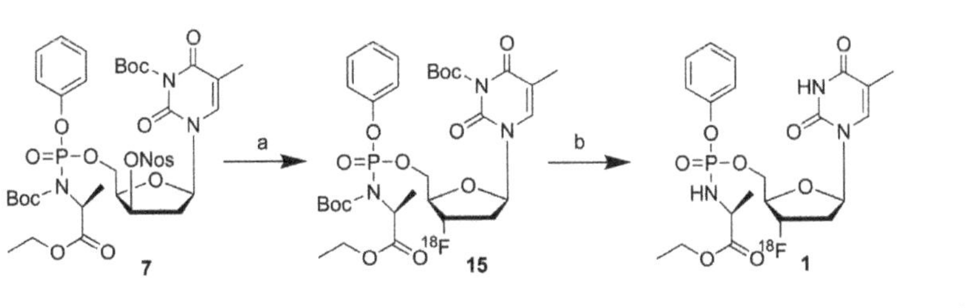

Scheme 4. ^{18}F-fluorination of the Boc protected nosyl derivative 7 and deprotection. Reagents and conditions: (a) ^{18}F⁻, Kryptofix, anh. CH₃CN, 90 °C, 30 min.; (b) 2 N HCl, 95 °C, 10 min.

The compound was then purified by semi-preparative HPLC (Phenomenex Synergi 4µ Hydro-RP 80, C-18, 10 × 250 mm) and was eluted after 35 min at a flow rate of 3.5 mL/min using 30% CH$_3$CN/70% H$_2$O as the mobile phase. To confirm the identity of the ^{18}F-product (**1**), an aliquot of the purified sample was analysed by HPLC (Phenomenex Synergi 4µ Hydro-RP 80, C-18, 4.6 × 250 mm) via co-elution with the cold standard. The ^{18}F-product showed a R$_t$ of 9.5 min and the cold standard co-injected eluted at a Rt of 9.2 min (Figure 2) confirming the identity of the [^{18}F]FLT ProTide (**1**). Radiochemical reactions were carried out using starting activities between 1.5–8 GBq, leading to final product activities of 300–580 MBq in a good radiochemical yield (RCY) of 15–30% (n = 5, decay-corrected from end of bombardment (EoB)). High radiochemical purities (≥97%) and molar activities of 56 GBq/µmol were obtained, and the total synthesis time was 130 min after the end of bombardment (EoB).

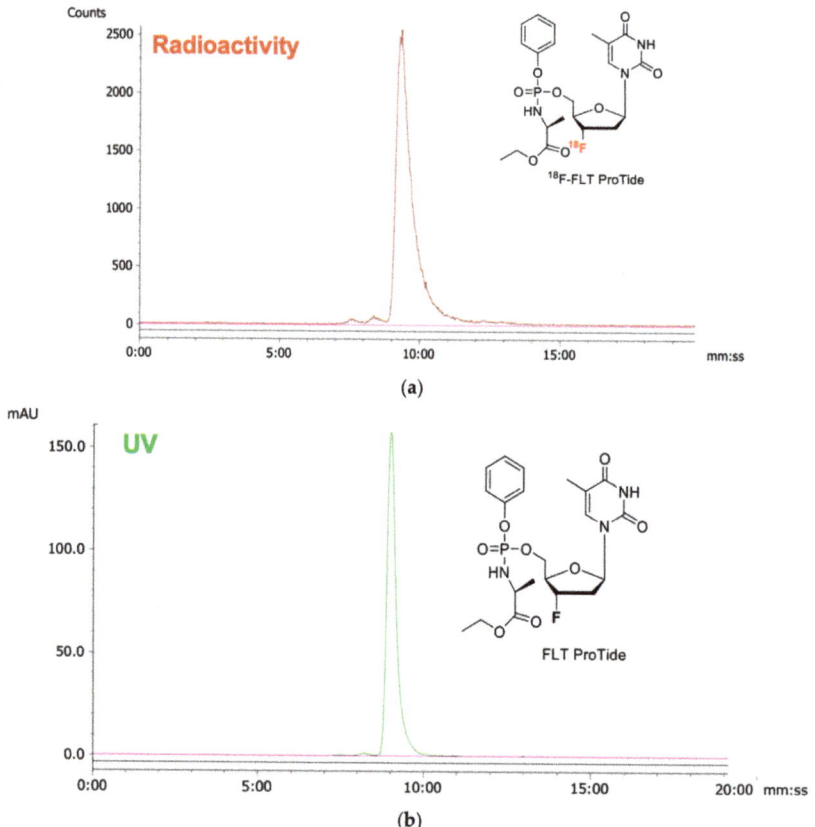

Figure 2. HPLC of the [^{18}F]FLT ProTide (**1**). (**a**) Radioactive chromatogram of the purified [^{18}F]FLT ProTide with R$_t$ of 9.5 min; (**b**) UV chromatogram of the reaction mixture co-spiked with the cold standard (R$_t$: 9.2 min). HPLC system: 90% H$_2$O/10% CH$_3$CN, to 50% H$_2$O/50% CH$_3$CN.

2.2. [^{18}F]FIAU - A Prototypical 2′-Fluorinated ProTide

2′-deoxy-2′-[^{18}F]fluoro-1-β-D-arabinofuranosyl-5-iodouracil ([^{18}F]FIAU (**2**), is a PET biomarker used for imaging HSV1-*tk* gene expression in biological processes including transcriptional regulation, lymphocyte migration and stem-cell tracking [23]. Building on previous developed radiosyntheses of this tracer for PET imaging, we decided to synthesise a ProTide of [^{18}F]FIAU (**2**) as a model of the class of the 2′-fluorinated ProTides [10] introducing ^{18}F early in the synthetic sequence as outlined in Scheme 5.

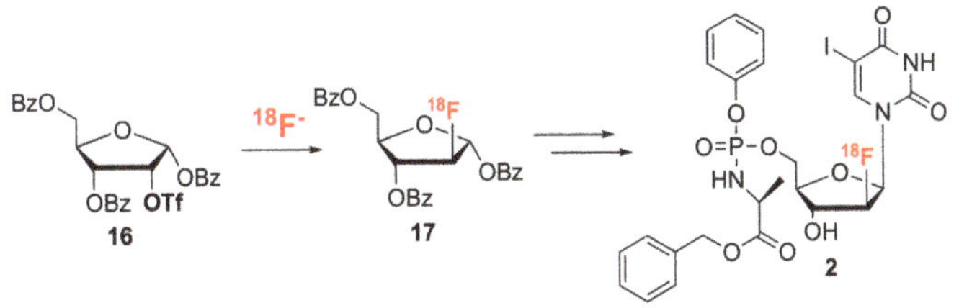

Scheme 5. Synthetic approach for the radiosynthesis of the [^{18}F]FIAU ProTide (**2**).

2.2.1. Synthesis of the Non-Radioactive FIAU ProTide Standard

A cold standard of FIAU ProTide (**21**) was synthesised following the synthesis outlined in Scheme 6. The commercially available compound **18** was firstly iodinated at the C-5 position upon reaction with iodine and cerium ammonium nitrate to give compound **19** under previously reported conditions [24]. The ProTide **21** was synthesised using methodology described above with the exception that the L-alanine ethyl ester was here replaced by a benzyl ester (using compound **20**). The phosphoramidate **21** was obtained as a diastereoisomeric mixture for co-injection with the radiolabelled counterpart, the [^{18}F]FIAU ProTide, to confirm its identity by HPLC.

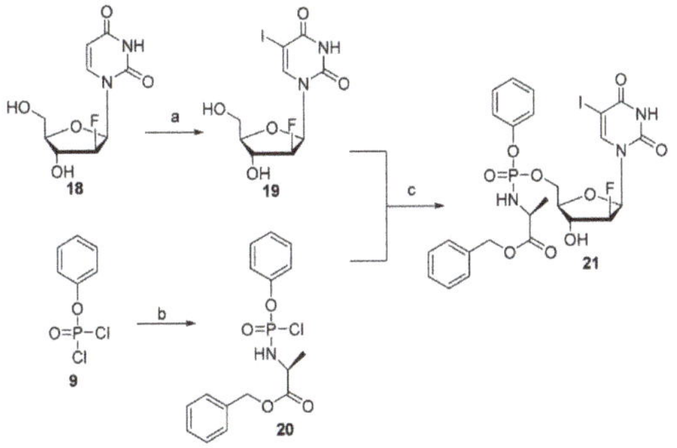

Scheme 6. Synthesis of the non-radioactive standard FIAU ProTide (**21**). Reagents and conditions: (a) I$_2$, Ceric ammonium nitrate, ACN, 75 °C, 1 h, 60%; (b) L-alanine benzyl ester hydrochloride salt, Et$_3$N, −75 °C to rt, anh. CH$_2$Cl$_2$, 3 h, 88%; (c) NMI, anh. THF, 0 °C to rt, 16 h, 10%.

2.2.2. Radiochemical Synthesis of the [^{18}F]FIAU ProTide

The first step consisted of the radioactive fluorination of the commercially available sugar **16** bearing a triflate as leaving group according to literature precedent ([^{18}F]fluoride, Kryptofix, anh. CH$_3$CN, 95 °C) [25]. The reaction was again carried out using the E&Z modular lab. After purification with an alumina sep-pak [26], the radiolabelled sugar (**17**) was used for the next step without further purification. When an aliquot of the radioactive mixture was co-spiked with a cold standard (Sigma Aldrich), it showed the same retention time at around 3 min (Figure S10).

The second step consisted in the protection of the base moiety **22** with hexamethyldisilazane and the catalyst trimethylsilyl trifluoromethanesulfonate (TMSOTf) [27] to obtain compound **23** that was coupled with the radiolabelled sugar (**17**) without further purification. The glycosylation reaction led to the formation of two anomers following removal of the TMS groups under basic conditions, the β-anomer ([^{18}F]FIAU) (**24**) and the α-anomer in a ratio 2:1 (Figure S10). Attempts to increase the speed of the reaction by either reducing the time or using a combination of catalysts (TMSOTf and SnCl$_4$) led to incomplete conversion into the final product or favoured the formation of the α-anomer (ratio β:α = 1:1.3), as shown in Table 1 [27,28]. Therefore, based on these attempts to optimise the reaction conditions, the synthetic pathway in Scheme 7 was established as the most suitable for synthesis of [^{18}F]FIAU **24**.

Table 1. Tentative optimization of the glycosylation reaction.

Solvent	Temperature	Time	Catalyst	Ratio β:α
CH$_3$CN	85 °C	30 min	TMSOTf	Incomplete
CH$_3$CN	85 °C	45 min	TMSOTf	Incomplete
CH$_3$CN	85 °C	1 h	TMSOTf	2:1
CH$_3$CN	95 °C	30 min	TMSOTf	Incomplete
CH$_3$CN	85 °C	15 min	TMSOTf+SnCl$_4$ [24]	1:1.3

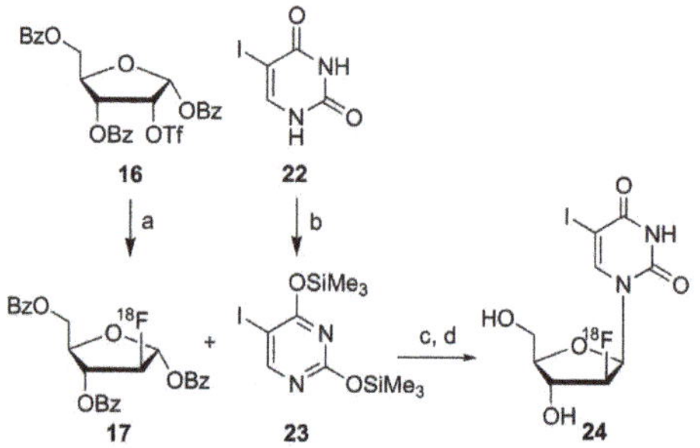

Scheme 7. Radiochemical synthesis of [^{18}F]FIAU (**24**). Reagents and conditions: (a) ^{18}F$^-$, Kryptofix, anh. CH$_3$CN, 95 °C, 30 min; (b) Hexamethyldisilazane, TMSOTf, anh. dichloroethane, 85 °C, 2 h; (c) anh. CH$_3$CN, 85 °C, 1 h; (d) NaOCH$_3$/CH$_3$OH, 85 °C, 10 min.

Finally the last step consisted of the coupling between the [^{18}F]FIAU (**24**) and the appropriate phosphorochloridate previously synthesised according to the standard NMI promoted procedure [19]. The phosphoramidate reaction between phosphorochloridate and nucleoside under non-radioactive conditions is reported in the literature as a room temperature reaction over 16h [19]. However, this procedure would not be suitable for a reaction as time sensitive as one involving the short-lived radionuclide fluorine-18 ($t_{1/2}$ = 109.7 min.). For this reason, we developed an assay to observe the progress of the phosphoramidate bond formation. The substrate of this assay was the non-radioactive FIAU and the reaction was monitored via ^{31}P NMR spectroscopy and HPLC chromatography. Surprisingly we observed almost complete conversion into the ProTide after 15 min when the reaction was conducted at mild temperatures (50 °C) to then reach a steady state at around 30 min (Figure 3).

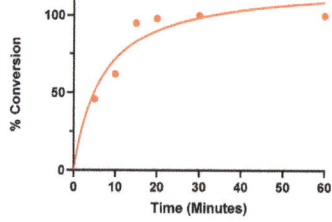

Figure 3. Conversion rate of FIAU (**19**) into FIAU ProTide (**21**): % conversion to FIAU ProTide during the coupling reaction was calculated over time at 50 °C using ^{31}P NMR spectroscopy and analytical HPLC chromatography.

We therefore applied the same conditions for the radioactive reaction and observed formation of the final compound (**2**) at 50 °C after 15–20 min. (Scheme 8).

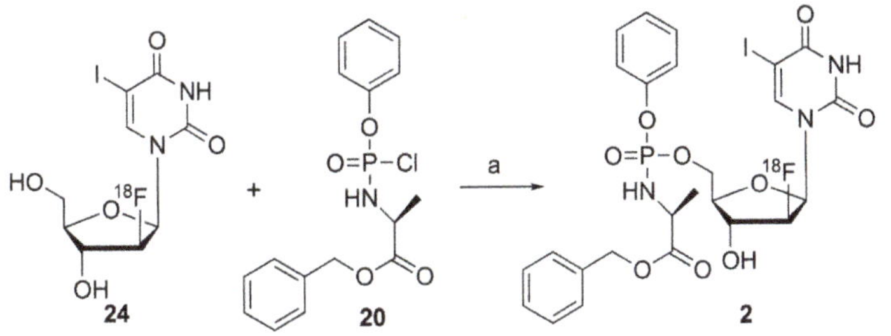

Scheme 8. Synthesis of the [^{18}F]FIAU ProTide. Reagents and conditions: (a) NMI, anh. THF, 50 °C, 20 min.

Satisfyingly, when an aliquot was taken to perform an analytical HPLC evaluation, [^{18}F]FIAU ProTide (**2**) was observed to be the main product of the reaction (Figure S11). The product was then isolated via semi preparative HPLC and was eluted after 23 min at a flow rate of 3.5 mL/min using 50% CH$_3$CN/50% H$_2$O as the mobile phase. An aliquot of the purified sample was analysed by analytical HPLC via co-elution with the non-radioactive standard (Figure 4).

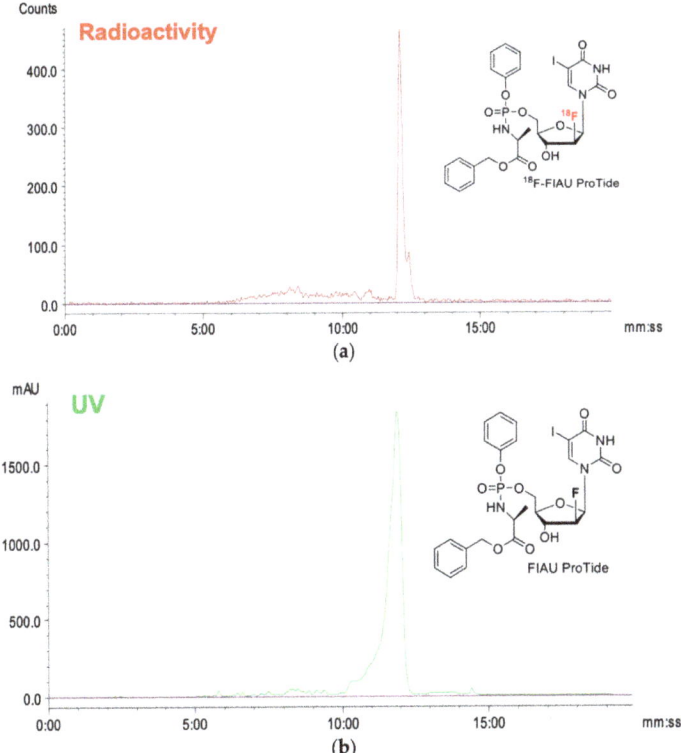

Figure 4. HPLC of [^{18}F]FIAU ProTide. (**a**) Radioactive chromatogram of the purified [^{18}F]FIAU ProTide with R$_t$ of 12.3 min; (**b**) UV chromatogram of the [^{18}F]FIAU ProTide co-spiked with the non-radioactive standard. (R$_t$ of non-radioactive standard: 12.2 min). HPLC system: 90% H$_2$O/10% CH$_3$CN to 50% H$_2$O/50% CH$_3$CN.

Radiochemical reactions were carried out using starting activities between 7–15 GBq, leading to final product activities of 8–46 MBq in RCY of 1–5% (*n* = 7, decay-corrected from end of bombardment

(EoB)), with high radiochemical purities (98%) and molar activities of 53 GBq/µmol. The total synthesis time was 240 min after the end of bombardment (EoB), within the acceptable range of just over two half-lives for future pre-clinical/clinical applications.

3. Materials and Methods

3.1. General Non-Radioactive Chemistry: Reagents and Analytical Methods

All the reagents and anhydrous solvents were purchased from Sigma-Aldrich. FLT was purchased from Carbosynth Ltd. (Berkshire, UK). Fluka silica gel (35–70 mm) was used as stationary phase for column chromatography. ^{1}H NMR spectra were acquired for all known compounds whereas for novel compounds ^{1}H NMR, ^{31}P NMR, ^{13}C NMR, MS and HPLC data were acquired. ^{1}H NMR were measured using a Bruker Advance Ultra Shield spectrometer (500 MHz) at ambient temperature. Data were recorded as follows: chemical shift in δ ppm from internal standard tetramethylsilane; multiplicity (s = singlet; d = doublet; t = triplet; m = multiplet); coupling constant (Hz); integration and assignment. ^{13}C NMR spectra were measured using a Bruker Advance Ultra Shield spectrometer (125 MHz) at ambient temperature. Chemical shifts were recorded in ppm from the solvent resonance used as the internal standard (e.g., CDCl$_3$ at 77.00 ppm). ^{31}P NMR spectra were recorded on a Bruker Advance Ultra Shield spectrometer (202 MHz) at ambient temperature. ^{19}F NMR spectra were recorded on a Bruker Advance Ultra Shield (474 MHz) spectrometer at ambient temperature. High-performance liquid chromatography (HPLC) analysis was conducted on an Agilent Technology 1200 Series System at the PET imaging centre in Cardiff (PETIC) with an analytical reversed phase column (Phenomenex Synergi 4µ Hydro-RP 80, C-18, 4.6 × 250 mm). Thin-layer chromatography (TLC) was conducted on pre-coated silica gel 60 GF$_{254}$ plates. Mass spectrometry analysis (LC-ESI-MS) was performed on a Bruker micro-TOF and with an Agilent 6430 T-Quadrupole spectrometer. High-resolution mass spectrometry (ESI-HRMS) was determined at the EPSRC National Mass Spectrometry facility at Swansea University (Swansea, UK).

3.2. General Radiochemistry: Source, Equipment and Analytical Methods

[^{18}F]Fluoride was produced in an IBA Cyclon 18/9 cyclotron using the ^{18}O(p,n)^{18}F nuclear reaction. ^{18}O-Enriched water (enrichment grade 98%, 2.2 mL, Nukem GmbH, Alzenau, Germany) was irradiated with 18 MeV protons. Radiofluorinations were performed on an Eckert & Ziegler module system. The drying procedures were performed with a vacuum pump N820 (Neuberger, Freiburg, Germany). Semi-prep HPLC (Phenomenex Synergi 4µ Hydro-RP 80, C-18, 10 × 250 mm) with a smartline pump 100–126 connected to the Eckert & Ziegler module system was used for the purification of the radiolabelled products. QMA and Al sep-pak (Waters corp., Milford, MA, USA) were used for purification of fluorine-18 intermediates. HPLC analytical evaluation was conducted on an Agilent Technology 1200 Series System with an analytical reversed phase column (Phenomenex Synergi 4µ Hydro-RP 80, C-18, 4.6 × 250 mm) coupled with a RAM/RAM Model 4 detector (Lablogic System, Ltd., Sheffield, UK) for radio-HPLC purposes.

3.3. Procedures and Spectroscopic Data for the Synthesis of the [^{18}F]FLT-ProTide

Synthesis of (2S)-ethyl-2-((chloro(phenyl)phosphoryl)amino)propanoate (10). C$_{11}$H$_{15}$ClNO$_4$P; MW: 291.6. Compound 10 was synthesized according to standard procedure [19]. Anhydrous triethylamine (2 eq; 0.662 mL; 0.480 g; 4.74 mmol) was added to the phenyldichlorophosphate (9) (1 eq; 0.354 mL; 0.500 g; 2.37 mmol) and L-alanine ethyl ester hydrochloride salt (1 eq; 0.364 g; 2.37 mmol) in anhydrous CH$_2$Cl$_2$ (5 mL) to obtain the final product 10 as a yellowish oil that was used without further purification. Yield: 92%. ^{1}H NMR (500 MHz, CDCl$_3$): δ 7.35–7.41 (m, 2H, Ar-H), 7.21–7.30 (m, 3H, Ar-H), 4.53 (m, 1H, NH), 4.21 (m, 1H, CH), 3.95 (m, 2H, CH$_2$), 1.51 (m, 3H, CH$_3$), 1.23 (m, 3H, CH$_3$). ^{31}P NMR (202 MHz, CDCl$_3$): δ 7.71, 8.05. Spectroscopic data in agreement with literature [29–31].

Synthesis of *(2S)-ethyl2-(((((2R,3S,5R)-3-fluoro-5-(5-methyl-2,4-dioxo-3,4-dihydropyrimidin-1(2H)-yl) tetrahydrofuran-2-yl)methoxy)(phenoxy)phosphoryl)amino)propanoate* **(11)**. $C_{21}H_{27}FN_3O_8P$, MW: 499.4. FLT (**8**) (1 eq; 0.100 g; 0.41 mmol) in anhydrous THF, was reacted with ^tBuMgCl (1.5 eq; 0.08 mL) under nitrogen atmosphere. The reaction mixture was stirred at rt for 30 min. A solution of the phosphorochloridate **10** (2 eq; 0.239 g; 0.82 mmol) in anhydrous THF was then added dropwise and the reaction mixture was left to stir overnight. The solvent was then evaporated under reduced pressure and the residue was purified by silica gel column chromatography (CH_2Cl_2/CH_3OH from 100% CH_2Cl_2 to 95% CH_2Cl_2) to give the FLT ProTide **(11)** as a yellowish oil. Yield: 23.5%. Rf: 0.44 in 90% CH_2Cl_2/10% CH_3OH TLC system. ^1H NMR (500 MHz, $CDCl_3$): δ 8.60 (br s, 1H, NH), 7.55 (s, 1H, H-6), 7.33–7.41 (m, 2H, Ar-H), 7.20–7.29 (m, 3H, Ar-H), 6.23–6.30 (m, 1H, H-1′), 5.33 (m, 1H, H-3′), 4.43–4.53 (m, 2H, NH, H-4′), 4.21 (m, 1H, CH), 3.95 (m, 2H, CH_2), 3.35-3.51 (m, 2H, H-5′, H-5″), 2.48–2.51 (m, 1H, H-2″), 2.55–2.33 (m, 1H, H-2′), 1.93–1.74 (m, 3H, CH_3, thy), 1.30–1.18 (d, J = 6.9, CH_3-ala). ^{19}F NMR (479 MHz, $CDCl_3$): δ −173.70, −175.20. ^{31}P NMR (202 MHz, $CDCl_3$): δ 4.34, 4.12. MS (ESI$^+$): 498.1 [M − H$^+$]. HPLC: Rt: 10.8 min; Purity > 96%; [Gradient: (0′) 95%H_2O/5% CH_3CN − (5′) 50% H_2O/50% CH_3CN − (15′) 50% H_2O/50% CH_3CN − (20′) 95% H_2O/5% CH_3CN]. Spectroscopic data in agreement with literature [17].

Synthesis of *(3R,5R)-3-(hydroxymethyl)-8-methyl-2,3-dihydro-5H,9H-2,5-methanopyrimido[2,1-b][1,5,3] dioxazepin-9-one* **(12)**. MF: $C_{10}H_{12}N_2O_4$; MW: 224.22. Thymidine (**3**) (1 eq, 0.250g, 1.32mmol) and triphenylphosphine (Ph_3P) (2 eq, 0.541 g, 2.64 mmol) were suspended in anhydrous acetonitrile (20 mL) and cooled down to −15 °C. Diisopropylazadicarboxylate (DIAD) (2 eq, 0.406 mL, 0.417 g, 2.64 mmol) was then added dropwise maintaining the temperature below −5 °C with vigorous stirring. The reaction was allowed to stir for 5h at 0 °C and then again cooled down to −20 °C. Cold ethyl acetate (20 mL) was added and the reaction was stirred for a further 15 min. A white precipitate was formed and was collected by Buchner filtration. The filtrate was washed with cold ethyl acetate and evaporated to dryness. The resulting crude compound was purified by silica gel column chromatography using 90% CH_2Cl_2/10% CH_3OH as eluent to obtain the product **12** as a white solid. Yield: 52%. Rf: 0.5. ^1H NMR (500 MHz, DMSO-d_6): δ 7.55 (d, J = 1.2, 1H, ArH), 5.80 (d, J = 3.9, 1H, H-1′), 5.23 (brs, 1H, H-3′), 5.01 (t, 1H, 5′-OH), 4.20 (m, 1H, H-4′), 3.51 (m, 2H, H-5′, H-5″), 2.55 (d, J = 1.2, H-2′,1H), 2.47 (ddd, $J^{1,8}$ = 19.0, $J^{1,4}$ = 6.7, $J^{1,2}$ = 3.0, 1H, H-2″), 1.76 (d, J = 1.1, 3H, CH_3). Spectroscopic data in agreement with literature [32].

Synthesis of *1-((2R,4R,5R)-4-hydroxy-5-(hydroxymethyl)tetrahydrofuran-2-yl)-5-methylpyrimidine-2,4(1H,3H)-dione* **(13)**. MF: $C_{10}H_{14}N_2O_5$; MW: 242. 3′-anhydrothymidine (**12**) (0.200 g; 0.892 mmol) in aq. 1.5 M NaOH (3.33 mL) was stirred in methanol (30 mL) under reflux for 3 h. Upon heating, the solution changed colour from clear to golden-brown. The reaction was monitored by TLC chromatography. When the conversion into the final product was confirmed, the solvent was evaporated under reduced pressure. The resulting crude compound was purified by silica gel column chromatography (CH_2Cl_2/CH_3OH gradient from 100% to 90% of CH_2Cl_2) to obtain the final product (**13**) as a white powder. Yield: 64%. Rf: 0.4 in 90% CH_2Cl_2/10% CH_3OH TLC system. ^1H NMR (500 MHz, DMSO-d_6): δ 11.24 (s, 1H, NH), 7.78 (s, 1H, H-6), 6.07 (dd, J = 8.5, 2.44, 1H, H-1′), 5.25 (d, J = 3.35, 1H, 3′-OH), 4.67 (t, J = 5.49, 1H, 5′-OH), 4.23 (m, 1H, H-3′), 3.60–3.84 (m, 3H, H-4′, H-5′ and H-5″), 2.55–2.59 (m, 1H, H-2″), 1.84 (dd, J = 14.95, J = 2.14, 1H, H-2′), 1.76 (s, 3H, CH_3). Spectroscopic data in agreement with literature [32].

Synthesis of *((2S)-ethyl-2-(((((2R,3R,5R)-3-hydroxy-5-(5-methyl-2,4-dioxo-3,4-dihydropyrimidin-1(2H)-yl) tetrahydrofuran-2-yl)methoxy)(phenoxy)phosphoryl)amino)propanoate* **(14)**. MF: $C_{21}H_{28}N_3O_9P$; MW: 497.4. Compound **14** was synthesised according to standard procedure [16]. 1-(2-deoxy-β-lyxofuranoxyl thymidine) (**13**) (1 eq; 0.175 g; 0.721 mmol) was reacted with NMI (5 eq; 0.287 mL; 0.297 g, 3.62 mmol) and ethyl-(2-chloro(phenyl)phosphorylamino)propanoate (**10**) (3 eq; 0.633 g; 2.17 mmol) to obtain a crude residue that was purified by silica gel column chromatography (97% CH_2Cl_2/3% CH_3OH) to afford the final compound **14** as a white solid. Yield: 18.4%. Rf: 0.4. ^{1}H NMR (500 MHz, CDCl$_3$): δ 7.58 (d, J = 7.0, 1H, H-6), 7.35–7.43 (m, 2H, Ar-H), 7.20–7.29 (m, 3H, Ar-H), 6.25–6.32 (m, 1H, H-1'), 5.22 (m, 1H, 3'-OH), 4.94 (m, 1H, H-3'), 4.31–4.52 (m, 2H, NH, H-4'), 3.95–4.03 (m, 1H, CH), 3.68 (d, J = 7.0, 2H, CH$_2$-ester), 3.35 (d, J = 16.0, 2H, H-5', H-5''), 2.48–2.51 (m, 1H, H-2''), 2.09–2.25 (m, 1H, H-2'), 1.85 (d, J = 10.1, 3H, CH$_3$-thy), 1.23 (m, 3H, CH$_3$-ala), 1.16 (d, J = 7.2, 3H, CH$_3$-ester). ^{31}P NMR (200 MHz, CDCl$_3$): δ 5.34, 4.98.

Synthesis of *(2S)-ethyl-2-(((((2R,3R,5R)-5-(5-methyl-2,4-dioxo-3,4-dihydropyrimidin-1(2H)-yl)-3-((methylsulfonyl) oxy)tetrahydrofuran-2-yl)methoxy)(phenoxy)phosphoryl)amino)propanoate* (**4**). MF: C$_{22}$H$_{30}$N$_3$O$_{11}$PS; MW: 575.5. Triethylamine (10 eq; 1.06 mL; 0.769 g; 7.6 mmol) and mesyl chloride (4 eq; 0.235 mL; 0.348 g; 3.04 mmol) were reacted with a solution of compound **14** (1 eq; 0.378 g; 0.76 mmol) in anhydrous CH$_2$Cl$_2$ (20 mL) at 0 °C. The reaction mixture was stirred at 0 °C for 10 min and then warmed to rt and stirred for 1.5 h. The crude mixture was diluted with sat. NaHCO$_3$ solution and extracted with CH$_2$Cl$_2$. After drying over Na$_2$SO$_4$, the solution was reduced under reduced pressure and the resulting crude compound was purified by silica gel column chromatography (CH$_2$Cl$_2$/CH$_3$OH gradient from 100% CH$_2$Cl$_2$ to 95% CH$_2$Cl$_2$) to give the product **4** as a white solid. Yield: 29.5%. Rf: 0.45 in 90% CH$_2$Cl$_2$/10% CH$_3$OH TLC system. ^{1}H NMR (500 MHz, CDCl$_3$): δ 8.93–8.96 (s, 1H, NH, thy), 7.33–7.32 (d, J = 7.0, 1H, H-6), 7.24–7.28 (m, 2H, Ar-H), 7.10–7.15 (m, 3H, Ar-H), 6.23–6.21 (m, 2H, H-1'), 5.19–5.15 (s, 1H, NH-ala), 3.84–4.37 (m, 7H, H-3', H-4', CH, H-5', H-5'', CH$_2$-ethyl), 2.96–3.01 (s, 3H, SO$_2$CH$_3$), 2.72–2.75 (m, 1H, H-2''), 2.38–2.42 (m, 1H, H-2'), 1.87 (d, J = 1.2, 3H, CH$_3$-thy), 1.28–1.33 (m, 3H, CH$_3$-ala), 1.19–1.16 (m, 3H, CH$_3$-ethyl). ^{13}C NMR (125 MHz, CDCl$_3$) δ 173.7–173.4 (C=O, acetyl), 163.6 (C-2), 150.42 (C-1), 135.1 (C-4), 129.8 (C-2; C-6Ar), 125.2 (C-4Ar), 120.2 (C-3; C-5Ar), 111.6 (C-3), 83.5 (C-1'), 79.9 (C-3'), 77.3 (C-4'), 63.7 (C-5'), 61.7 (CH$_2$-ethyl), 50.8 (CH-ala), 39.2 (C-2'), 38.8 (CH$_3$-mesyl), 21.34 (CH$_3$-ethyl), 14.04 (CH$_3$-ala), 12.76 (CH$_3$-thy). ^{31}P NMR (202 MHz, CDCl$_3$): δ 2.92, 2.63. MS(ESI$^+$): 576.2 [M + H$^+$]. HPLC: Rt: 13.2 min; Purity > 98%; [Gradient: (0') 95% H$_2$O/5% CH$_3$CN – (5') 50% H$_2$O/50% CH$_3$CN – (15') 50% H$_2$O/50% CH$_3$CN – (20') 95% H$_2$O/5% CH$_3$CN].

Synthesis of *(2S)-ethyl-2-(((((2R,3R,5R)-5-(5-methyl-2,4-dioxo-3,4-dihydropyrimidin-1(2H)-yl)-3-(tosyloxy) tetrahydrofuran-2-yl)methoxy)(phenoxy)phosphoryl)amino)propanoate* (**5**). MF: C$_{28}$H$_{34}$N$_3$O$_{11}$PS; MW: 651.6. To a solution of compound **14** (1 eq; 0.181 g; 0.364 mmol) in pyridine (5 mL), tosyl chloride (2 eq; 0.138 g; 0.727 mmol) and silver trifluoromethanesulfonate (AgOTf) (2 eq; 0.186 g; 0.727 mmol) were added at 0 °C. The reaction was stirred for 1 h and then slowly allowed to warm to rt and stirred for another 2 h. The reaction mixture was then diluted with EtOAc, filtered, and the filtrate was washed with H$_2$O and brine. The organic layer was dried over anhydrous Na$_2$SO$_4$ and the solvent was evaporated under reduced pressure. The crude residue was purified by silica gel column chromatography (95% CH$_2$Cl$_2$/5% CH$_3$OH) to furnish the tosylated compound **5** as a yellowish solid. Yield: 35%. Rf: 0.55. ^{1}H NMR (500 MHz, CDCl$_3$) δ 7.76–7.75 (d, J = 1.9, 1H, H6 Ar), 7.34 (dd, J = 15.7, 8.6, 4H, Ar-H tosyl), 7.27–7.17 (m, 5H, Ar), 6.19–5.20 (td, J = 7.8, 2.9, 1H, H-1'), 4.45–3.74 (m, 10H), 2.72–2.61 (m, 1H, CH-ala), 2.46 (s, 3H, CH$_3$, tosyl), 1.85 (s, 3H, CH$_3$, thy), 1.39 (t, J = 7.2 Hz, 3H, CH$_3$, ethyl), 1.31 (m, 3H, CH$_3$-ala). ^{13}C NMR (126 MHz, CDCl$_3$) δ 173.65, 173.59 (C-ala), 163.52 (C1-thy), 150.58, 150.53 (C3-thy), 150.25, 150.19 (C1-tosyl), 145.99, 145.98 (C1-phenyl), 135.04, 134.95 (CH-thy), 133.05, 132.90 (C4-tosyl), 130.27, 130.21 (CH, C2, C6-tosyl), 129.75, 129.70 (CH, C2, C6-phenyl), 127.64, 127.58 (CH, C4-phenyl), 125.10, 120.35 (CH, C3, C5-phenyl), 120.31, 120.21 (CH, C3, C5-tosyl), 111.15, 110.98 (C3-thy), 84.22, 83.99 (CH, C1'), 80.96, 80.90 (CH, C3'), 80.72, 80.66 (CH, C4'), 63.89, 63.85 (CH$_2$, C5'), 63.33, 63.29 (CH$_2$, ethyl), 50.39, 50.38 (CH, ala), 39.03 (CH$_2$, C3'), 21.69, 21.00 (CH$_3$-ethyl), 20.96, 20.95 (CH$_3$-tosyl), 14.12 (CH$_3$, ala), 12.49, 12.44 (CH$_3$-thy). ^{31}P NMR (202 MHz, CDCl$_3$) δ 2.78, 2.66. MS (ESI)$^+$: 652.2 [M + H$^+$]; 674.1 [M + Na$^+$]. HPLC: Rt: 16.03 min; Purity > 98%; [Gradient: (0') 95% H$_2$O/5% CH$_3$CN – (5') 50% H$_2$O/50% CH$_3$CN – (15') 50% H$_2$O/50% CH$_3$CN – (20') 95% H$_2$O/5% CH$_3$CN].

Synthesis of *(S)-ethyl-2-(((((2R,3R,5R)-5-(5-methyl-2,4-dioxo-3,4-dihydropyrimidin-1(2H)-yl)-3-(((4-nitrophenyl) sulfonyl)oxy)tetrahydrofuran-2yl)methoxy)(phenoxy)phosphoryl)amino)propanoate* (**6**). MF: C$_{27}$H$_{31}$N$_4$O$_{13}$PS;

MW: 682.5. The ProTide **14** (1 eq; 1.16 g; 2.34 mmol) was dissolved in pyridine (20 mL) at 0 °C. 4-nitrobenzenesulfonylchloride (nosyl chloride) (2 eq; 1.06 g; 4.79 mmol) and silver trifluoromethanesulfonate (AgOTf) (2 eq; 1.23 g; 4.79 mmol) were added and the reaction mixture was stirred at 0 °C. After 1h the reaction mixture was allowed to slowly warm to rt and stirred for another 2 h. The reaction mixture was then diluted with EtOAc, filtered, and the filtrate was washed with H_2O and brine. The organic layer was dried over anhydrous Na_2SO_4 and the solvent was evaporated under reduced pressure. Purification of the crude residue was accomplished by silica gel column chromatography (95% CH_2Cl_2/5% CH_3OH) to give the desired compound **6** as a yellowish solid. Yield: 60%. Rf: 0.6. ^1H NMR (500 MHz, $CDCl_3$) δ 8.77–8.67 (s, 1H, NH, thy), 8.41–8.39 (d, J = 2.2, 2H-Ar, nosyl), 8.14–8.08 (m, 2H-Ar, nosyl), 7.71 (ddd, J = 7.6, 4.7, 1.7, 1H, H6), 7.42–7.30 (m, 5H-Ar), 6.28–5.28 (m, 1H-H-1'), 4.51–3.75 (m, 7H), 2.79–2.46 (m, 1H, CH-ala), 1.96–1.86 (m, 2H, H-2', H-2''), 1.37 (t, J = 7.6, 3H, CH_3 ester), 1.31–1.25 (m, 6H, CH_3-thy, CH_3-ala). ^{13}C NMR (126 MHz, $CDCl_3$) δ 173.66, 173.36 (C-ala), 163.48 (C2, thy), 151.13, 151.10 (C1, thy), 150.29, 150.25 (C1, nosyl), 141.45, 141.35 (C1, phenyl), 134.73, 134.67 (CH, thy), 129.86, 129.81 (CH, C2-6, nosyl), 129.15, 129.13 (CH, C2, C6, phenyl), 125.28 (CH, C4, phenyl), 124.84, 124.77 (CH, C3, C5, phenyl), 120.15, 120.11 (CH, C3, C5, nosyl), 120.06, 120.02, 111.36, 111.23 (C, C3, thy), 84.14, 84.00 (CH, C1'), 80.58, 80.52 (CH, C3'), 80.25, 80.18 (CH, C4'), 63.17, 63.14 (CH_2, C5'), 62.82, 62.79 (CH_2, ethyl), 50.41, 50.20 (CH, ala), 39.18, 39.16 (CH_2, C2'), 20.94, 20.90 (CH_3, ethyl), 14.12, 14.11 (CH_3, ala), 12.58, 12.56 (CH_3, thy). ^{31}P NMR (202 MHz, $CDCl_3$) δ 2.75, 2.48. MS (ESI)$^+$: 705.1 [M + Na$^+$]. HPLC: Rt: 15.88 min; Purity > 99%; [Gradient: (0') 95% H_2O/5% CH_3CN – (5') 50% H_2O/50% CH_3CN – (15') 50% H_2O/50% CH_3CN – (20') 95% H_2O/5% CH_3CN].

Synthesis of *tert-butyl-3-((2R,4R,5R)-5-(((((tert-butoxycarbonyl)((S)-1-ethoxy-1-oxopropan-2-yl)amino)(phenoxy) phosphoryl)oxy)methyl)-4-(((4-nitrophenyl)sulfonyl)oxy)tetrahydrofuran-2-yl)-5-methyl-2,6-dioxo-3,6-dihydropyrimidine-1(2H)-carboxylate* (**7**). MF: $C_{37}H_{47}N_4O_{17}PS$. MW: 882.83. The nosylated ProTide (**6**) (1 eq, 0.050 g, 0.073 mmol) was dissolved in pyridine (6 mL) at rt under nitrogen atmosphere. To the stirring solution, di(tert-butyl)dicarbonate (Boc$_2$O) (1.3 eq, 0.021 mL, 0.020 g, 0.095 mmol) was added dropwise and the reaction was stirred for 16h. The crude mixture was evaporated under reduced pressure and was purified by silica gel column chromatography (CH_2Cl_2/CH_3OH gradient from 100% CH_2Cl_2 to 95% CH_2Cl_2) to give the final di-protected nosylated derivate **7** as a yellowish oil. Yield: 56%. Rf: 0.67 in 90% CH_2Cl_2/10% CH_3OH as TLC system. ^1H NMR (500 MHz, $CDCl_3$) δ 8.47–8.36 (d, 2H, J = 2.2, Ar, nosyl), 8.17–8.08 (m, 2H, Ar, nosyl), 7.38 (m, 1H, Ar), 7.32–7.24 (m, 4H), 7.17 (s, 1H, thy), 6.31–6.24 (m, 1H-H-1'), 4.51–3.75 (m, 7H), 2.78–2.43 (m, 1H, CH-Ala), 2.31–2.23 (m, 1H-H-2'), 2.01 (m, 3H, CH_3, thy), 1.53–1.49 (m, 9H, CH_3, tert-butyl), 1.44 (m, 9H, CH_3, tert-butyl), 1.38 (t, J = 7.6, 3H, CH_3, ester), 1.31 (m, 3H, CH_3, ala). ^{13}C NMR (126 MHz, $CDCl_3$) δ 175.50, 174.29 (C, ala), 161.32 (C2, thy), 150.13, 150.08 (C1, thy), 150.02, 150.00 (C1, nosyl) 143.51, 142.21 (C1, phenyl), 132.71, 132.23 (CH, thy), 130.68, 129.99 (CH, C2–C6, nosyl), 129.34, 129.5 (CH, C2, C6, phenyl), 126.28 (CH, C4, phenyl), 125.79, 124.85 (CH, C3, C5, phenyl), 120.15, 120.13 (CH, C3,C5, nosyl), 120.06, 120.02, 111.36, 111.23 (C, C3, thy), 84.13, 84.10 (CH, C1'), 80.78–80.77 (C-tert-butyl), 80.58, 80.52 (CH, C3'), 80.25, 80.18 (CH, C4'), 63.21, 63.20 (CH_2, C5'), 62.79, 62.77 (CH_2, ethyl), 50.39, 50.30 (CH, ala), 39.18, 39.15 (CH_2, C2'), 28.41–28.23 (CH_3, tert-butyl), 20.94, 20.90 (CH_3, ethyl), 14.12, 14.11 (CH_3, ala), 12.58, 12.56 (CH_3, thy). ^{31}P NMR: (202 MHz, $CDCl_3$): δ 2.61, 2.53. MS (ESI$^+$): 905.83 [M + Na$^+$]. HPLC: Rt: 17.1 min; Purity > 99%; [Gradient: (0') 95% H_2O/5% CH_3CN – (5') 50% H_2O/50% CH_3CN – (15') 50% H_2O/50% CH_3CN – (20') 95% H_2O/5% CH_3CN].

Synthesis of *tert-butyl-3-((2R,4S,5R)-5-((((N-((S)-1-ethoxy-1-oxopropan-2-yl)-3,3-dimethylbutanamido)(phenoxy) phosphoryl)oxy)methyl)-4-[^{18}F]fluoro-tetrahydrofuran-2-yl)-5-methyl-2,6-dioxo-3,6-dihydropyrimidine-1(2H)-carboxylate* (**15**). MF: $C_{32}H_{45}^{18}FN_3O_{11}P$; MW: 696.70. Aqueous [$^{18}F$]fluoride (2–8 GBq) produced by the cyclotron was trapped in a QMA cartridge and was then eluted through the cartridge by an aqueous solution of $KHCO_3$ and Kryptofix in CH_3CN. The resulting [^{18}F]F$^-$/$KHCO_3$/Kryptofix complex was dried by an azeotropic distillation with anhydrous CH_3CN (2 × 1 mL) under reduced pressure and a stream of nitrogen. A solution of the precursor **7** (10 mg) in anhydrous CH_3CN (1 mL) was added and the reaction was stirred for 30 min at 95 °C. The resulting reaction mixture was passed through an alumina cartridge to obtain the radiolabelled product **15**. The reaction mixture was analysed by analytical radio HPLC: Rt: 15 min (analytical HPLC: (0′) 95% H_2O/5% CH_3CN − (5′) 50% H_2O/50% CH_3CN − (15′) 50% H_2O/50% CH_3CN − (20′) 95% H_2O/5% CH_3CN).

Synthesis of *ethyl ((((2R,3S,5R)-3-[^{18}F]fluoro-5-(5-methyl-2,4-dioxo-3,4-dihydropyrimidin-1(2H)-yl)tetrahydrofuran-2-yl)methoxy)(phenoxy)phosphoryl)-L-alaninate* (**1**). MF: $C_{21}H_{27}{}^{18}FN_3O_8P$; MW: 498.43. To the Boc protected [^{18}F]-radiolabelled ProTide (**15**), a solution of 2 N HCl (1 mL) was added and the reaction was stirred for 10 min. The solution was then neutralised with 2 N NaOH (1 mL). The crude mixture was then purified by semi-preparative HPLC and the desired compound was eluted after 35 min at a flow rate of 3.5 mL/min using 70% H_2O/30% CH_3CN as mobile phase. The compound was then dried under a stream of nitrogen, taken up in saline and flushed through a sterility filter to obtain the aqueous solution of [^{18}F]FLT ProTide (**1**). RCY of 15–30% (n = 5, decay-corrected from end of bombardment (EoB)), with high radiochemical purities (97%) and molar activities of 56 GBq/μmol. The total synthesis time was 130 min after the end of bombardment (EoB). The reaction mixture was analyzed by analytical radio-HPLC. Analytical HPLC: (0′) 95% H_2O/5% CH_3CN − (5′) 50% H_2O/50% CH_3CN − (15′) 50% H_2O/50% CH_3CN − (20′) 95% H_2O/5% CH_3CN); Rt: 9.5 min.

3.4. Procedures and Analytical Data for the Synthesis of the [^{18}F]FIAU ProTide

Synthesis of *2′-deoxy-2′-α-fluoro-5-iodouridine* (**19**). MW: 372.1; MF: $C_9H_{10}FIN_2O_4$. Iodine (1.2 eq; 1.24 g; 4.87 mmol) and ceric ammonium nitrate (CAN) (1 eq; 2.23 g; 4.062 mmol) were added to a stirring solution of 2′-β-fluoro-2′-deoxyuridine (**18**) (2 eq; 2.0 g; 8.12 mmol) in anhydrous acetonitrile (50 mL). The mixture was stirred at 75 °C for 1 h and was then quenched with a saturated solution of $Na_2S_2O_3$ and concentrated under reduced pressure. The residue was then re-dissolved in ethyl acetate and washed twice with saturated NaCl. The organic layer was dried over $MgSO_4$, filtered and concentrated to give compound **19** as a pale yellow solid. Yield: 60%. HPLC: Rt: 2.3 min; Purity > 95% (98% H_2O/2% CH_3CN). ^{1}H NMR (500 MHz, DMSO-d_6): δ 11.69 (s, 1H, NH), 8.53 (s, 1H, 6-CH), 5.86 (d, *J* = 15.8, 1H, 1′-CH), 5.60 (d, *J* = 6.4, 1H, 3′-OH), 5.39 (t, *J* = 4.5, 1H, 3′-CH), 5.04 (dd, *J* = 5.2, 4.1, 1H, 2′-CH), 4.18 (ddd, *J* = 23.4, 11.4, 7.2, 1H, 4′-CH), 3.90 (d, *J* = 8.2, 1H, 5′-OH), 3.85–3.79 (m, 1H, 5′-CH), 3.63–3.58 (m, 1H, 5′-CH). ^{13}C NMR (125 MHz, DMSO-d_6): δ 167.88 (C=O), 165.01 (C=O), 145.02 (C-6), 125.19 (CH, C-2′), 121.26 (CH, C-1′), 115.81 (CH, C-4′), 61.11 (CH, C-5), 57.30 (CH, C-3′), 45.87 (CH, C-5′). ^{19}F NMR (470 MHz, DMSO-d_6): δ −202.09. Spectroscopic data in agreement with literature [10].

Synthesis of *benzyl(chloro(phenoxy)phosphoryl)-L-alaninate* (**20**). MW: 353.73; MF: $C_{16}H_{17}ClNO_4P$. Compound **20** was synthesised according to the standard procedure [16]. Anhydrous triethylamine (2 eq; 1.26 mL; 0.918 g; 9.08 mmol), phenyl dichlorophosphate (1 eq; 0.678 mL; 0.958 g; 4.54 mmol) and L-alanine benzyl ester hydrochloride salt (1 eq; 1.50 g; 4.54 mmol) were reacted to give compound **20** as a yellowish oil. Yield: 88%. ^{1}H NMR (500 MHz, CDCl$_3$): δ 7.54–7.47 (m, 7H, Ar-H), 7.46–7.40 (m, 3H, Ar-H), 5.27 (d, *J* = 8.4, 2H, CH$_2$-ester), 4.69 (d, *J* = 9.9, 1H, NH), 4.13 (dd, *J* = 34.4, 29.8, 1H, CH-ala), 1.52 (m, 3H, CH$_3$-ala). ^{31}P NMR (202 MHz, CDCl$_3$): δ 8.03, 7.75. Spectroscopic data in agreement with literature [29–31].

Synthesis of *benzyl((((2R,3R,4S,5R)-4-fluoro-3-hydroxy-5-(5-iodo-2,4-dioxo-3,4-dihydropyrimidin-1(2H)-yl)tetrahydrofuran-2-yl)methoxy)(phenoxy)phosphoryl)-L-alaninate* (**21**). MF: $C_{25}H_{26}FIN_3O_9P$; MW: 689.3. Compound **21** was prepared according to the standard procedure [16]. 1-(3-Fluoro-4-hydroxy-5-(hydromethyl)tetrahydrofuran-2-yl)-5-iodopyrimidine-2,4(1H,3H)-dione (**19**) (1 eq; 0.400 g; 1.07 mmol) and NMI (5 eq; 0.424 mL; 0.439 g; 5.35 mmol) were reacted with benzyl 2-(chloro(phenoxy)phosphorylamino)propanoate (3 eq; 1.05 g; 3.22 mmol) (**20**). The crude mixture was purified by silica gel column chromatography (CH_2Cl_2/CH_3OH from 100% CH_2Cl_2 to 95% CH_2Cl_2) to obtain the product

21 as a yellowish oil. Yield: 10%. ^{1}H NMR (500 MHz, CDCl$_3$): δ 10.59 (s, 1H, 3-NH), 7.89 (s, 1H, 6-CH), 7.53–7.48 (m, 2H, CH-phenyl), 7.45 (t, J = 8.0, 2H, CH-phenyl), 7.41–7.38 (m, 2H, CH-benz), 7.17–7.15 (m, 2H, CH-benz), 7.13 (t, J = 8.0, 1H, CH-benzyl), 7.12 (t, J = 7.4, 1H, CH-phenyl), 5.99 (m, 1H, 1'-CH), 5.79 (dd, J = 47.6, 4.6, 1H, 2'-CH), 5.14 (m, 2H, CH$_2$-benz), 4.90 (m, 2H, 5'-CH$_2$), 4.39 (m, 1H, 4'-CH), 4.27 (m, 1H, 3'-CH), 4.19 (m, 1H, 3'-OH), 4.02 (m, 1H, NH-ala), 3.99 (m, 1H, CH-ala), 1.29 (d, J = 7.0, 3H, CH$_3$-ala). ^{13}C NMR (125 MHz, CDCl$_3$): δ 171.9 (C=O, ala), 168.12 (C=O, C-4), 144.18 (C=O, C-2), 149.0 (C-phenyl), 147.91 (C-benzyl), 145.5, 145.1 (CH, C-6), 128.31–121.48 (CH, Ar-C), 94.3 (C-5), 92.10 (CH, C-4'), 89.13 (CH, C-2'), 83.6, 82.97 (CH, C-1'), 81.08, 80.97 (CH$_2$, C-ester), 69.61, 68.14 (CH, C-3'), 67.23 (CH, C-5'), 52.8 (CH-ala), 21.98 (CH$_3$-ala). ^{19}F NMR (470 MHz, CDCl$_3$): δ -200.91, -201.15. ^{31}P NMR (202 MHz, CDCl$_3$): δ 3.90, 3.86. MS (ESI)$^+$: 690.3 [M + H$^+$]. HPLC: Rt: 13.4 min; purity > 97%; [Gradient: (0') 95% H$_2$O/5% CH$_3$CN – (5') 50% H$_2$O/50% CH$_3$CN – (15') 50% H$_2$O/50% CH$_3$CN – (20') 95% H$_2$O/5% CH$_3$CN].

Synthesis of *(2R,3S,4R,5R)-5-((benzoyloxy)methyl)-3-[^{18}F]fluoro-tetrahydrofuran-2,4-diyl-dibenzoate* (**17**). MF: C$_{26}$H$_{21}$^{18}FO$_7$ MW: 463.4 Aqueous [^{18}F]fluoride (4.11 GBq), produced by the cyclotron was trapped in a QMA cartridge before it was eluted with an aqueous solution of KHCO$_3$ and Kryptofix in anhydrous CH$_3$CN. The [^{18}F]F$^-$/KHCO$_3$/Kryptofix complex was dried by co-evaporation with anhydrous CH$_3$CN (2 × 1 mL) under reduced pressure and a stream of nitrogen. A solution of the triflate precursor (**16**) (10 mg) in anhydrous CH$_3$CN (1 mL) was added to the reaction vial and the reaction was stirred for 30 min at 95 °C. The mixture was passed through an alumina cartridge to obtain the radiolabelled product **17** that was used for next step without further purification. The reaction mixture was analysed by analytical HPLC. Rt: 8.3 min (100% H$_2$O).

Synthesis of *5-iodo-2,4-bis((trimethylsilyl)oxy)pyrimidine* (**23**). MF: C$_{10}$H$_{19}$IN$_2$O$_2$Si$_2$; MW: 382.3. To a solution of 5-iodouracil (**22**) (1 eq; 10 mg; 0.042 mmol) in dichloroethane (500 µL), hexamethyldisilazane (11.4 eq; 100 µL; 0.0774 mg; 0.479 mmol) and TMSOTf (13.1 eq; 100 µL, 0.123 mg; 0.549 mmol) were added. The mixture was stirred for 2 h at 85 °C and was used for next step without further purification. The purity of the compound was assessed by analytical HPLC (Rt = 6.9 min; 88% H$_2$O/12% CH$_3$CN) and LC-MS ([M + H$^+$]: 383.2). Spectroscopic data was in agreement with literature [10].

Synthesis of *1-((2R,3S,4R,5R)-3-[^{18}F]fluoro-4-hydroxy-5-(hydroxymethyl)tetrahydrofuran-2-yl)-5-iodopyrimidine-2,4(1H,3H)-dione ([^{18}F]FIAU)* (**24**). MF: C$_9$H$_{10}$^{18}FN$_2$O$_5$ MW: 371.0. Compound **17** was delivered to a vial containing 2-4-bis(trimethylsilyl)-5-iodouracil (**23**). The mixture was then heated at 85 °C for 60 min. To this mixture, 0.5 M of NaOCH$_3$ in CH$_3$OH (1 mL) was then added and the reaction was stirred at 85 °C for another 5 min. The precipitate was then reconstituted in water (1 mL) and neutralised with 6 N HCl. The reaction mixture was analyzed by analytical HPLC showing the formation of the 2 anomers (α and β) of the 2'-deoxy-2'-fluoro-5-iodouridine. Analytical HPLC: 98% H$_2$O/2% CH$_3$CN; Rt: α anomer 2.1 min; β anomer 2.9 min). The anomeric mixture was then purified by semi-preparative HPLC and the target compound (**24**) was eluted after 7.3 min at a flow rate of 3.5 mL/min using 20% CH$_3$CN/80% H$_2$O as mobile phase to obtain the final compound. HPLC: Rt: 2.1 min; 98% H$_2$O/2% CH$_3$CN. Data in agreement with literature [10].

Synthesis of *[^{18}F]FIAU ProTide* (**2**). MF: C$_{25}$H$_{26}$^{18}FIN$_3$O$_9$P; MW: 688.3. To [^{18}F]FIAU (**24**), NMI (0.1 mL) and a solution of benzyl-2-(chloro(benzyloxy)phosphorylamino)propanoate (**20**) (0.050 g) in anhydrous THF (0.5 mL) were added together under nitrogen atmosphere. The reaction mixture was stirred at 50 °C for 20 min and then dried under a flow of nitrogen, re-dissolved in CH$_3$CN and purified via semi-preparative HPLC. The compound **2** was eluted after 23 min at a flow rate of 3.5 mL/min using 50% CH$_3$CN/50% H$_2$O as the mobile phase. The solvent was then removed from the mixture under a stream of nitrogen. The final product was then re-formulated in saline and flushed through a sterility filter to furnish a clean sterile aqueous solution of [^{18}F]FIAU ProTide (**2**). Radiochemical reactions were carried out using starting activities between 4–15 GBq, leading to RCY's of 1–5% (n = 7, decay-corrected from end of bombardment (EoB)), with high radiochemical purities (98%) and molar activities of 53 GBq/μmol. The total synthesis time was 240 min after the end of bombardment (EoB). Analytical HPLC: (0′) 95% H$_2$O/5% CH$_3$CN - (5′) 50% H$_2$O/50% CH$_3$CN- (15′) 50% H$_2$O/50% CH$_3$CN- (20′) 95% H$_2$O/5% CH$_3$CN); Rt: 12.3 min.

4. Conclusions

Phosphoramidate ProTide technology is a successful prodrug strategy to deliver nucleosides to their target sites, reducing toxicity issues and improving the potency of their parent nucleosides. Several fluorinated ProTides are currently being evaluated as anticancer and antiviral agents at different stages of clinical trials. PET imaging has the potential to provide the pharmacokinetic profile of certain drug candidates directly in vivo and therefore to predict the response to therapy. In this study we have developed the first radiochemical synthesis of the [^{18}F]FLT ProTide (**1**) chosen as a model standard of the class of 3′-fluorinated ProTides. An automated late stage [^{18}F]fluorination was tested on four different precursors with the best yields obtained when using a di-Boc protected nosyl derivative (**7**). The late stage fluorination and easy purification make this tracer a good candidate as a PET imaging probe with substantial potential for clinical application.

[^{18}F]FIAU ProTide (**2**) was synthesised as a model of the class of 2-′fluorinated ProTides. Despite the early stage introduction of the fluorine-18, we have optimized the following steps involving the formation of the phosphoramidate bond. This optimization reduced the overall reaction time whilst maintaining a reasonable yield and high purity of the final compound.

To our knowledge, this is the first time that [^{18}F] radiolabelled ProTides have been synthesised. These radiotracers have the potential in preclinical models to further elucidate the in vivo mechanism of biodistribution and metabolism, as well as to be clinically translated for diagnostic and therapeutic evaluation purposes.

Supplementary Materials: The following are available online. Figure S1. Internalization and metabolism of ProTides, bypassing the first-rate limiting step of the nucleoside analogues phosphorylation cascade; Figure S2. ^{31}P NMR stability study; Figure S3: E&Z modular lab sketch; Figure S4: Representative analytical HPLC chromatogram for the fluorination of the mesyl precursor **4**; Figure S5: Representative analytical HPLC chromatogram for the fluorination of the tosyl precursor **5**; Figure S6: Representative analytical HPLC chromatogram for the fluorination of the nosyl unprotected precursor **6**; Figure S7: Representative analytical HPLC chromatogram of the fluorination of the nosyl protected precursor **7**; Figure S8: Representative analytical HPLC chromatogram for the deprotection of the precursor **15** before purification; Figure S9: Representative analytical HPLC chromatogram for the fluorination of the sugar; Figure S10: Representative analytical HPLC chromatogram of the glycosylation reaction; Figure S11: Representative analytical HPLC chromatogram of the coupling reaction; Table S1: Radiolabelling attempts for the mesyl precursor (compound **4**); Table S2: Radiolabelling attempts for the tosyl precursor (compound **5**); Table S3: Radiolabelling attempts for the unprotected nosyl precursor (compound **6**).

Author Contributions: Conceptualization, C.M. (Christopher McGuigan) and A.D.W.; methodology, K.C.P., S.J.P.; investigation, A.C., A.K.H.D.; resources, C.M. (Christopher Marshall), F.A.; writing – original draft preparation, A.C., A.D.W.; writing – review and editing, A.C., A.D.W.; supervision, F.A., C.M. (Christopher McGuigan), A.D.W.; funding acquisition, C.M. (Christopher McGuigan), A.D.W. All authors have read and agreed to the published version of the manuscript.

Funding: This research was funded by Life Science Research Network of Wales, in collaboration with Cardiff University (U.K.).

Acknowledgments: We acknowledge Cardiff University (U.K.), the PET imaging centre in Cardiff (PETIC) and their staff for technical support, and the WBIC at Cambridge University (U.K.).

Conflicts of Interest: The authors declare no conflict of interest

References

1. Shelton, J.; Lu, X.; Hollenbaugh, J.A.; Cho, J.H.; Amblard, F.; Schinazi, R.F. Metabolism, biochemical actions, and chemical synthesis of anticancer nucleosides, nucleotides, and base analogs. *Chem. Rev.* **2016**, *116*, 14379–14455. [CrossRef] [PubMed]
2. Seley-Radtke, K.L.; Yates, M.K. The evolution of nucleoside analogue antivirals: A review for chemists and non-chemists. Part 1: Early structural modifications to the nucleoside scaffold. *Antivir. Res.* **2018**, *154*, 66–86. [CrossRef] [PubMed]

3. Damaraju, V.L.; Damaraju, S.; Young, J.D.; Baldwin, S.A.; Mackey, J.; Sawyer, M.B.; Cass, C.E. Nucleoside anticancer drugs: The role of nucleoside transporters in resistance to cancer chemotherapy. *Oncogene* **2003**, *22*, 7524–7536. [CrossRef] [PubMed]
4. Van Rompay, A.R.; Johansson, M.; Karlsson, A. Phosphorylation of nucleosides and nucleoside analogs by mammalian nucleoside monophosphate kinases. *Pharmacol. Ther.* **2000**, *87*, 189–198. [CrossRef]
5. Petrelli, F.; Coinu, A.; Borgonovo, K.; Cabiddu, M.; Ghilardi, M.; Barni, S. Polychemotherapy or gemcitabine in advanced pancreatic cancer: A meta-analysis. *Dig. Liver Dis.* **2014**, *46*, 452–459. [CrossRef] [PubMed]
6. Mehellou, Y.; Rattan, H.S.; Balzarini, J. The ProTide prodrug technology: From concept to the clinic. *J. Med. Chem.* **2018**, *61*, 2211–2226. [CrossRef]
7. McQuaid, T.; Savini, C. Sofosbuvir, a significant paradigm change in HCV treatment. *J. Clin. Transl. Hepatol.* **2015**, *3*, 27–35.
8. Palmer, D.H.; Ross, P.J.; Silcocks, P.; Greenhalf, W.; Faluyi, O.O.; Ma, Y.T.; Wadsley, J.; Rawcliffe, C.L.; Valle, J.W.; Neoptolemos, J.P. ACELARATE: A phase III, open label, multicentre randomised clinical study comparing Acelarin (NUC-1031) with gemcitabine in patients with metastatic pancreatic carcinoma. *J. Clin. Oncol.* **2018**, *36*, TPS537. [CrossRef]
9. Vaquero, J.J.; Kinahan, P. Positron Emission Tomography: Current challenges and opportunities for technological advances in clinical and preclinical imaging systems. *Annu. Rev. Biomed. Eng.* **2015**, *17*, 385–414. [CrossRef]
10. Anderson, H.; Pillarsetty, N.; Cantorias, M.; Lewis, J.S. Improved synthesis of 2′-deoxy-2′-[18F]-fluoro-1-β-d-arabinofuranosyl-5-iodouracil ([18F]-FIAU). *Nucl. Med. Biol.* **2010**, *37*, 439–442. [CrossRef]
11. Jacobson, O.; Kiesewetter, D.O.; Chen, X. Fluorine-18 radiochemistry, labeling strategies and synthetic routes. *Bioconjug. Chem.* **2015**, *26*, 1–18. [CrossRef]
12. Cavaliere, A.; Probst, K.C.; Westwell, A.D.; Slusarczyk, M. Fluorinated nucleosides as an important class of anticancer and antiviral agents. *Future Med. Chem.* **2017**, *9*, 1809–1833. [CrossRef]
13. Surasi, D.S.; Bhambhvani, P.; Baldwin, J.A.; Almodovar, S.E.; O'Malley, J.P. 18F-FDG PET and PET/CT patient preparation: A review of the literature. *J. Nucl. Med. Technol.* **2014**, *42*, 5–13. [CrossRef]
14. Meyer, J.P.; Probst, K.C.; Trist, I.M.L.; McGuigan, C.; Westwell, A.D. A Novel radiochemical approach to 1-(2′-deoxy-2′-[^{18}F]fluoro-beta-D-arabinofuranosyl)cytosine (F-18-FAC). *J. Label. Cpds. Radiopharm.* **2014**, *57*, 637–644. [CrossRef] [PubMed]
15. Meyer, J.P.; Probst, K.C.; Westwell, A.D. Radiochemical synthesis of 2′-[F-18]-labelled and 3′-[F-18]-labelled nucleosides for positron emission tomography imaging. *J. Label. Cpds. Radiopharm.* **2014**, *57*, 333–337. [CrossRef]
16. Peck, M.; Pollack, H.A.; Friesen, A.; Muzi, M.; Shoner, S.C.; Shankland, E.G.; Fink, J.R.; Armstrong, J.O.; Link, J.M.; Krohn, K.A. Applications of PET imaging with the proliferation marker [^{18}F]-FLT. *Q. J. Nucl. Med. Mol. Imaging* **2015**, *59*.
17. Velanguparackel, W.; Hamon, N.; Balzarini, J.; McGuigan, C.; Westwell, A.D. Synthesis, anti-HIV and cytostatic evaluation of 3′-deoxy-3′-fluorothymidine (FLT) pro-nucleotides. *Bioorg. Med. Chem. Lett.* **2014**, *24*, 2240–2243. [CrossRef] [PubMed]
18. Swamy, K.C.K.; Kumar, N.N.B.; Balaraman, E.; Kumar, K.V.P.P. Mitsunobu and related reactions: Advances and applications. *Chem. Rev.* **2009**, *109*, 2551–2651. [CrossRef] [PubMed]
19. Serpi, M.; Madela, K.; Pertusati, F.; Slusarczyk, M. Synthesis of phosphoramidate prodrugs: ProTide approach. In *Current Protocols in Nucleic Acid Chemistry*; John Wiley & Sons, Inc.: Hoboken, NJ, USA, 2013; Vol. Chapter 15; pp. 15.5.1–15.5.15. ISBN 9780471142706.
20. Yun, M.; Oh, S.J.; Ha, H.-J.; Ryu, J.S.; Moon, D.H. High radiochemical yield synthesis of 3′-deoxy-3′-[^{18}F] fluorothymidine using (5′-O-dimethoxytrityl-2′-deoxy-3′-O-nosyl-β-D-*threo*- pentofuranosyl)-thymidine and its 3-*N*-Boc-protected analogue as a labeling precursor. *Nucl. Med. Biol.* **2003**, *30*, 151–157. [CrossRef]
21. Buckingham, F.; Gouverneur, V. Asymmetric 18F-fluorination for applications in positron emission tomography. *Chem. Sci.* **2016**, *7*, 1645. [CrossRef]
22. Cole, E.L.; Stewart, M.N.; Littich, R.; Hoareau, R.; Scott, P.J.H. Radiosyntheses using fluorine-18: The art and science of late stage fluorination. *Curr. Top Med. Chem.* **2014**, *14*, 875–900. [CrossRef]
23. Alauddin, M.M. Nucleoside-based probes for imaging tumor proliferation using positron emission tomography. *J. Label. Comp. Radiopharm.* **2013**, *56*, 237–243. [CrossRef]

24. Sridharan, V.; Menéndez, J.C. Cerium(IV) ammonium nitrate as a catalyst in organic synthesis. *Chem. Rev.* **2010**, *110*, 3805–3849. [CrossRef]
25. Lepore, S.D.; Mondal, D. Recent advances in heterolytic nucleofugal leaving groups. *Tetrahedron* **2007**, *63*, 5103–5122. [CrossRef]
26. Casella, V.; Ido, T.; Wolf, A.P.; Fowler, J.S.; MacGregor, R.R.; Ruth, T.J. Anhydrous F-18 labeled elemental fluorine for radiopharmaceutical preparation. *J. Nucl. Med.* **1980**, *21*, 750–757.
27. Zhang, H.; Cantorias, M.V.; Pillarsetty, N.; Burnazi, E.M.; Cai, S.; Lewis, J.S. An improved strategy for the synthesis of [^{18}F]-labeled arabinofuranosyl nucleosides. *Nucl. Med. Biol.* **2012**, *39*, 1182–1188. [CrossRef]
28. Alauddin, M.M. Positron emission tomography (PET) imaging with (18)F-based radiotracers. *Am. J. Nucl. Med. Mol. Imaging* **2012**, *2*, 55–76. [PubMed]
29. Slusarczyk, M.; Lopez, M.H.; Balzarini, J.; Mason, M.; Jiang, W.G.; Blagden, S.; Thompson, E.; Ghazaly, E.; McGuigan, C. Application of ProTide technology to gemcitabine: A successful approach to overcome the key cancer resistance mechanisms leads to a new agent (NUC-1031) in clinical development. *J. Med. Chem.* **2014**, *57*, 1531–1542. [CrossRef] [PubMed]
30. McGuigan, C.; Murziani, P.; Slusarczyk, M.; Gonczy, B.; Vande Voorde, J.; Liekens, S.; Balzarini, J. Phosphoramidate ProTides of the anticancer agent FUDR successfully deliver the preformed bioactive monophosphate in cells and confer advantage over the parent nucleoside. *J. Med. Chem.* **2011**, *54*, 7247–7258. [CrossRef] [PubMed]
31. Derudas, M.; Quintiliani, M.; Brancale, A.; Graciela, A.; Snoeck, R.; Balzarini, J.; McGuigan, C. Evaluation of novel phosphoramidate ProTides of the 2'-fluoro derivatives of a potent anti-varicella zoster virus bicyclic nucleoside analogue. *Antivir. Chem. Chemother.* **2010**, *21*, 15–31. [CrossRef] [PubMed]
32. Tang, G.; Tang, X.; Wen, F.; Wang, M.; Li, B. A facile and rapid automated synthesis of 3'-deoxy-3'-[18F]fluorothymidine. *Appl. Radiat. Isot.* **2010**, *68*, 1734–1739. [CrossRef] [PubMed]

Sample Availability: Samples of intermediate (non-radioactive) compounds are available from the authors.

© 2020 by the authors. Licensee MDPI, Basel, Switzerland. This article is an open access article distributed under the terms and conditions of the Creative Commons Attribution (CC BY) license (http://creativecommons.org/licenses/by/4.0/).

Article

The Radiolabeling of a Gly-Sar Dipeptide Derivative with Flourine-18 and Its Use as a Potential Peptide Transporter PET Imaging Agent

Andrei Molotkov [1], John W. Castrillon [1], Sreevidya Santha [1], Paul E. Harris [2], David K. Leung [1], Akiva Mintz [1,*] and Patrick Carberry [1,*]

[1] Department of Radiology, Columbia University Medical Center, 722 W. 168th St., Room B05, New York, NY 10032, USA; am3355@cumc.columbia.edu (A.M.); jc944@cumc.columbia.edu (J.W.C.); sreevidya.gp@gmail.com (S.S.); davidleungmd@gmail.com (D.K.L.)
[2] Medicine Endocrinology, Columbia University Medical Center, 722 W. 168th St., Room B05, New York, NY 10032, USA; peh1@cumc.columbia.edu
* Correspondence: am4754@cumc.columbia.edu (A.M.); pc2545@cumc.columbia.edu (P.C.); Tel.: 1-(212) 305-8815 (P.C.)

Academic Editor: Svend Borup Jensen
Received: 13 December 2019; Accepted: 30 January 2020; Published: 2 February 2020

Abstract: We have developed a novel fluorine-18 radiotracer, dipeptide **1**, radiolabeled in two steps from mesylate **3**. The initial radiolabeling is achieved in a short reaction time (10 min) and purified through solid-phase extraction (SPE) with modest radiochemical yields (rcy = 10 ± 2%, $n = 5$) in excellent radiochemical purity (rcp > 99%, $n = 5$). The de-protection of the *tert*-butyloxycarbonyl (Boc) and trityl group was achieved with mild heating under acidic conditions to provide ^{18}F-tagged dipeptide **1**. Preliminary analysis of ^{18}F-dipeptide **1** was performed to confirm uptake by peptide transporters (PepTs) in human pancreatic carcinoma cell lines Panc1, BxPC3, and ASpc1, which are reported to express the peptide transporter 1 (PepT1). Furthermore, we confirmed in vivo uptake of ^{18}F-dipeptide tracer **1** using microPET/CT in mice harboring subcutaneous flank Panc1, BxPC3, and Aspc1 tumors. In conclusion, we have established the radiolabeling of dipeptide **1** with fluoride-18, and demonstrated its potential as an imaging agent which may have clinical applications for the diagnosis of pancreatic carcinomas.

Keywords: fluorine-18; peptide transporters; positron emission tomography; radiotracer

1. Introduction

The proton-coupled oligopeptide transport (POT), also called the peptide transport (PTR), comprises a family of transporters which are found in animal, plant, yeast, archaea, and both Gram-negative and Gram-positive bacterial cells [1–3]. In mammalian cells, the POT family is comprised of four members (PepT1, PepT2, PHT1, and PHT2) [4,5]. These elements are responsible for coordinating the intracellular transport of small peptides across membranes by coupling to an inwardly directed proton gradient and negative trans-membrane electrical potential.

In humans, the normal tissue distribution of PepT1 is predominantly in the apical plasma membrane of enterocytes in the small intestine where it helps in the absorption of nutrients and small peptides [6,7]. It is also found in renal proximal tubular cells and in bile duct epithelial cells. PepT2 is found in epithelial cells of the kidney tubule, lung, and cerebral cortex. PepT2 function is the reabsorption of di/tri-peptides and peptide-like drugs from the glomerular filtrate [8]. The expression of PepT1 and PepT2 has also been studied in tumor cells [9]. Immunohistochemistry, Western blotting, and gene expression studies revealed that PepT1 and PepT2 are over-expressed and regulated by different kinases in a variety of human cancer cell lines like bile duct epithelial cells and colon, prostate, and pancreatic cancer cell lines [10–13]. These results suggest that PepT1 and PepT2 may be clinically useful to target as biomarkers of specific epithelial cancers. H$^+$/Peptide transporters have been shown to be responsible for peptide transport activity in cancer cells, including pancreatic carcinoma [9,11,14].

The dipeptide PET radiotracer, [^{11}C]glycylsarcosine **2** (Gly-Sar) [15] has been shown to be specific in the imaging of PepT2, with a higher specificity to distinguish between tumors versus inflammatory tissue with respect to [^{18}F]FDG [9,11]. The design of our dipeptide is a modification from the work performed by Nabulsi et al. [15], in which the dipeptide Gly-Sar was radiolabeled with carbon-11 on the internal amine. We propose here the incorporation of fluorine-18 into a dipeptide Gly-Sar derivative. Brandsch et al. [16] lists criteria that are essential for dipeptides high-affinity substrate-receptor binding for PepT1 and PepT2. Following these suggestions, we designed a dipeptide that possesses a free amine and carboxyl-terminal, a bulky peptide side chain, and an overall hydrophobicity that follows the criteria to increase PepTs binding (Figure 1).

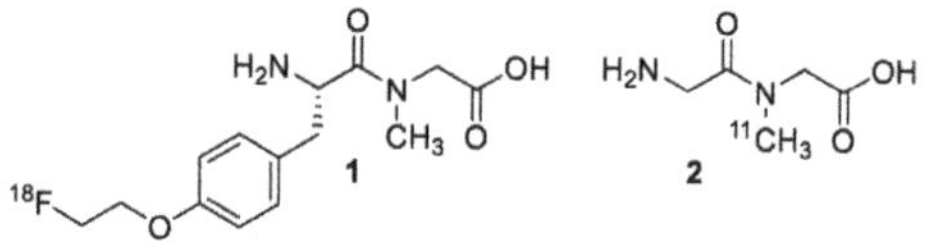

Figure 1. Structure of [^{18}F]FEPPG **1** and previously reported [^{11}C]Gly-Sar **2**.

Herein we report the synthesis of a novel ^{18}F-dipeptide **1**, (S)-N-(2-amino-3-(4-(2-(fluoro-^{18}F)ethoxy)phenyl)propanoyl)-N-methylglycine ([^{18}F]FEPPG), that meets the criteria as a suitable PepT binding substrate. We also demonstrate preliminary data confirming cell uptake of the radiolabeled dipeptide by several human pancreatic carcinoma cell lines and uptake of the [^{18}F]FEPPG **1** by μ-PET imaging on subcutaneous human xenograft Panc1, BxPC3, and ASpc1 tumor models in mice. This suggests that [^{18}F]FEPPG **1** is suitable for further study to explore its specificity as a PET tracer.

2. Results and Discussion

2.1. Gly-Sar Dipeptide Derivative Radiolabeled with Fluorine-18

In previous studies [^{11}C]Gly-Sar (Figure 1) has been reported [17] as a PET radiotracer for mouse heterotopic pancreatic cancer. It was suggested that [^{11}C]Gly-Sar had an advantage over [^{18}F]FDG by not binding to inflammation sites allowing to differentiate between inflammatory response and tumor. We propose the use of fluorine-18 nuclide to radiolabel a Gly-Sar dipeptide derivative since fluorine-18 has several important advantages over carbon-11 labeled compounds. Common advantages of fluorine-18 include a relatively high amount of theoretical molar activity (1.71×10^3 Ci/μmol), high resolution with a low positron energy (0.64 MeV), and a half-life of 109.8 min that allows for easier commercial distribution and longer scan times as compared to carbon-11. [18,19].

The synthesis of dipeptide **1** was a two-step procedure in which the fluorine-18 was first incorporated in the molecule, followed by the deprotection of the *tert*-butyloxycarbonyl (Boc) and *tert*-butyl group. Initially, solvents were screened (dimethyl sulfoxide, dimethylformamide, and acetonitrile) at a constant temperature, time and concentration to determine the suitable reaction media. From these tests (data not shown), deprotection performed in dimethyl sulfoxide result in least amount of side products, in the highest overall yield; it was therefore used as the solvent in all other subsequent reactions for the optimization of intermediate **4**. Next, a series of temperatures were studied, with 110 °C leading to the highest overall yield with the use of dimethyl sulfoxide as the solvent.

The optimization of the first step was further evaluated by adjusting the amount mesylate precursor, compound **3**, used for the formation of intermediate **4** (Table 1).

Table 1. Varying amount of starting material to produce intermediate **4**.

Entry	Amount of 3 (mg)	% Conversion to 4
1	0.5	8
2	1.0	12
3	2.0	23
4	4.0	18

An aliquot (10 µL) of reaction media was removed after 10 min and quenched in 1 mL of 50/50 (v/v) water/acetonitrile. This sample was evaluated by high-performance liquid chromatography (HPLC) to determine the amount of product formed relative to unreactive fluoride-18 complex and other impurities identified by the radiation detector. The use of 2.0 mg of compound **3** (entry 3, Table 1) provided radiolabeling of intermediate **4** in a 23% conversion. Doubling the amount of precursor from 1.0 mg (entry 2, Table 1) to 2.0 mg led to a 50% increase in the overall conversion of intermediate **4**. However, when 4.0 mg of precursor **3** (entry 3, Table 1) was used, the conversion to radiolabeled compound **4** had no significant increase in yield.

With the amount of starting material, solvent, and temperature determined, our attention turned to optimizing the reaction time for the radiolabeling of intermediate **4**. For this study, the initial labeling reaction was set-up and monitored using an analytical HPLC system. Aliquots were removed at set time points (Table 2) and the formation of desired radiolabeled species **4** was determined relative to other peaks observed in the radio-trace. As shown in Table 2, a 19% conversion of the desired product was attained in 10 min (entry 4). Since additional impurities were detected after 20 min in the radio-trace, 10 min was used as the optimized time for the reaction of mesylate **3** with dry flouride-18 complex.

Table 2. Kinetic study of the labeling of mesylate **3**.

Entry	Time (min)	% Conversion to 4
1	0	ND
2	2	5
3	5	14
4	10	19
5	20	14
6	60	5

For purification of ^{18}F-intermediate **4**, we tested HPLC with fraction collection as well as solid-phase extraction (SPE) cartridges. We found no detectable differences from the overall decay corrected yield or radiochemical purity achieved with the use of either purification method (data not shown). However, an important advantage of the SPE (t-C18) cartridge was the shortening of the synthesis time, by approximately 30 min, making it an optimal choice.

To obtain the de-protection of both the Boc and *tert*-butyl group in one-step, we screened several reaction conditions. We found that using hydrogen chloride solution (4.0 M in dioxane) and tetrahydrofuran with mild heating (see Method A) led to the final production of [^{18}F]FEPPG **1** with the minimal side products [20]. The use of 6 N hydrochloric acid in dimethyl sulfoxide (see Method B) also provided the desired radioligand (Scheme 1). When trifluoroacetic acid or higher temperatures (hydrochloric acid, 90 °C) was used for the de-protection, a partial decomposition of product was observed or sluggish reactions conditions resulted in elongated reaction times with only mono-de-protection of the final product.

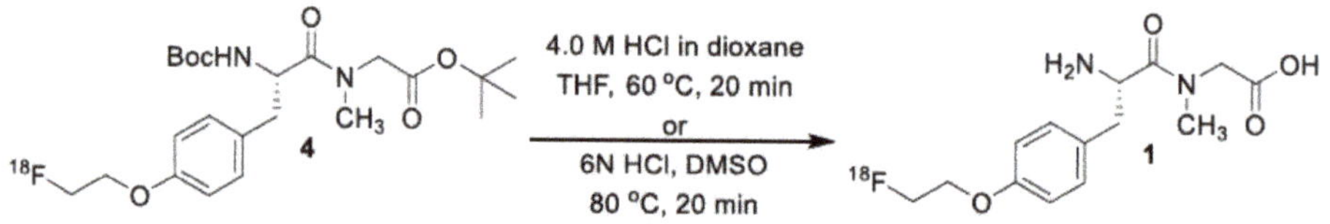

Scheme 1. Formation of desired [^{18}F]FEPPG 1.

Purification of [^{18}F]FEPPG 1 was achieved with the use of reverse phase HPLC (Figure 2). The fractions collected representing the desired product were pooled together and assayed to produce approximately 70% decay corrected yield of [^{18}F]FEPPG 1. However, after trapping and releasing (SPE) followed by the formulation process, less than 10% of radioligand was secured in >95% radiochemical purity. Current work is underway with the use of different SPE cartridges.

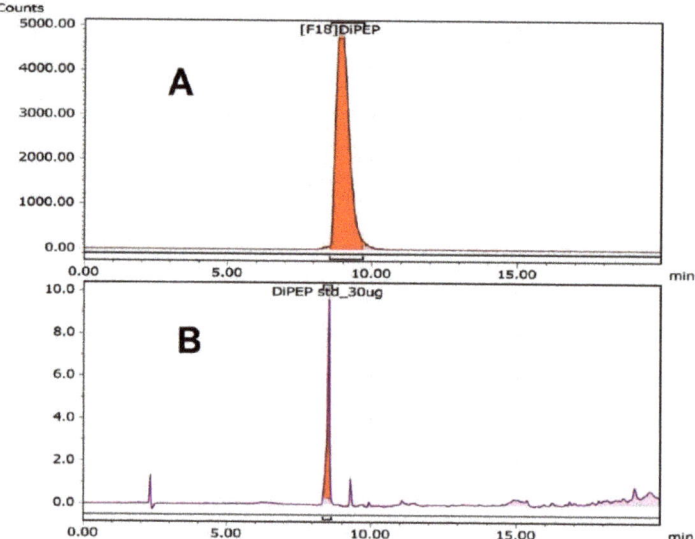

Figure 2. HPLC chromatogram of [^{18}F]FEPPG 1. (A) Radiotrace; (B) UV @ 280 nm. HPLC conditions: acetonitrile: 0.1% trifluoroacetic acid (aq) solution 10/90 ramp to 90/10 over 20 min; flow rate = 1.5 mL/min; Luna 5 u C18(2) 100 A, 250 × 4.60 nm column Agilent 1260 Infinity (Santa Clara, CA, USA).

The overall decay corrected yield for the formation of [^{18}F]FEPPG 1, starting from a raw solution of aqueous fluoride-18/[^{18}O]-enriched water (98%), was 19 ± 9% ($n = 5$) with an average total synthesis time of 117 min. The chemical purity of the final product, determined by HPLC analysis, was 33 ± 5% ($n = 5$) with a range of 0.5–4.8 µg/mL of the fluorine-19 isotope of FEPPG. The radiochemical purity determined by HPLC analysis of [^{18}F]FEPPG was determined to be 89 ± 5% ($n = 5$).

2.2. Initial Testing of [^{18}F]FEPPG in Pancreatic Carcinoma Cell Lines and Tumors

While the main goal of this work was to establish the chemical synthesis of [^{18}F]FEPPG **1**, we performed initial cell and tumor uptake studies to ensure that newly introduced fluorine-18 moiety did not negate cell and tumor binding previously reported in the carbon-11 analog. We therefore performed initial cell uptake studies on Panc1, BxPC3, and AsPC1 cells, which have been reported to express the PepTs, and found that all tested cell lines had uptake of [^{18}F]FEPPG **1** (Figure 3A).

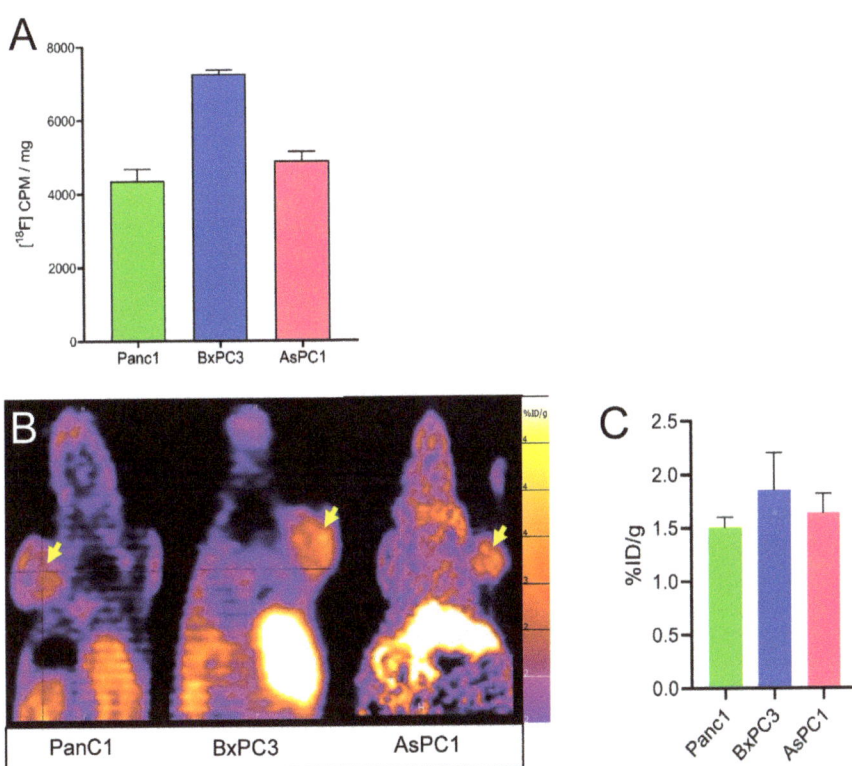

Figure 3. Cell uptake (**A**) and PET scanning (**B**,**C**) using [^{18}F]FEPPG **1**. (**A**) Panc1, BxPC3, and AsPC1 human pancreatic carcinoma cell uptake of [^{18}F]FEPPG **1**; (**B**,**C**) PET scanning using [^{18}F]FEPPG **1** tracer of mice (n = 3) with heterotrophic sc tumors originated from Panc1, BxPC3, and AsPC1 cells. Yellow arrow, sq tumor labeled by [^{18}F]FEPPG.

Furthermore, we found similar [^{18}F]FEPPG **1** uptake on PET imaging 30 min after injection of 3.7–5.5 MBq (100–150 µCi) of [^{18}F]FEPPG **1** (Figure 3B,C) in subcutaneous xenograft models of all pancreatic cancer cell lines tested (Panc1, BxPC3, and AsPC1). As expected, higher amounts of [^{18}F]FEPPG **1** uptake were observed in the liver and kidneys, in line with the transporter expression and excretion [21–23] (Figure 3B and data not shown).

In conclusion, we have a feasible synthesis pathway and initial indication of tumor uptake, justifying further study of [^{18}F]FEPPG **1**. Our future focus is on examining receptor specificity using unlabeled substrates of PepTs to demonstrate competition as well as defining radioligand binding to specific PEPT transporters by transfecting either PepT1 or PepT2 into non expressing cells and quantitating uptake of [^{18}F]FEPPG **1** caused by each transporter.

3. Conclusions

We have reported a new fluorine-18 labeled dipeptide **1**, which is successfully synthesized in two steps. The intermediate was radiolabeled in short reaction time (10 min) and could be easily purified by way of solid-phase extraction (*t*-C_{18} plus cartridge), eluted with acetonitrile, dried, and used directly in the next synthetic sequence without any further purification. The desired final product was formulated into an isotonic injectable solution. Of significance, the final injectable product demonstrated excellent radiochemical purity. We also performed preliminary cell and tumor uptake studies to demonstrate retained in vitro and in vivo uptake of our new molecular entity that has an additional fluorine-18 moiety compared to the originally reported carbon-11 tracer, justifying its further study to confirm PEPT specificity.

4. Material and Methods

4.1. General Information

Mesylate precursor **3**, intermediate **4**, and final product standard **1** were purchased from Syngene International Ltd. (Bangalore, India). Hydrogen chloride solution was purchased from Sigma-Aldrich (St. Louis, MO, USA) and used without further purification. Oxygen-18 enriched water (min. 98%) was purchased from Advanced Accelerator Applications (Milburn, NJ, USA) and used for the cyclotron bombardments (2.4 mL per bombardment). The cell lines (AsPC1, Panc1, and BxPC3) and culture media were obtained from American Type Culture Collection (ATCC, Manassas, VA, USA) Company. Fetal bovine serum came from Atlanta Biologicals (Norcross, GA, USA). DMEM was obtained from Fisher Scientific (Pittsburg, PA, USA). Fetal bovine serum (FBS) was obtained from Atlanta Biologicals (Norcross, GA, USA). All other reagents not listed above were of the highest grade available from Sigma-Aldrich (St. Louis, MO, USA) and Fisher Scientific (Pittsburgh, PA, USA). Radiochemical yields (rcy) are reported as decay corrected to the end of cyclotron target bombardment.

4.2. Animals

CrTac:NCr-Foxn1nu mice (Taconic, Rensselaer, NY, USA) referred to in text as NCr were maintained on a normal mouse diet. All animal experiments were conducted according to protocols approved by the Institutional Animal Care and Use Committee of Columbia University Medical Center.

4.3. Radiochemistry

4.3.1. (*S*)-*tert*-Butyl 2-(2-((*tert*-butoxycarbonyl)amino)-3-(4-(2-[^{18}F]fluoro-ethoxy)phenyl)-*N*-methylpropanamido)acetate (**4**)

A typical procedure for the formation of radioactive intermediate **4** is as follows: [^{18}O]-Enriched water (98%) aqueous solution of fluoride-18, produced from a Siemens 111 RDS cyclotron (Knoxville, TN, USA), was trapped on a pre-conditioned QMA light (Waters Corporation, Milford, MA, USA) cartridge. A solution of 1.5 mL (1.0 mL acetonitrile, 0.5 mL water) containing cryptand 222 (15 mg of 4,7,13,16,21,24-hexaoxa-1,10-diazabicyclo[8.8.8]hexacosane) and 3.0 mg potassium carbonate, was passed through the QMA light cartridge into a reaction vessel to provide K[222]^{18}F complex. This complex was dried in a heat block at 110 °C with a flow of argon until most of the solvent was evaporated. The solution was further azeotroped with the use of acetonitrile (3 × 1 mL) at 110 °C over a stream of argon gas. A solution of mesylate **3** (2.0 mg, 3.8 µmol) in 0.5 mL of anhydrous dimethyl sulfoxide was added to the "dry" K[222]^{18}F complex at 110 °C. After 10 min, the reaction mixture was removed from the heat and allowed to cool. The reaction mixture was then taken up in 20 mL of water and passed through a *t*-C_{18} plus (Waters Corporation) cartridge (activated by flushing with 10 mL ethanol followed by 10 mL water). The cartridge was then washed with another 10 mL of water and eluted into a test tube with 2.0 mL of acetonitrile to provide radiolabeled intermediate **4** (dcy = 10 ± 2%, n = 5; rcp > 99%). Radiochemical purity of intermediate **4** was determined by HPLC. HPLC conditions: Luna 5 μ C18(2) 100 Å column, 250 × 4.60 mm, 5 μm; acetonitrile/0.1 M ammonium formate (aq) in 0.5% acetic acid solution 50/50 ramped to 80/20 over 20 min with a flow rate of 2.0 mL/min; r_t = 10.05 min. The newly formed intermediate was dried at 90 °C over a stream of argon. The dried compound was taken to the next step.

4.3.2. (S)-2-(2-Amino-3-(4-(2-[^{18}F]fluoroethoxy)phenyl)-N-methylpropanamido)acetic acid, [^{18}F]FEPPG (1)

Method A: A typical procedure for the formation dipeptide **1** is as follows: Tetrahydrofuran (167 μL) followed by hydrogen chloride solution (833 μL, 4.0 M in dioxane) was added to dry intermediate **4**. The reaction mixture was placed in a heat block set to 60 °C. After 20 min, remove from heat and allowed to cool. The reaction mixture was diluted with 9/1 (v/v) 0.1% trifluoroacetic acid (aq)/acetonitrile (1 mL) solution and purified by HPLC (Luna 10 μ C18(2) 100 Å column, new column 250 × 10 mm; acetonitrile/0.1% trifluoroacetic acid (aq) solution 10/90 ramped to 90/10 over 30 min as mobile phase with a flow rate of 4.0 mL/min; fractions collected from 10.00 to 10.80 min). The fractions were pooled together and diluted with 40 mL of sterile water for injection. The reaction media were then passed through a t-C_{18} light (Waters Corporation, Milford, MA, USA) cartridge (activated by flushing with 10 mL ethanol followed by 10 mL water). The t-C_{18} light cartridge was then washed with sterile water for injection (10 mL), eluted into sterile 10 mL final product vial with ethanol (0.45 mL) followed by 0.9% saline (2.55 mL). An aliquot was removed for quality control testing.

Method B: A typical procedure for the formation of dipeptide **1** is as follows:

Dimethyl sulfoxide (20 μL) followed by 6 N hydrogen chloride (200 μL) was added to intermediate **4**. The reaction mixture was placed in a heat block set to 80 °C. After 20 min, was removed from heat and allowed to cool. The reaction mixture was then taken up in 50 mL of 10% sodium ascorbate solution (pH = 7.5). An aliquot was removed for quality control testing.

Each batch of [^{18}F]FEPPG **1** passed all required quality control tests, which included radionuclidic purity (511 keV), chemical (HPLC, identity) and radiochemical purity (HPLC, rcp ≥ 85%), excipients present (acetonitrile < 410 ppm, ethanol < 150,000 ppm), pH (range 4–8), and half-life (110.0 ± 5.0 min). Radiochemical purity/identity of **1** was determined by analytical HPLC. Analytical HPLC conditions: Luna 5 μ C18(2) 100 Å column, 250 × 4.60 mm, 5 μm; acetonitrile/0.1% trifluoroacetic acid (aq) solution 10/90 ramped to 90/10 over 20 min with a flow rate of 1.5 mL/min; r_t = 8.68 min. Fluorine-19 derivative of **1** varied from 0.5–4.8 μg/mL, based off calibration curve (y = 1.762x, R^2 = 0.996, not forced through zero intercept) at 280 nm; molar activity (A_m) = 17,000 ± 7000 MBq/μmol (460 ± 190 mCi/μmol), where n = 5.

4.4. Cell Uptake

For cell-uptake studies, Panc1, BxPC3, and AsPC1 human pancreas carcinoma cells were plated at 1 × 10^5 cells per well in individual 24-well plates 48 h before assay in 10% heat-inactivated FBS in DMEM with 1/100 of anti-anti and 1/200 of glutamine. For non-specific binding control, some wells were left without cells. [^{18}F]FEPPG **1**, at ~1.5 MBq per well (~40 μCi), was added to cells for 30 min and washed four times with PBS. Cells were lysed with 0.1 M sodium hydroxide and the resulting gamma-activity was measured on Hidex gamma counter (Hidex, Turku, Finland). For protein concentration, cells were harvested using 100 μL of RIPA buffer; protein concentration was measured using colorimetric assay.

4.5. PET Experiments

CrTac:NCr-Foxn1nu mice (Taconic, Rensselaer, NY, USA) were injected sc with 1 × 10^6 of Panc1, BxPC3 or AsPC1 human pancreatic carcinoma cells (ATCC) in 200 μL of Matrigel®. After tumors reached 1 cm^3, mice were injected IV with 3.7–5.5 MBq (100–150 μCi) of [^{18}F]FEPPG **1**. Static 60 min PET was acquired 30 min after [^{18}F]FEPPG **1** injection using Inveon micro PET scanner (Siemens, Munich, Germany). PET images were reconstructed using the 3D-OSEM algorithm with three iterations in a 256 × 256 matrix, without attenuation correction (Inveon, Siemens, Munich, Germany) and analyzed using VivoQuant ver 4 (Invicro, Boston, MA, USA).

4.6. Quantification and Statistical Analysis

Statistical analysis was performed using Prism 8.0 (GraphPad Software, San Diego, CA, USA). All data are represented as mean ± standard error. Statistical *p*-values were calculated using two-tailed Student's *t*-test for unpaired samples.

Author Contributions: Conceptualization: D.K.L., P.C.; Methodology: P.C., P.E.H.; Formal Analysis: P.C., A.M. (Andrei Molotkov), A.M. (Akiva Mintz); Investigation: P.C., J.W.C., A.M. (Andrei Molotkov), S.S., P.E.H.; Resources: P.E.H., D.K.L.; Data Curation: P.C., A.M. (Andrei Molotkov); Writing—Original Draft Preparation: P.C., A.M. (Andrei Molotkov); Writing—Review and Editing: P.C., A.M. (Andrei Molotkov), A.M. (Akiva Mintz), D.K.L., P.E.H.; Visualization, P.C., A.M. (Andrei Molotkov); Supervision: P.C., P.E.H., A.M. (Andrei Molotkov); Project Administration: P.C.; Funding Acquisition: P.C. All authors have read and agreed to the published version of the manuscript.

Funding: This publication was supported by the National Center for Advancing Translational Sciences, National Institutes of Health, through Grant Number UL1TR001873 (Reilly) to P.C.

Conflicts of Interest: Authors declare no conflict of interest.

References

1. Fei, Y.J.; Kanai, Y.; Nussberger, S.; Ganapathy, V.; Leibach, F.H.; Romero, M.F.; Singh, S.K.; Boron, W.F.; Hediger, M.A. Expression cloning of a mammalian proton-coupled oligopeptide transporter. *Nature* **1994**, *368*, 563–566. [CrossRef]
2. Verri, T.; Maffia, M.; Danieli, A.; Herget, M.; Wenzel, U.; Daniel, H.; Storelli, C. Characterisation of the H(+)/peptide cotransporter of eel intestinal brush-border membranes. *J. Exp. Biol.* **2000**, *203*, 2991–3001.
3. Conrad, E.M.; Ahearn, G.A. 3H-L-histidine and 65Zn(2+) are cotransported by a dipeptide transport system in intestine of lobster Homarus americanus. *J. Exp. Biol.* **2005**, *208*, 287–296. [CrossRef]
4. Sleisenger, M.H.; Burston, D.; Dalrymple, J.A.; Wilkinson, S.; Mathews, D.M. Evidence for a single common carrier for uptake of a dipeptide and a tripeptide by hamster jejunum in vitro. *Gastroenterology* **1976**, *71*, 76–81. [CrossRef]
5. Shen, H.; Smith, D.E.; Brosius, F.C., 3rd. Developmental expression of PEPT1 and PEPT2 in rat small intestine, colon, and kidney. *Pediatric Res.* **2001**, *49*, 789–795. [CrossRef]
6. Liang, R.; Fei, Y.J.; Prasad, P.D.; Ramamoorthy, S.; Han, H.; Yang-Feng, T.L.; Hediger, M.A.; Ganapathy, V.; Leibach, F.H. Human intestinal H+/peptide cotransporter. Cloning, functional expression, and chromosomal localization. *J. Biol. Chem.* **1995**, *270*, 6456–6463. [CrossRef]
7. Knutter, I.; Hartrodt, B.; Toth, G.; Keresztes, A.; Kottra, G.; Mrestani-Klaus, C.; Born, I.; Daniel, H.; Neubert, K.; Brandsch, M. Synthesis and characterization of a new and radiolabeled high-affinity substrate for H+/peptide cotransporters. *Febs J.* **2007**, *274*, 5905–5914. [CrossRef]
8. Groneberg, D.A.; Eynott, P.R.; Doring, F.; Dinh, Q.T.; Oates, T.; Barnes, P.J.; Chung, K.F.; Daniel, H.; Fischer, A. Distribution and function of the peptide transporter PEPT2 in normal and cystic fibrosis human lung. *Thorax* **2002**, *57*, 55–60. [CrossRef]
9. Mitsuoka, K.; Miyoshi, S.; Kato, Y.; Murakami, Y.; Utsumi, R.; Kubo, Y.; Noda, A.; Nakamura, Y.; Nishimura, S.; Tsuji, A. Cancer detection using a PET tracer, 11C-glycylsarcosine, targeted to H+/peptide transporter. *J. Nucl. Med. Off. Publ. Soc. Nucl. Med.* **2008**, *49*, 615–622. [CrossRef]
10. Tai, W.; Chen, Z.; Cheng, K. Expression profile and functional activity of peptide transporters in prostate cancer cells. *Mol. Pharm.* **2013**, *10*, 477–487. [CrossRef]
11. Mitsuoka, K.; Kato, Y.; Miyoshi, S.; Murakami, Y.; Hiraiwa, M.; Kubo, Y.; Nishimura, S.; Tsuji, A. Inhibition of oligopeptide transporter suppress growth of human pancreatic cancer cells. *Eur. J. Pharm. Sci. Off. J. Eur. Fed. Pharm. Sci.* **2010**, *40*, 202–208. [CrossRef] [PubMed]
12. Knutter, I.; Rubio-Aliaga, I.; Boll, M.; Hause, G.; Daniel, H.; Neubert, K.; Brandsch, M. H+-peptide cotransport in the human bile duct epithelium cell line SK-ChA-1. *Am. J. Physiol Gastrointest Liver Physiol* **2002**, *283*, G222–G229. [CrossRef] [PubMed]

13. Gonzalez, D.E.; Covitz, K.M.; Sadee, W.; Mrsny, R.J. An oligopeptide transporter is expressed at high levels in the pancreatic carcinoma cell lines AsPc-1 and Capan-2. *Cancer Res.* **1998**, *58*, 519–525. [PubMed]
14. Nakanishi, T.; Tamai, I.; Sai, Y.; Sasaki, T.; Tsuji, A. Carrier-mediated transport of oligopeptides in the human fibrosarcoma cell line HT1080. *Cancer Res.* **1997**, *57*, 4118–4122. [PubMed]
15. Nabulsi, N.B.; Smith, D.E.; Kilbourn, M.R. [11C]Glycylsarcosine: Synthesis and in vivo evaluation as a PET tracer of PepT2 transporter function in kidney of PepT2 null and wild-type mice. *Bioorganic Med. Chem.* **2005**, *13*, 2993–3001. [CrossRef]
16. Brandsch, M.; Knutter, I.; Bosse-Doenecke, E. Pharmaceutical and pharmacological importance of peptide transporters. *J. Pharm. Pharmacol.* **2008**, *60*, 543–585. [CrossRef]
17. Cherry, S.R. Fundamentals of positron emission tomography and applications in preclinical drug development. *J. Clin. Pharmacol.* **2001**, *41*, 482–491. [CrossRef]
18. Schlyer, D.J. PET tracers and radiochemistry. *Ann. Acad Med. Singap.* **2004**, *33*, 146–154.
19. Tateoka, R.; Abe, H.; Miyauchi, S.; Shuto, S.; Matsuda, A.; Kobayashi, M.; Miyazaki, K.; Kamo, N. Significance of substrate hydrophobicity for recognition by an oligopeptide transporter (PEPT1). *Bioconjug. Chem.* **2001**, *12*, 485–492. [CrossRef]
20. Jappar, D.; Wu, S.P.; Hu, Y.; Smith, D.E. Significance and regional dependency of peptide transporter (PEPT) 1 in the intestinal permeability of glycylsarcosine: In situ single-pass perfusion studies in wild-type and Pept1 knockout mice. *Drug Metab. Dispos. Biol. Fate Chem.* **2010**, *38*, 1740–1746. [CrossRef]
21. Shen, H.; Smith, D.E.; Yang, T.; Huang, Y.G.; Schnermann, J.B.; Brosius, F.C., 3rd. Localization of PEPT1 and PEPT2 proton-coupled oligopeptide transporter mRNA and protein in rat kidney. *Am. J. Physiol.* **1999**, *276*, F658–F665. [CrossRef] [PubMed]
22. Herrera-Ruiz, D.; Wang, Q.; Gudmundsson, O.S.; Cook, T.J.; Smith, R.L.; Faria, T.N.; Knipp, G.T. Spatial expression patterns of peptide transporters in the human and rat gastrointestinal tracts, Caco-2 in vitro cell culture model, and multiple human tissues. *Aaps Pharmsci* **2001**, *3*, E9. [CrossRef] [PubMed]
23. Wang, Y.; Hu, Y.; Li, P.; Weng, Y.; Kamada, N.; Jiang, H.; Smith, D.E. Expression and regulation of proton-coupled oligopeptide transporters in colonic tissue and immune cells of mice. *Biochem. Pharmacol.* **2018**, *148*, 163–173. [CrossRef] [PubMed]

Sample Availability: Samples of the standards for compounds **1** and compounds **4** are available from the authors.

© 2020 by the authors. Licensee MDPI, Basel, Switzerland. This article is an open access article distributed under the terms and conditions of the Creative Commons Attribution (CC BY) license (http://creativecommons.org/licenses/by/4.0/).

Review

Huntington's Disease: A Review of the Known PET Imaging Biomarkers and Targeting Radiotracers

Klaudia Cybulska [1,2,*], Lars Perk [2], Jan Booij [1,3], Peter Laverman [1] and Mark Rijpkema [1]

[1] Department of Radiology and Nuclear Medicine, Radboud University Medical Center, Geert Grooteplein-Zuid 10, 6525 EZ Nijmegen, The Netherlands; jan.booij@radboudumc.nl (J.B.); peter.laverman@radboudumc.nl (P.L.); mark.rijpkema@radboudumc.nl (M.R.)
[2] Radboud Translational Medicine B.V., Radboud University Medical Center, Geert Grooteplein 21 (route 142), 6525 EZ Nijmegen, The Netherlands; lars.perk@radboudumc.nl
[3] Department of Radiology and Nuclear Medicine, Amsterdam University Medical Centers, Academic Medical Center, Meibergdreef 9, 1105 AZ Amsterdam, The Netherlands
* Correspondence: klaudia.cybulska@radboudumc.nl

Academic Editor: Svend Borup Jensen
Received: 18 December 2019; Accepted: 15 January 2020; Published: 23 January 2020

Abstract: Huntington's disease (HD) is a fatal neurodegenerative disease caused by a CAG expansion mutation in the *huntingtin* gene. As a result, intranuclear inclusions of mutant huntingtin protein are formed, which damage striatal medium spiny neurons (MSNs). A review of Positron Emission Tomography (PET) studies relating to HD was performed, including clinical and preclinical data. PET is a powerful tool for visualisation of the HD pathology by non-invasive imaging of specific radiopharmaceuticals, which provide a detailed molecular snapshot of complex mechanistic pathways within the brain. Nowadays, radiochemists are equipped with an impressive arsenal of radioligands to accurately recognise particular receptors of interest. These include key biomarkers of HD: adenosine, cannabinoid, dopaminergic and glutamateric receptors, microglial activation, phosphodiesterase 10 A and synaptic vesicle proteins. This review aims to provide a radiochemical picture of the recent developments in the field of HD PET, with significant attention devoted to radiosynthetic routes towards the tracers relevant to this disease.

Keywords: Huntington's disease; mutant huntingtin; Positron Emission Tomography; radiochemistry; fluorine-18; carbon-11; radiopharmaceuticals; [^{11}C]raclopride; [^{18}F]MNI-659

1. Introduction

The purpose of this review is to provide a radiochemistry focused summary of the recent advancements in Positron Emission Tomography (PET) imaging of a rare genetic condition, Huntington's disease (HD). The prevalence of HD is 5 to 10 cases per 100,000 people worldwide [1]. It progresses with fatal and devastating psychiatric, cognitive and motor impairments, caused by mutant huntingtin (mHTT) protein expression. Some excellent reviews on the molecular imaging of HD have been published, however focusing primarily on clinical data [2–5]. This review is divided according to HD biomarkers thought to be most affected by the pathology. For each biomarker, a suitable PET tracer and its radiosynthesis are provided, together with the most important PET data, both clinical and preclinical, if available. The authors' intention is to highlight the importance of radiochemistry and design of novel and highly specific radioligands for PET imaging, which will further our understanding of the changes orchestrated by mutant huntingtin and may eventually lead to the invention of disease-modifying treatments.

1.1. Positron Emission Tomography

PET plays a pivotal part in the diagnosis and understanding of neurological pathophysiologies. It is a non-invasive molecular imaging technique employing radiopharmaceuticals, which, after crossing the blood–brain barrier, bind to a specific molecular target, such as a transporter or receptor, and enable accurate tracking of changes in their function. Nowadays, PET is equipped with an array of radiolabelled biomarkers for neuroimaging in psychiatry and neurodegenerative pathologies, such as Parkinson's disease (PD), Alzheimer's disease (AD) and HD.

1.2. Huntington's Disease

Although considerable progress has been made towards identifying some of the mechanisms involved in the HD pathogenesis, there are currently no disease-modifying strategies available [6]. The uncovering of the huntingtin (*htt*) mutation in 1993 has enabled intensified research efforts with the hope to slow down or stop progressive neuronal damage [4]. HD is an autosomal dominant disease of the central nervous system (CNS) caused by an expansion of the CAG sequence in the *huntingtin* gene (*HTT*), located on chromosome 4 [3,7]. As a result, the expressed huntingtin protein is a mutant (mHTT) with an expanded polyglutamine tract. It has been proposed that a flawed proteostasis network results from the aggregation of mHTT, initiating a cascade of devastating consequences for synchronised neuroreceptors [8]. Individuals with 7-12 CAG tracts are usually considered healthy, whilst those with 35 suffer from HD [4]. Additionally, full penetrance and consequently, rapid progression of the disease, is associated with more than 40 repeats [9]. There is an inverse correlation between the number of CAG repeats and the age of HD onset, usually mid-life, but juvenile forms have also been reported [10,11]. Aggregated mHTT, also called inclusion bodies (IBs), manifests itself primarily in a brain region called the striatum, where it causes dysfunction of medium spiny neurons (MSNs) and eventually their death [8]. HD-affected subjects suffer from different types of impairments: motor (chorea, loss of coordination and involuntary movements), cognitive, in the form of widely understood dementia, and psychiatric (depression, anxiety and personality changes).

2. Indicators of Huntington's Disease: From A to Z

Within the complex brain networks, several molecular mechanisms have been recognised for their involvement in the HD pathogenesis. These include glucose metabolism, dopaminergic system, phosphodiesterases and neuroinflammation. PET imaging of these biomarkers provides useful insight into the disease progression by quantifying, for example, receptor expression and density. To the best of our knowledge, no PET tracer targeting mHTT has been proposed to date. The emergence of such a radioligand could aid validation of novel disease-modifying therapies aimed at lowering levels of this neurotoxic aggregate. This review summarises the key biochemical targets within the central nervous system which could be relevant for HD and the corresponding PET radioligands, along with their radiosynthetic routes.

2.1. Adenosine Receptors

Somewhat less studied in the context of HD, adenosine receptors have been considered potential biomarkers of the pathology due to their involvement in neurotransmission [12,13]. Adenosine is a vital dopaminergic and glutamatergic modulator [14]. It acts as an inhibitory neurotransmitter by monitoring energy levels and usage. It exerts its action through four G protein-coupled receptors, A_1, A_{2A}, A_{2B} and A_3. A_1 receptors are expressed on dynorphinergic MSNs which also co-express dopamine D_1 receptors (for a more detailed explanation of D_1 receptor involvement in HD, see Section 2.3). Enkephalinergic MSNs co-express A_{2A} and dopamine D_2 receptors. The latter are found abundantly in the striatum and significantly less in other brain regions [15–17]. Both types have been investigated, although not considerably in HD, with PET imaging in preclinical and clinical settings.

2.1.1. [^{18}F]CPFPX

[^{18}F]CPFPX is a potent xanthine-based antagonist of the adenosine A$_1$ receptor. Its binding selectivity over A$_{2A}$ is 1200-fold [18]. The radioligand is synthesised via nucleophilic aliphatic substitution at the tosylate leaving group in the presence of Kryptofix 222 and potassium carbonate, followed by deprotection of the pivaloyloxymethyl group with an aqueous solution of sodium hydroxide (Scheme 1). The average radiochemical yield (RCY) is 45 ± 7%, with molar activity values exceeding 270 GBq/µmol. Molar activity expresses the measured radioactivity per mole of compound and is commonly reported in GBq/µmol [19]. Radiochemical purity (RCP) exceeded 98%. In vivo rodent experiments by Holschbach et al. and later by Bauer and co-workers, confirmed the suitability of the tracer for A$_1$ receptor studies, owing to favourable kinetics and behaviour in the presence of a standard A$_1$ desfluorinated antagonist, DPCPX [18,20].

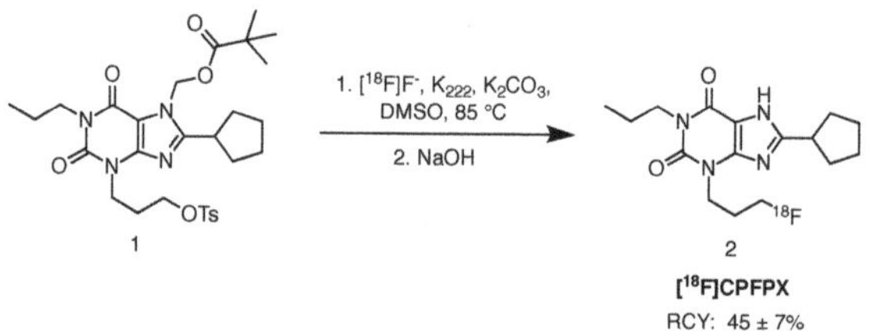

Scheme 1. Radiolabelling of [^{18}F]CPFPX, starting from the tosyloxy precursor 3 [18].

The tracer was used by Matusch et al. in premanifest and manifest HD individuals in an effort to elucidate the role of adenosine A$_1$ receptors in the pathology [21]. The former group was divided into far and near, based on the number of years until the calculated clinical onset. It was discovered that [^{18}F]CPFPX binding was globally higher for the far premanifest subjects than for the healthy controls, then levelling up for the near-to-onset subset and finally, decreasing in the caudate nucleus (part of the striatum), frontal cortex and amygdala for the symptomatic cohort. These results, although preliminary, suggest potential discriminatory power of adenosine A$_1$ biomarker for HD phenoconversion.

2.1.2. [^{11}C]KF18446

Ishiwata et al. investigated the adequacy of adenosine A$_{2A}$ receptor imaging with [^{11}C]KF18446, another xanthine-based ligand, based on evidence that there is a marked reduction in the receptor density of HD patients, particularly in the caudate nucleus, putamen and external globus pallidus [22,23]. Like the caudate nucleus, putamen is part of the striatum and together, the three brain areas are part of the so-called basal ganglia—responsible for the coordination of motor functioning. The study was performed with the aforementioned radioligand due to its promising imaging properties. The tracer was accessed through ^{11}C-methylation with [^{11}C]methyl iodide ([^{11}C]CH$_3$I) in the presence of caesium carbonate in DMF at 120 °C (Scheme 2). Radiochemical yields ranged between 25% and 46% with molar activities of 10–72 GBq/µmol. Radiochemical purity (RCP) was higher than 99%.

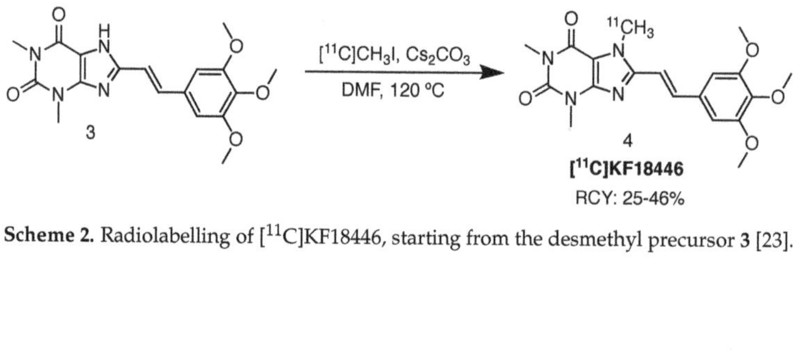

Scheme 2. Radiolabelling of [^{11}C]KF18446, starting from the desmethyl precursor 3 [23].

In 2002, Ishiwata and colleagues published an interesting PET study in which rats had been injected with quinolinic acid, an excitotoxin, to mimic HD pathology [23]. The compound causes damage to the dopaminergic neurons in the striatum. PET data, co-registered with Magnetic Resonance Imaging (MRI), were accompanied by in vitro and ex vivo autoradiography. The images displayed clear degeneration of not only A_{2A}, but also D_1 and D_2 receptors, calculated as a difference in the binding of the tracer between the non-lesioned and lesioned sides [23].

2.2. Cannabinoid Receptors

Gaining better understanding of the implication of the endocannabinoid system in synaptic transmission could bring us closer to unravelling the mechanism of HD and perhaps, further down the line, designing an efficient therapeutic against it. Cannabinoid receptor type 1, CB_1, is one of the major cannabinoid receptors, belonging to the G protein-coupled family, expressed abundantly in striatal MSNs, and overlooking the complex process of inhibitory neurotransmission [3,24]. The marked loss of CB_1 receptors in post mortem human brains, described by Glass et al., was one of the first receptor pathologies documented in HD [25]. It was most pronounced in the caudate nucleus, putamen and globus pallidus externus, even in the very early stage of the disease [26].

2.2.1. [^{18}F]MK-9470

PET allows for non-invasive in vivo imaging of various CNS receptors, including CB_1. Van Laere and colleagues administered a selective inverse agonist, [^{18}F]MK-9470, to 20 manifest HD patients [27]. Loss of PET signal was detected in the caudate and putamen, as well as the grey matter of cerebrum, brain stem and cerebellum. The HD cohort showed a significant loss of cannabinoid CB_1 receptors compared to the healthy controls. No relationship was elucidated between receptor binding and clinical HD scales or the number of CAG repeats. With this investigation, the researchers pioneered the use of CB_1 as a biomarker for HD PET imaging.

[^{18}F]MK-9470 can be radiolabelled by aliphatic nucleophilic substitution with [^{18}F]F$^-$ using the corresponding tosyloxy precursor. Thomae and co-workers published an automated Good Manufacturing Practice (GMP) compliant radiosynthetic route towards this ligand (Scheme 3) [28]. The authors reported an average RCY of $30 \pm 12\%$ ($n = 12$), RCP and molar activity exceeding 95% and 370 GBq/μmol, respectively.

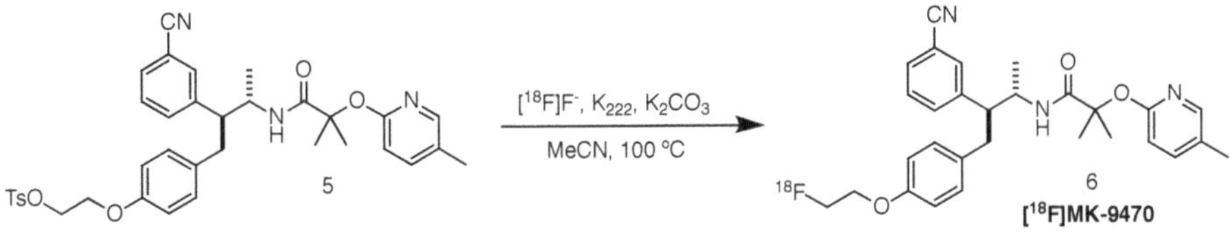

Scheme 3. Radiolabelling of [^{18}C]MK-9470, starting from the tosyloxy precursor 5 [28].

2.3. Dopaminergic Receptors

Dopaminergic receptors D_1 and D_2 constitute another part of the complex neurotransmission system which is affected by mHTT, mainly postsynaptically. Dysfunction of and fluctuations in the availability of these receptors have been one of the hallmarks of HD. Approximately half of MSNs express D_1 implicated in the "direct pathway", while the other half expresses mainly D_2 receptors implicated in the "indirect pathway" [29]. Postsynaptic nerve terminals which express D_1 and D_2 receptors, are severely affected by MSN degradation, caused by the presence of neurotoxic mHTT. PET radioligands which visualise density changes of these receptors can also provide insight into the pathology.

2.3.1. [^{11}C]SCH-23390

SCH-23390 is a potent halobenzazepine-based D_1 receptor antagonist [30,31]. The active form of this molecule is the *R*-enantiomer. Interestingly, in 1985, Friedman et al. began their evaluation of a ^{76}Br-brominated version of the molecule using small animal PET [31]. In 1989, DeJesus and colleagues evaluated the carbon-11 version of the ligand, which contains a ^{11}C-tagged *N*-methyl moiety, as a potential tracer for the imaging of CNS D_1 levels [32]. Nowadays, [^{11}C]SCH-23390 is still frequently used in preclinical and clinical PET studies of the D_1 receptor [33].

Typical radiolabelling via ^{11}C-alkylation of the secondary amine is achieved with [^{11}C]methyl iodide in the presence of a mild base, such as sodium bicarbonate, in DMF at 50 °C [34]. DeJesus et al. described the wet method, starting with lithium aluminium hydride-mediated reduction of cyclotron-acquired [^{11}C]CO$_2$, followed by iodination with hydroiodic acid to yield [^{11}C]methyl iodide. Radiolabelling was achieved by bubbling the resulting gaseous ^{11}C-methylating agent into a solution of the corresponding amine precursor and sodium hydrogen carbonate (Scheme 4). The authors reported RCY of 72% over 30 runs (based on [^{11}C]methyl iodide). The product obtained was radiochemically pure; however, contaminated with approximately 5% of the desmethyl precursor. Molar activities ranged from 370 to 8695 GBq/mmol. Halldin et al. also achieved successful radiolabelling of [^{11}C]SCH-23390, starting from the free base desmethyl precursor in acetone or DMSO-DMF at elevated temperatures [35]. Radiolabelling in acetone yielded the desired tracer in 80% RCY (based on [^{11}C]methyl iodide) and RCP higher than 99%. Publications that followed thereafter described radiolabelling of this tracer based on either of the protocols.

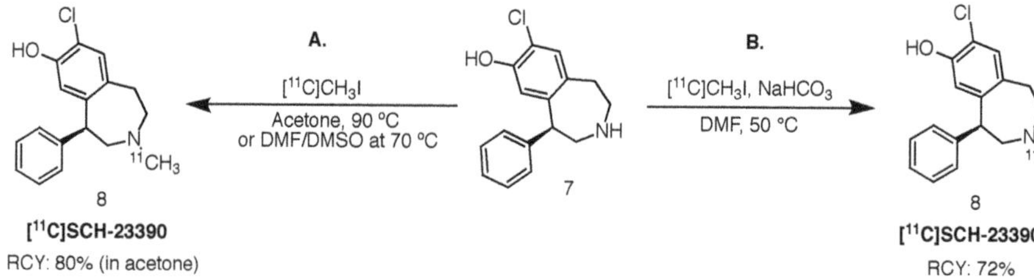

Scheme 4. A. Radiolabelling of [^{11}C]SCH-23390, starting from the desmethyl precursor **7**, described by Halldin and colleagues [35]; **B.** Method from DeJesus and colleagues [34].

Sedvall and co-workers investigated the performance of the radioligand in HD-affected individuals [36]. Five clinically diagnosed patients with motor dysfunction and one asymptomatic gene carrier were chosen alongside five healthy male volunteers. The authors reported a 50% decrease in D_1 receptor density in the putamen in comparison to the healthy subjects. For the sole asymptomatic HD individual, this value lied in the lower boundary of that of the healthy ones. Andrews et al. studied the rate of dopamine D_1 and D_2 receptor loss over 40 months in a larger group including 9 asymptomatic, 4 symptomatic and 3 patients at risk, complemented by 7 healthy controls [37]. In the first group [^{11}C]SCH-23390 signal was lost at a mean rate of 2% annually, with some patients progressing actively with a mean yearly loss of 4.5%. For the manifest individuals, a mean annual decrease of radioligand binding of 5.0% was reported.

2.3.2. [^{11}C]NNC-112

In the aforementioned publication, the study of longitudinal D_1 receptor changes was performed with a carbon-11 based ligand, [^{11}C]NNC-112 [38]. It is a derivative of [^{11}C]SCH-22390 with a benzofuran substituent in the 5-position of the central tetrahydrobenzazepine. The authors reported a 28% difference in striatal binding values between zQ175 (HD animal model) and wildtype mice at 6 months of age, increasing further to 34% three months later. Interestingly, the diseased animals expressed less D_1 receptors in the cortex and hippocampus than their healthy counterparts, hence expanding the potential of D_1 PET imaging in HD.

[^{11}C]NNC-112 is administered in the active S-enantiomeric form. It is accessed in a high-yielding (50–60%, calculated from the end of bombardment) N-^{11}C-methylation of the enantionmerically pure precursor with [^{11}C]methyl triflate in acetone at room temperature (Scheme 5) [39]. Molar activity was established at 110 GBq/μmol and RCP at more than 99%.

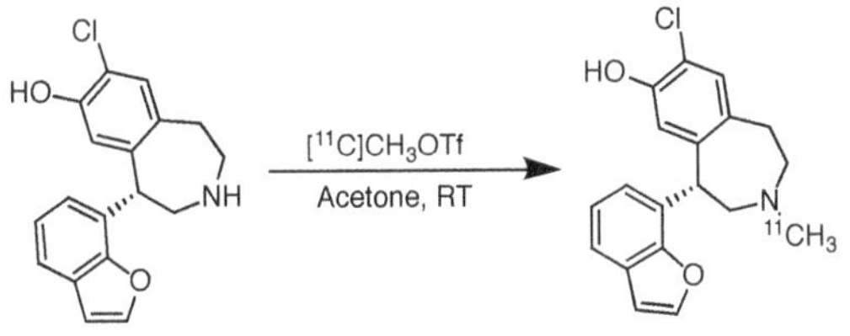

Scheme 5. Radiolabelling of [^{11}C]NNC-112, starting from the enantiomerically pure desmethyl precursor **9** [39]. RT: room temperature.

2.3.3. [^{11}C]Raclopride

[^{11}C]Raclopride is often used in conjunction with [^{11}C]SCH-23390 to get a more in-depth image of the dopaminergic system. It is a well known antagonist of $D_{2/3}$ receptors, used clinically, for instance, in parkinsonian patients [40].

Ehrin et al. were the first to publish the radiolabelling protocol of [^{11}C]raclopride in 1985 [41]. The precursor was N-[^{11}C]ethylated with [^{11}C]ethyl iodide in the presence of 2,2,6,6-tetramethylpiperidine in DMF at room temperature. Presently, the tracer is produced routinely for clinical purposes by O-methylation of the hydroxyl group with [^{11}C]methyl iodide or [^{11}C]methyl triflate ([^{11}C]CH$_3$OTf) in the presence of an inorganic base. Langer and co-workers reported RCYs in the range of 55–65% and molar activities of 56–74 GBq/μmol [42]. Several groups published various optimisations to the protocol ever since, including base-free synthesis, microfluidics and captive solvent methods, among many others [43–45]. A typical [^{11}C]raclopride synthesis is shown in Scheme 6 below.

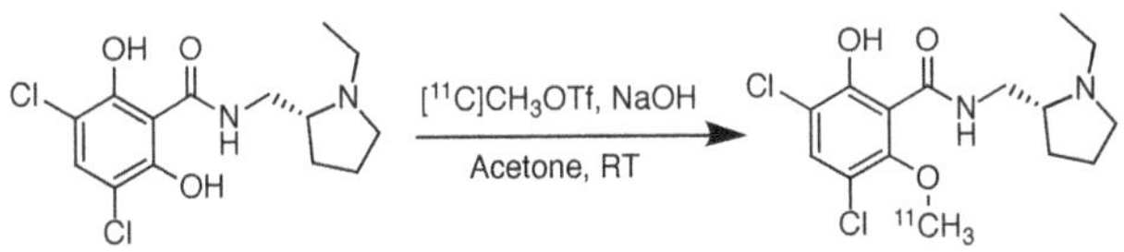

Scheme 6. Radiolabelling of [^{11}C]raclopride, starting from the desmethyl precursor **11**, published by Langer and co-workers. RCY is based on [^{11}C]methyl triflate [42]. RT: room temperature.

Antonini and co-workers examined 8 symptomatic and 10 asymptomatic HD mutation carriers with [^{11}C]raclopride PET, along with a 10-member healthy control group, in an effort to verify the correlation between CAG repeat lengths and striatal degeneration and age [46]. The authors unveiled a positive relationship between the two. Pavese et al. recruited 27 HD gene carriers with a minimum of 39 CAG repeats—16 symptomatic and 11 asymptomatic—for D_2 receptor studies with [^{11}C]raclopride PET [47]. Cortical reductions in $D_{2/3}$ binding were analysed in conjunction with neuropsychological tests, such as verbal fluency, Rivermead Behavioural Memory and Boston Naming

Tests. Manifest HD subjects with decreased cortical binding of the radioligand scored lower on the tests than those with unimpaired cortical dopaminergic system.

[^{11}C]Raclopride was also employed in a longitudinal cross-sectional HD biomarker study in zQ175 mice by Häggkvist and colleagues, in order to pinpoint the most powerful PET tracer to detect subtle receptor changes [38]. This mouse model was engineered to express motor and cognitive impairment, as well as decreased body weight with disease progression. There was a pronounced difference in tracer uptake between heterozygous and WT animals at both 6 and 9 months of age (Figure 1). For the earlier timepoint, this difference was 40%, then increasing further to 44% at 9 months of age. In addition, both genotypes exhibited a decline in receptor density due to age progression, a natural factor.

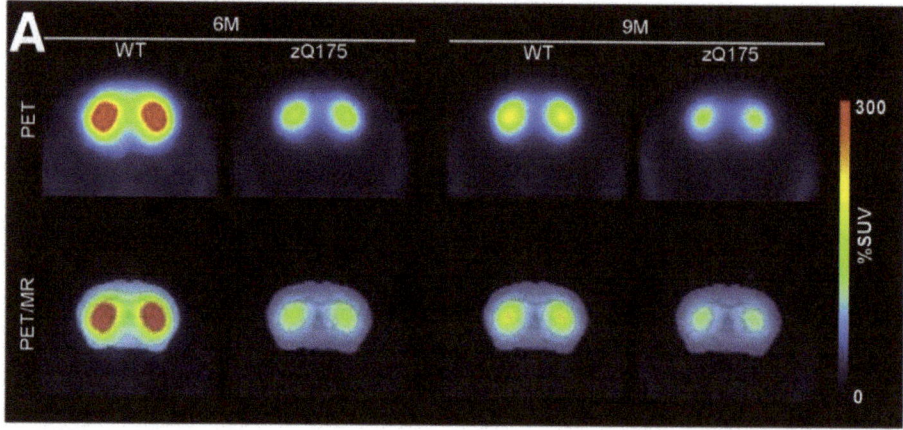

Figure 1. Average %SUV images of mice at 6 and 9 months of age acquired after [^{11}C]raclopride injection. Lower receptor binding was observed for zQ175 mice at both timepoints. Top row: Averaged PET measurement, 15–63 min post-injection. Bottom row: PET co-registered to template MRI. Reprinted from Häggkvist and colleagues [38].

2.4. GABA Receptors

Alterations of GABA, or *gamma*-amino butyric acid, receptor expression has also been studied in the context of HD. GABAergic neurotransmission is severely impaired in HD - GABAergic striatal medium spiny projection neurons are particularly targeted during the course of the disease [48,49].

2.4.1. [^{11}C]Flumazenil

Künig et al. studied changes in GABA receptor density in 23 HD gene carriers (10 manifest and 13 premanifest) using the PET tracer [^{11}C]flumazenil, a potent benzodiazepine-based GABA antagonist [50]. The authors reported reduced [^{11}C]flumazenil binding in the caudate nucleus, corresponding to the loss of projection neurons and consequently, GABA receptors. Interestingly, there was no difference in tracer binding in the putamen of the symptomatic HD subjects and healthy controls. It was proposed that an upregulatory GABA compensation mechanism was initiated, which was absent in the premanifest cohort, where neuronal loss was much less pronounced. The study was accompanied by the use of a $D_{2/3}$ receptor radioligand, [^{11}C]raclopride (for more about this tracer, see Section 2.3.3), and a glucose metabolism biomarker, [^{18}F]FDG. In summary, it was reported that reduced binding of these two had taken place before that of [^{11}C]flumazenil, and that this was only evident in manifest HD gene carriers.

An efficient and rapid [^{11}C]flumazenil radiolabelling protocol was published by Cleij et al. [51]. The typical ^{11}C-methylation, achieved by trapping [^{11}C]methyl iodide in a solution of the precursor, was replaced by a captive solvent method. The precursor solution was first adsorbed onto a stainless steel-packed support (inside an empty HPLC guard column), through which gaseous [^{11}C]methyl iodide was subsequently fed. The precursor solution was hence efficiently dispersed, owing to the large surface area of the metallic powder. Using only 40 μg of the precursor, the authors were able to obtain the desired product with RCYs reaching 80%. Trapping efficiency of [^{11}C]methyl iodide reached 90%. Very high molar activities were reported, amounting to 600 GBq/μmol, typically hard to achieve with the [^{11}C]CO$_2$ synthon. Due to the small size of this improvised "reactor" and reaction mixture volumes, HPLC purification was shortened significantly, arriving at an injectable solution of [^{11}C]flumazenil in 20 min, counted from the end of bombardment of the cyclotron target. The synthetic route towards the tracer is shown in Scheme 7.

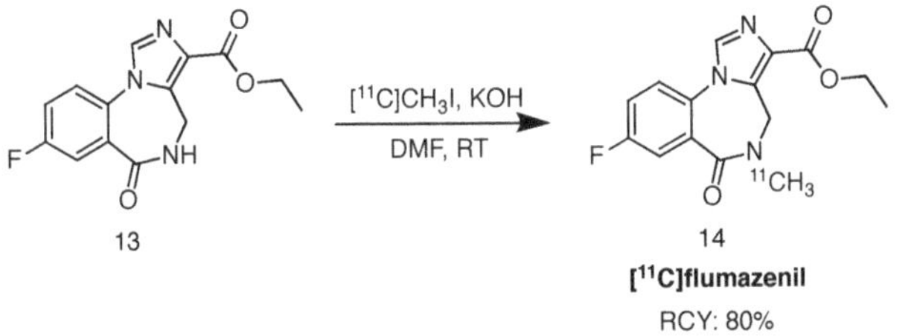

Scheme 7. Radiolabelling of [^{11}C]flumazenil, starting from the desmethyl precursor **13**, described by Cleij and colleagues [51]. A solution of the precursor in DMF is exposed to [^{11}C]methyl iodide in a stream of helium, in a 'micro-reactor' column.

2.5. Glucose Metabolism

A great deal of publications have been devoted to the study of brain metabolism in the context of HD. Metabolic glucose changes can be traced with the use of [^{18}F]FDG PET imaging. The last decade has witnessed major developments in the radiosynthetic route towards this PET tracer with regard to cassette-based synthesis using automated modules produced by companies such as Trasis and General Electric. This has enabled research institutions and hospitals to produce it in quantitative radiochemical yields and activities allowing for scanning dozens of patients per batch. In this review, however, we decided to focus on novel radiopharmaceuticals in order to highlight great advancements which have been made in the field of PET radiochemistry and the ever-expanding pool of the highly specialised molecular arsenal for HD.

2.6. Glutamatergic Receptors

Although the role of glutamate receptors in HD has been highlighted in numerous publications, only a few studies with glutamate-targeting PET tracers have been reported [52]. Ribeiro et al. demonstrated altered metabotropic glutamate receptor (mGluR) signalling, particularly that of group I, in presymptomatic HD mice [53]. mGluRs are members of the G protein coupled receptor family, receiving signal from glutamate, one of the main neurotransmitters. mGluR5, a member of the group I mGluR subset, is highly expressed in striatal MSNs [54].

2.6.1. [^{11}C]ABP-688

[^{11}C]ABP-688 is a high affinity oxime-based structural analogue of the prototypic mGluR5 antagonist MPEP. Ametamey et al. described radiolabelling and preclinical evaluation of the tracer in 2006, followed by healthy volunteer studies in 2007 [55,56]. [^{11}C]ABP-688 was accessed via ^{11}C-methylation of the sodium salt of the desmethyl precursor with [^{11}C]methyl iodide at 90 °C in DMF. The tracer was obtained in 35 ± 8% RCY and RCP greater than 95% (Scheme 8). Molar activity values ranged from 100 to 200 GBq/μmol. The authors also performed an analogous investigation with the ligand labelled with an NMR active carbon-13 nuclide. Not only was formation of the O-methylated product confirmed, but also the E/Z isomeric ratio was established. It was demonstrated that a high ratio of 10:1 favouring the more potent (E)-[^{11}C]ABP-688 can be achieved by deprotonating the precursor with sodium hydride, followed by treatment with [^{11}C]methyl iodide. An alternative method involves trapping of the more reactive methylating agent, [^{11}C]methyl triflate, in an acetone solution of the precursor with sodium hydroxide at room temperature. Average RCY of 25 ± 5%, at the end of synthesis, and molar activity values of 148 ± 56 GBq/µmol were obtained [57].

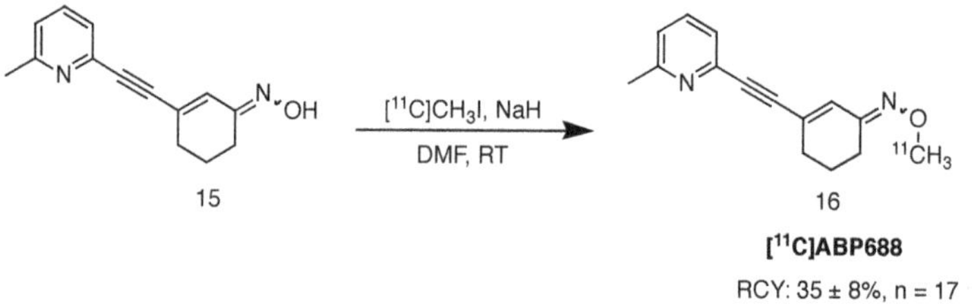

Scheme 8. Radiolabelling of [^{11}C]ABP-688, starting from the desmethyl precursor **15**, described by Amatamey and colleagues [55].

Bertoglio and co-workers employed [^{11}C]ABP-688 in a longitudinal investigation of mGlu5 receptor changes in the Q175 mouse model of HD using PET [58]. The choice of their study was motivated by the potential of mGluR5 targeting to increase cognitive performance in an effort to find a disease-modifying treatment for HD. Animals were scanned at 6, 9 and 13 months of age. A clear loss of signal was reported in the striatum and cortex of the heterozygous mice in comparison to the wildtypes. The Q175 mice displayed a significant reduction in the nondisplaceable binding potential (BP$_{ND}$) values between 6 and 13 months of age. The authors were cautious about drawing preliminary conclusions about the therapeutic potential of mGluR5 targeting, due to the dual role mGluR5 signalling is thought to play—neuronal activation or toxicity. In addition, they highlighted the importance of investigating age-related decline of mGluR5 availability as well as circadian-related variations. Further investigation is required to gain a better understanding of this intricate mechanism.

2.7. Microglia

Activation of microglia in neurodegenerative diseases has been described by various groups, yet the exact role of this mechanism is still unclear [59]. Microglia are another crucial component of the CNS, accounting for 5–15% of the entire cellular content [60]. They are heavily involved in neuroimmunity and one of their key roles is to maintain homeostasis and protect the system against pathogenesis [61]. They could also act as a trigger of damage as well as initiating compensatory action against mHTT damage. Imaging of microglia activation in HD can be performed using PET tracers binding to the 18-kDa translocator protein (TSPO). This transmembrane domain protein, formerly known as the peripheral benzodiazepine receptor (PBR), has a modest expression pattern in a healthy brain, in contrast to peripheral organs such as kidneys, heart and lungs. Increased TSPO expression has been reported in various neurodegenerative disorders, such as AD, PD and HD [60]. Consequently, the transformation from the quiescent to the activated state of the microglia has been considered a potential biomarker of HD using TSPO ligands, such as [^{11}C]PK11195.

2.7.1. [^{11}C]PK11195

[^{11}C]PK11195 is a first generation PET ligand targeting the TSPO protein. Radiolabelling, described by Toyama et al. in 2008, was achieved with [^{11}C]methyl iodide and sodium hydroxide in DMSO at 100 °C [62]. RCYs of 20 ± 6%, with high RCP (over 98%) and moderate molar activity (68.2 ± 18.1 GBq/µmol) at the end of synthesis were obtained (Scheme 9). In 2015, Alves and co-workers suggested an optimised radiolabelling protocol in which the tracer was performed using a captive solvent method, in an HPLC injection loop, with [^{11}C]methyl iodide at room temperature [63].

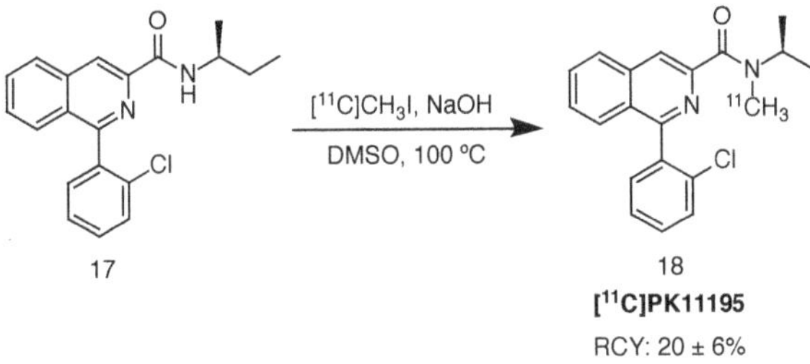

Scheme 9. Radiolabelling of [^{11}C]PK11195, starting from the desmethyl precursor **17** [62].

The tracer was employed in a few studies with HD patients. Tai et al. recruited presymptomatic HD gene carriers who underwent [^{11}C]PK11195 and [^{11}C]raclopride scans [64]. Striatal and cortical binding values for the premanifest subjects were higher than those of the healthy volunteers. The results pointed towards early stage microglia activation in the course of HD progression, although no prognostic conclusion could be drawn about the age of symptom onset and [^{11}C]PK11195 binding. The researchers supplemented this study with follow-up multimodal imaging with PET and MRI in a group of manifest HD individuals [59]. The results were divided regionally into brain regions of HD manifestations (motor, cognitive and psychiatric) and levels of microglia activation, D$_2$ receptor and neuronal loss (measured as decreases in volume using MRI volume-of-interest analysis) for each of the subject groups (normal brain, premanifest and manifest HD brains). An increase in radioligand binding was observed in the sensorimotor striatum, globus pallidus, substantia nigra and red nucleus, all the brain areas involved in motor functioning. Pathological changes in microglia activation were reported for the first two regions for the premanifest and manifest subjects. The same uptake pattern of the tracer was found in brain regions responsible for cognitive and psychiatric functions (e.g., associative striatum, insula, amygdala, hypothalamus). These results provide further evidence that activated microglia could orchestrate crucial mechanisms behind neuronal loss in HD, although the precise mode of action is not yet well understood.

2.7.2. [^{18}F]PBR06

In 2018, this radioligand was recognised by Simmons et al. as a potential biomarker for monitoring the therapeutic effect of an inflammation reducing agent LM11A-31 in a mouse model of HD [65]. [^{18}F]PBR06, a second generation TSPO ligand, was previously investigated in humans and non-human primates by Imaizumi et al. and Fujimura et al. [66,67]. Simmons highlighted the advantage of this [^{18}F]fluorinated tracer over [^{11}C]PK11195 with respect to half-life, affinity for the TSPO protein and better signal-to-noise ratio. The authors used PET imaging after administration of LM11A-31 to R6/2 and BACHD mice, both expressing mutant huntingtin protein. The vehicle-treated transgenic animals exhibited higher uptake of the tracer in the striatum, cortex and hippocampus, compared to wildtypes and LM11A-31 treated counterparts. These findings correlate with data acquired with [^{11}C]PK11195 in manifest and premanifest HD subjects.

Synthesis of [^{18}F]PBR06, reported by Wang et al. in 2011, was achieved on an in-house automatic synthesis module by heating the corresponding tosyloxy precursor in the presence of potassium carbonate and Kryptofix 222 in DMSO at 140 °C (Scheme 10) [68]. RCYs ranged from 30% to 60%, with molar activity values reaching 222 GBq/μmol. This route offers an improvement with regards to previously reported protocols by Briard et al., who performed radiolabelling with the bromo precursor, achieving lower RCYs and molar activity values [69]. Inferior performance of the latter can be attributed to the worse leaving group character of the bromide, in addition to problematic separation of the tracer from precursor using high-performance liquid chromatography.

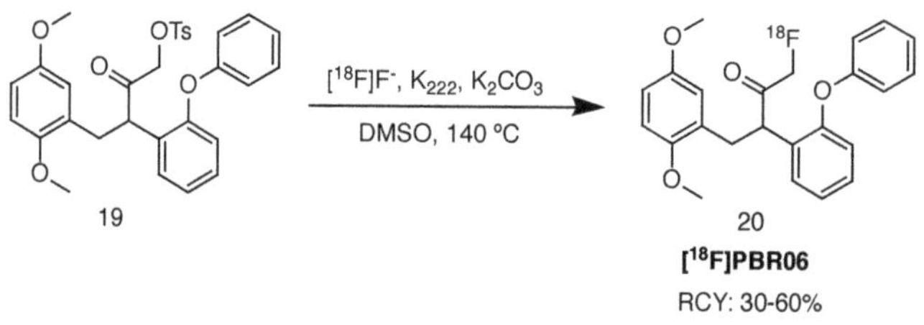

Scheme 10. Radiolabelling of [^{18}F]PBR06, starting from the tosyloxy precursor **19** [68].

2.8. Phosphodiesterase 10A

Phosphodiesterase 10A (PDE10A), a member of phosphodiesterases (PDEs), is an enzyme involved in deactivation of cyclic adenosine monophosphate and cyclic guanosine monophosphate (cAMP and cGMP, respectively). It is highly expressed in MSNs, where it regulates their excitability [70]. As mentioned earlier, MSNs are severely deregulated in various neurodegenerative pathologies, including HD. Hebb et al. reported that reduction of striatal PDE10A expression in transgenic R6/1 and R6/2 mice models of HD preceded motor impairment [71]. Several PET tracers targeting the enzyme have been proposed and studied preclinically and in humans. They were described by Boscutti et al. in their exhaustive review article in 2019 [72].

2.8.1. [^{18}F]JNJ42249152

Radiolabelling of this PDE10A inhibitor was first described by Andrés et al. in 2011 (Scheme 11) [73]. The precursor was synthesised through an 8-step route starting from 4-hydroxybenzoic acid methyl ester. Stability of the O-mesyl protected precursor proved to be a bottleneck in the process. Although isolation of the compound was possible, optimum results were reached when the precursor was synthesised the day before, having been passed through a Sep-Pak C18 cartridge and dried in vacuo overnight. It was then introduced to [^{18}F]fluoride and [^{18}F]fluorination proceeded with 17% RCY (based on [^{18}F]F$^-$, n = 8), with RCP higher than 97% after reverse-phase HPLC purification. The average molar activity (n = 8) was established at 167 GBq/µmol. In vitro PDE10A inhibition assays provided favourable pIC$_{50}$ and lipophilicity (clogP) values of 8.8 and 3.66, respectively, suggesting a potent ligand capable of crossing the blood–brain barrier. The authors also performed biodistribution studies in male Wistar rats and observed a time-dependent increased striatal accumulation and slow 2-to-30 min washout, compared to the hippocampus, cortex and cerebellum, in line with PDE10A expression levels. The evaluation of [^{18}F]JNJ42249152 as a PET ligand for PDE10A imaging continued with in-human studies performed by Van Laere and co-workers in collaboration with Janssen Pharmaceuticals [74,75]. Six healthy male volunteers were subjected to a whole body PET/CT scan. Unsurprisingly, the radiopharmaceutical exhibited rapid uptake in the striatum, particularly the putamen, reaching its peak at 12–15 min, and a high clearance rate. The test-retest evaluation in a concomitant human study with healthy male subjects was characterised by low intra-subject variability. The presence of two plasma metabolites, resulting from the cleavage of the 3,5-dimethylpyridine motif and yielding the radioactive phenol, is one of the limitations of this tracer. Ahmad and co-workers also tested it in an additional cohort of 5 HD sufferers, determined to unveil the fate of PDE10A in the presence of mHTT [76]. This quest was fuelled by contradictory data about enzyme depletion in a post-mortem striatal analysis of HD-affected individuals versus overexpression in a mouse model of the disease. Results of this in vivo experiment provided further evidence towards a reduction of phosphodiesterase 10 A in the presence of mHTT, however no clear link to clinical rating scales was demonstrated (Figure 2). The authors also presented alternative scenarios, according to which PDE10A levels were correlated with dropping cAMP levels.

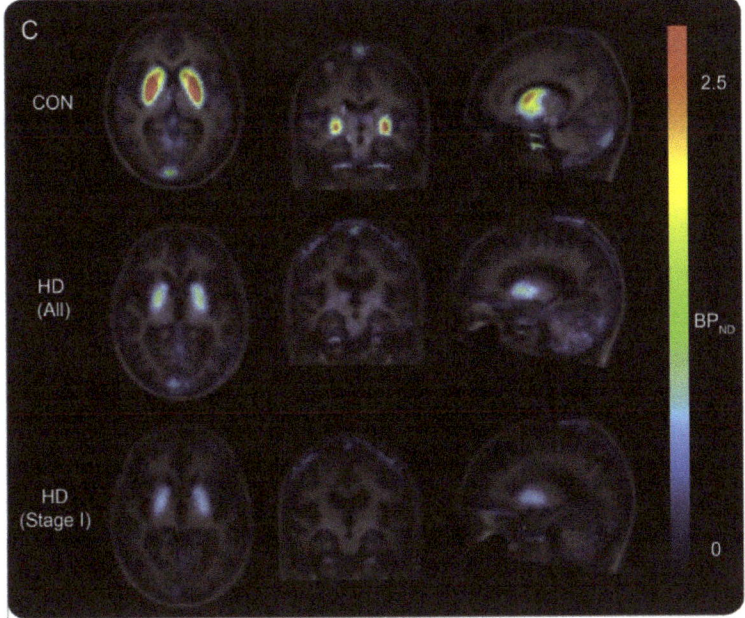

Figure 2. Mean PDE10A BP$_{ND}$ maps, obtained with [^{18}F]JNJ-42249152, projected on the mean group T1 MRI for 11 healthy controls, 5 manifest and 2 premanifest HD participants (stage I). Partial volume correction was performed. Loss of striatal PDE10A is clearly visualised in HD patients using this ligand. Reprinted from Ahmad and colleagues [76].

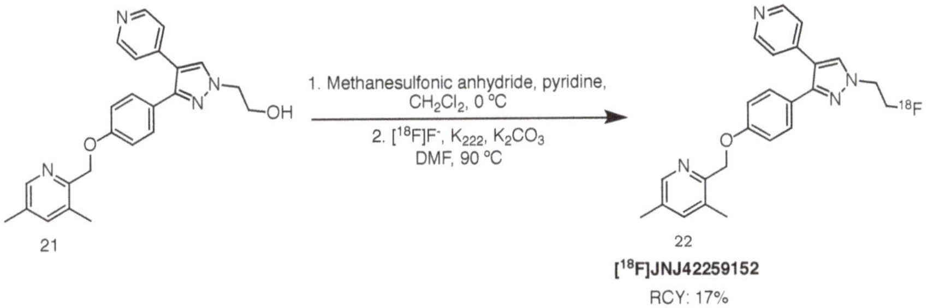

Scheme 11. Most efficient radiolabelling of [^{18}F]JNJ42249152, starting from the hydroxy building block **21**. The precursor for labelling, the mesylate derivative of compound **21**, is unstable, hence it must be produced shortly before the reaction. It is introduced into a reactor vial with dried [^{18}F]F$^-$ after solid phase extraction the day before [73].

2.8.2. [^{18}F]MNI-659

[^{18}F]MNI-659 and its derivative [^{18}F]MNI-654 were proposed as promising PDE10A radioligands by Barret et al. in 2012 [77]. Studies in non-human primates exhibited the highest update in the putamen and globus pallidus. Cerebellum, with lowest tracer accumulation, was used as a reference region for BP$_{ND}$ quantification. In their follow-up investigation in humans, [^{18}F]MNI-654 was disregarded as a potential PDE10A tracer due to slow kinetics. Its analogue, however, displayed a highly favourable in vivo profile.

[^{18}F]MNI-659 was accessed via S$_N$2 substitution of the tosylate leaving group with [^{18}F]fluoride, in the presence of Kryptofix 222 and potassium carbonate. The reaction was performed in DMF at 100 °C using a TRACERlab FX$_{FN}$ (GE Healthcare) automated unit. Reverse-phase HPLC purification on a C18 column yielded the injectable solution of the radioligand in 21 ± 5% RCY (n = 50). RCP was over 99%, with molar activity values exceeding 185 GBq/µmol. The reaction is shown in Scheme 12.

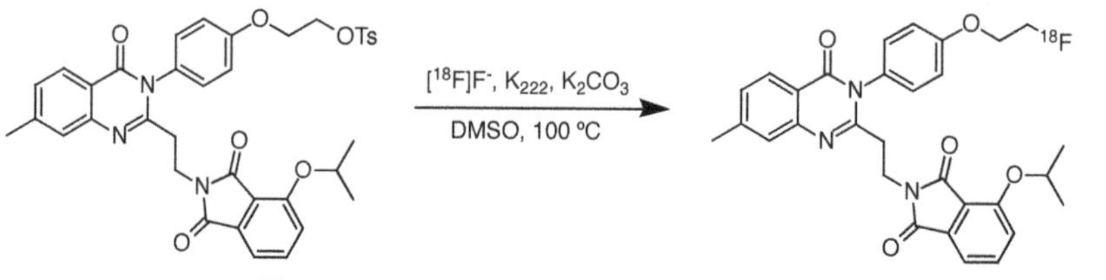

Scheme 12. Radiolabelling of [^{18}F]MNI-659, starting from the tosylate precursor 23 [77].

Plasma analysis revealed a moderate metabolic profile of the tracer, with 20% intact compound at 2 h post-injection. The presence of radioactive metabolites did not raise concern as they were more polar than the tracer itself, hence unlikely to cross the blood–brain barrier. The highest [^{18}F]MNI-659 uptake was recorded in the putamen, globus pallidus and caudate, in line with PDE10A expression patterns. Peak accumulation in these bran regions was observed 10–20 min post-injection, while cerebellum exhibited much faster washout, making it a suitable candidate for the reference region (to assess non-specific binding).

Russell et al. evaluated the potential of [^{18}F]MNI-659 as a PET tracer for HD imaging [78]. The cohort included nine healthy volunteers and 11 HD sufferers, of which three were ranked as premanifest and the remaining eight as manifest. PET images revealed a clear lack of striatal PDE10A binding among HD-affected individuals that correlated with disease severity. The same group published a follow-up study two years later [79]. Along with the healthy controls, eight HD patients were recruited, 2 premanifest and 6 manifest with early stage disease rating. The subjects were scanned twice, with a one year difference in between. Loss of radioligand binding in the putamen, caudate nucleus and globus pallidus was pronounced among the HD patients, 16.6%, 6.9% and 5.8%, respectively, with only 1% decline for the healthy volunteers.

2.8.3. [^{11}C]IMA-107

Another potential PET tracer for the imaging of HD was developed in collaboration with Imanova (now Invicro). A series a potent PDE10A inhibitors was proposed and characterised in pigs, baboons and humans [80]. The ligands all contained a central pyrazolo[1,5-a]pyrimidine unit. The choice was narrowed down to three tracers after the results of the in vivo evaluation in pigs, where [^{11}C]IMA-107 displayed specific binding to PDE10A during a blocking study. The highest uptake was noted in the striatum and kinetics were suitable for PET application. The next step involved testing the radioligands in baboons, where kinetics were generally much slower than in the porcine brain, yet [^{11}C]IMA-107 proved to be the most specific tracer in the series. Studies in healthy humans showed a similar uptake profile, with the highest standardised uptake values (SUVs) for the putamen and globus pallidus. Reversible kinetics and favourable washout confirmed its potential as a PET tracer for the imaging of PDE10A alterations.

Radiosynthesis of the tracer, also reported by Plisson et al., proceeded with an estimated RCY of 30% and the tracer can be produced in clinically useful quantities. [^{11}C]methylation of the amine moiety with [^{11}C]methyl iodide was achieved in the presence of tetrabutylammonium hydroxide in DMF at 80 °C (Scheme 13). The total synthesis time, from the end of cyclotron bombardment to obtain [^{11}C]carbon dioxide, and arriving at the formulated injectable product, was approximately 40 min. RCY was approximately 30% and RCP exceeded 95%. Moderate to good molar activity values for human studies were obtained, 62–287 GBq/µmol. Radiolabelling with carbon-11 is particularly beneficial during preclinical development, where a higher throughput can be achieved due to its short half-life and consequently, the possibility to use automated modules to produce ^{11}C-labelled radioligands more than once within a day. Despite the presence of a fluorine atom, radiolabelling of the scaffold with fluorine-18 was not attempted.

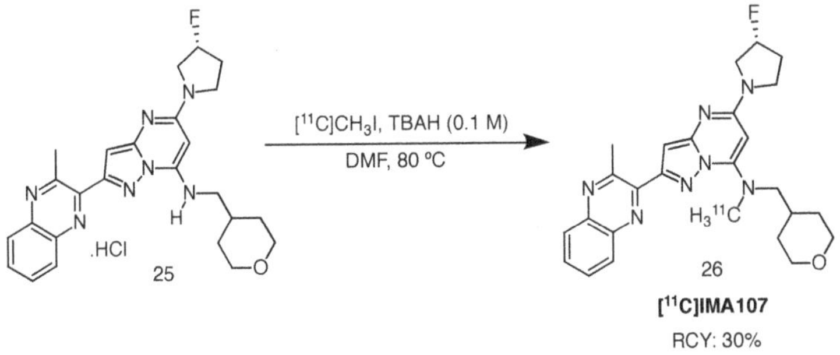

Scheme 13. Radiolabelling of [^{11}C]IMA-107, starting from precursor **25** [80].

Niccolini et al. assessed the discriminatory power of this radioligand for the imaging of PDE10A alterations in 12 early premanifest carriers of the *mHTT* mutation [81]. Individuals were chosen such that the symptom onset would not appear earlier than 25 ± 6.9 years with 90% probability. The resulting PET data were compared to those of the healthy controls. Interestingly, the tracer was able to detect differences in PDE10A expression years before the appearance of the first clinical symptoms of HD. Binding was reduced in the striatum and globus pallidus, with the opposite effect in the motor thalamic nuclei and similar uptake in substantia nigra and ventral striatum. Changes of phosphodiesterase 10A expression affected the dorsolateral striatum primarily, sparing the limbic and cognitive parts. The authors speculated that the increased [^{11}C]IMA-107 binding in motor thalamic structures may be a compensatory mechanism, which eventually collapses, ending in symptom manifestation.

2.9. Synaptic Vesicle Protein 2A

Synaptic vesicles play a crucial yet cryptic part of the intricate neurotransmission process. Synaptic vesicle glycoprotein 2A, or SV2A, is expressed extensively in synaptic vesicles of the central nervous system [82]. Levetiracetam, also known by its brand name Keppra, is an anti-epileptic drug which binds to SV2A and impedes the action of voltage-dependent Ca^{2+} channels, hence reducing neurotransmission [83]. Defective SV2A expression has been reported in various neuropathologies, such as Alzheimer's disease and pertinent to this review, HD. Several research articles highlighted the potential use of SV2A as a biomarker of synaptopathies with PET [82,84].

2.9.1. [^{11}C]UCB-J

[^{11}C]UCB-J is a more structurally complex derivative of levetiracetam, with a central pyrrolidinone motif. After a thorough screening, UCB Pharma proposed three potential SV2A tracers, which exhibited favourable in vitro pharmacokinetics and were predicted to cross the blood-brain barrier [84,85].

[^{11}C]UCB-A (Figure 3, left) evaluation in six epileptic individuals and two healthy controls revealed slow plasma and brain kinetics, a challenge for kinetic modelling. [^{18}F]UCB-H (Figure 3, middle) performed better in rats and non-human primates. Further clinical evaluation of [^{18}F]UCB-H was positive—the tracer accumulated readily in all relevant brain regions. The lack of an accurate reference region imposed the need for invasive arterial blood sampling in order to allow accurate quantification, however, an image-derived input function was later proposed by Bahri and co-workers as a good estimate of the arterial input function [86]. [^{11}C]UCB-J (Figure 3, right) outperformed [^{11}C]UCB-A and [^{18}F]UCB-H in non-human primate and rat studies. It exhibited high uptake and very fast kinetics, allowing convenient scanning, with the short half-life of carbon-11 in mind. Clinically, the tracer accumulated rapidly in the brain, following the SV2A expression pattern, with washout starting approximately 20 min post-injection.

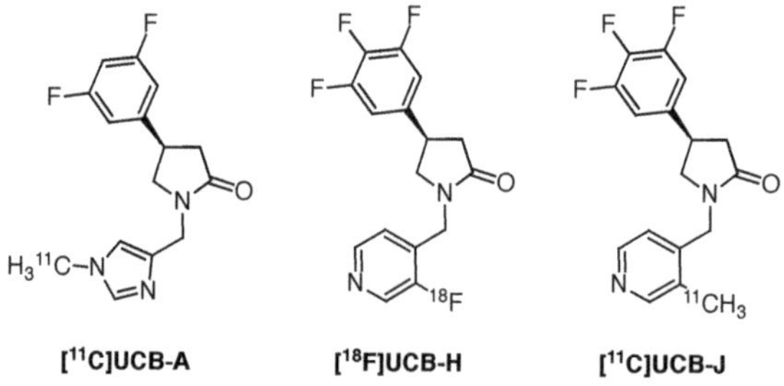

Figure 3. Radioligands for SV2A imaging proposed by UCB Pharma [84].

Radiolabelling of [^{11}C]UCB-J proceeds via palladium-mediated coupling of the trifluoroborate precursor with [^{11}C]methyl iodide, also known as Suzuki-Miyaura coupling (Scheme 14). The precursor is usually enriched with 2–5% of the boronic acid counterpart in order to ensure efficient radiolabelling, although the exact need for this is not entirely understood [84,87]. Reaction success is dependent on the purity of the precursor as well as proper handling of the Pd$_2$(dba)$_3$ catalyst and degassing the reaction mixture with inert gas prior to use. Nabulsi et al. reported 11 ± 4% RCY and molar activities reaching over 566.1 GBq/μmol [87]. The scheme is presented below.

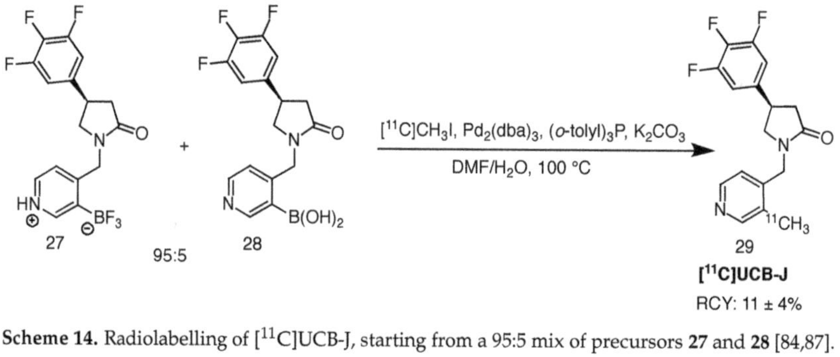

Scheme 14. Radiolabelling of [^{11}C]UCB-J, starting from a 95:5 mix of precursors **27** and **28** [84,87].

Recently, DiFilippo and co-workers published an improved protocol for [^{11}C]UCB-J synthesis, having struggled to reproduce the yields provided by Nabulsi and colleagues [88]. The authors introduced a hydrolysis step of the precursor prior to ^{11}C-methylation, in order to generate the corresponding boronic acid, which was then dissolved in DMF and exposed to [^{11}C]methyl iodide. The authors then followed the protocol of Nabulsi et al., except for the temperature—they heated the reaction mixture at 135 °C for 10 min. [^{11}C]UCB-J was obtained in 56 ± 7% RCY, with RCP exceeding 99% and a high molar activity of 477.3 ± 133.2 GBq/μmol.

To the best of our knowledge, Bertoglio and co-workers were the first ones to employ [^{11}C]UCB-J in a study with a HD animal model [89]. The group described kinetic modelling of the radioligand in 8-month-old Q175DN and wildtype mice. The tracer was synthesised on an automated module unit (Comecer) using the previously described protocol with a 95:5 mixture of the trifluoroborate and boronic acid precursors (Scheme 14). No RCYs were reported, but RCP over 99% and moderate molar activity of 78 ± 23 GBq/μmol were obtained. Along with baseline scans, a blocking study with leviteracetam was performed. The authors reported dose-dependent blocking of the radioligand, in line with SV2A expression. Relatively low non-specific binding was revealed. This work serves as premise for further evaluation of the tracer in different HD animal models, with the use of non-invasive quantification based on the one tissue compartment model (1TCM) and an image-derived input function.

3. Conclusions

Alterations in expressions of adenosine A_1 and A_2, cannabinoid CB_1, dopaminergic D_1 and D_2 and glutamatergic mGluR5 receptors, together with activation of microglia, PDE10A and SV2A protein dysfunction have been considered as promising biomarkers of HD. Advanced PET imaging using potent and specific radioligands tagged with carbon-11 or fluorine-18 enable visualisation of biochemical changes and intricate mechanisms related to these targets. As of now, no single PET tracer binding to mutant huntingtin protein has been reported. None of the molecular targets presented in this review can act as a stand-alone tool for HD progression monitoring, but the knowledge gathered since the mapping of the mHTT mutation in 1993, has brought researchers even closer to understanding the pathology and eventually, finding the highly sought after disease-modifying treatment.

Funding: This research received no external funding

Conflicts of Interest: The authors declare no conflict of interest.

Abbreviations

The following abbreviations are used in this manuscript:

A_1 and A_{2A}	Adenosine receptors
AD	Alzheimer's disease
BP_{ND}	Binding potential
Bq	Becquerel
CAG	Cytosine-adenine-guanine
CB_1	Cannabinoid receptor type 1
CNS	Central nervous system
D_1, D_2 and D_3	Dopamine receptors
DMF	Dimethyl formamide
DMSO	Dimethyl sulfoxide
GABA	*gamma*-Aminobutyric acid
GMP	Good manufacturing practice
HD	Huntington's disease
HPLC	High performance liquid chromatography
IB	Inclusion body
K_{222}	Kryptofix 222
MeCN	Acetonitrile
mHTT	Mutant huntingtin
MRI	Magnetic Resonance Imaging
MSN	Medium spiny neuron
OTf	Trifluoromethanesulfonate, triflate
PET	Positron Emission Tomography
PD	Parkinson's disease
RCP	Radiochemical purity; defined as the absence of other radiochemical compounds/species
RCY	Radiochemical yield; defined as the amount of activity in the product expressed as the percentage (%) of starting activity used in the considered process, all RCYs presented in this review are decay corrected
RT	Room temperature
SUV	Standardised uptake value
WT	Wildtype

References

1. Schapira, A.H.V.; Olanow, C.W.; Greenamyre, J.T.; Bezard, E. Slowing of neurodegeneration in Parkinson's disease and Huntington's disease: Future therapeutic perspectives. *Lancet* **2014**, *384*, 545–555. [CrossRef]
2. Wilson, H.; De Micco, R.; Niccolini, F.; Politis, M. Molecular Imaging Markers to Track Huntington's Disease Pathology. *Front. Neurol.* **2017**, *8*, 11. [CrossRef]
3. Pagano, G.; Niccolini, F.; Politis, M. Current status of PET imaging in Huntington's disease. *Eur. J. Nucl. Med. Mol. Imaging* **2016**, *43*, 1171–1182. [CrossRef]
4. Fazio, P.; Paucar, M.; Svenningsson, P.; Varrone, A. Novel Imaging Biomarkers for Huntington's Disease and Other Hereditary Choreas. *Curr. Neurol. Neurosci. Rep.* **2018**, *18*, 85. [CrossRef]
5. Roussakis, A.A.; Piccini, P. PET Imaging in Huntington's Disease. *J. Huntingtons. Dis.* **2015**, *4*, 287–296. [CrossRef] [PubMed]
6. Lee, J.M.; Correia, K.; Loupe, J.; Kim, K.H.; Barker, D.; Hong, E.P.; Chao, M.J.; Long, J.D.; Lucente, D.; Vonsattel, J.P.G.; et al. CAG Repeat Not Polyglutamine Length Determines Timing of Huntington's Disease Onset. *Cell* **2019**, *178*, 887–900.e14. [CrossRef] [PubMed]
7. Mahalingam, S.; Levy, L. Genetics of Huntington Disease. *Am. J. Neuroradiol.* **2014**, *35*, 1070–1072. [CrossRef] [PubMed]
8. Soares, T.R.; Reis, S.D.; Pinho, B.R.; Duchen, M.R.; Oliveira, J.M. Targeting the proteostasis network in Huntington's disease. *Ageing Res. Rev.* **2019**, *49*, 92–103. [CrossRef] [PubMed]
9. Neto, J.L.; Lee, J.M.; Afridi, A.; Gillis, T.; Guide, J.R.; Dempsey, S.; Lager, B.; Alonso, I.; Wheeler, V.C.; Pinto, R.M. Genetic contributors to intergenerational CAG repeat instability in Huntington's disease knock-in mice. *Genetics* **2017**, *205*, 503–516. [CrossRef] [PubMed]
10. Scherzinger, E.; Sittler, A.; Schweiger, K.; Heiser, V.; Lurz, R.; Hasenbank, R.; Bates, G.P.; Lehrach, H.; Wanker, E.E. Self-assembly of polyglutamine-containing huntingtin fragments into amyloid-like fibrils: Implications for Huntington's disease pathology. *Proc. Natl. Acad. Sci. USA* **1999**, *96*, 4604–4609. [CrossRef]
11. Chugani, H.T. Positron Emission Tomography in Pediatric Neurodegenerative Disorders. *Pediatr. Neurol.* **2019**, *100*, 12–25. [CrossRef] [PubMed]
12. Ehrlich, M.E. Huntington's Disease and the Striatal Medium Spiny Neuron: Cell-Autonomous and Non-Cell-Autonomous Mechanisms of Disease. *Neurotherapeutics* **2012**, *9*, 270–284. [CrossRef] [PubMed]
13. Vuorimaa, A.; Rissanen, E.; Airas, L. In Vivo PET Imaging of Adenosine 2A Receptors in Neuroinflammatory and Neurodegenerative Disease. *Contrast Media Mol. Imaging* **2017**, *2017*, 6975841. [CrossRef]
14. Blum, D.; Chern, Y.; Domenici, M.R.; Buée, L.; Lin, C.Y.; Rea, W.; Ferré, S.; Popoli, P. The Role of Adenosine Tone and Adenosine Receptors in Huntington's Disease. *J. Caffeine Adenosine Res.* **2018**, *8*, 43–58. [CrossRef] [PubMed]
15. Morigaki, R.; Goto, S. Striatal Vulnerability in Huntington's Disease: Neuroprotection Versus Neurotoxicity. *Brain Sci.* **2017**, *7*, 63. [CrossRef] [PubMed]
16. Gomes, C.V.; Kaster, M.P.; Tomé, A.R.; Agostinho, P.M.; Cunha, R.A. Adenosine receptors and brain diseases: Neuroprotection and neurodegeneration. *Biochim. Biophys. Acta Biomembr.* **2011**, *1808*, 1380–1399. [CrossRef]
17. Blum, D.; Hourez, R.; Galas, M.C.; Popoli, P.; Schiffmann, S.N. Adenosine receptors and Huntington's disease: implications for pathogenesis and therapeutics. *Lancet Neurol.* **2003**, *2*, 366–374. [CrossRef]
18. Holschbach, M.H.; Olsson, R.A.; Bier, D.; Wutz, W.; Sihver, W.; Schüller, M.; Palm, B.; Coenen, H.H. Synthesis and Evaluation of No-Carrier-Added 8-Cyclopentyl-3-(3-[^{18}F]fluoropropyl)-1-propylxanthine ([^{18}F]CPFPX): A Potent and Selective A_1-Adenosine Receptor Antagonist for in Vivo Imaging. *J. Med. Chem.* **2002**, *45*, 5150–5156. [CrossRef]
19. Coenen, H.H.; Gee, A.D.; Adam, M.; Antoni, G.; Cutler, C.S.; Fujibayashi, Y.; Jeong, J.M.; Mach, R.H.; Mindt, T.L.; Pike, V.W.; et al. Consensus nomenclature rules for radiopharmaceutical chemistry—Setting the record straight. *Nucl. Med. Biol.* **2017**, *55*, v–xi. [CrossRef]
20. Bauer, A.; Holschbach, M.H.; Cremer, M.; Weber, S.; Boy, C.; Shah, N.J.; Olsson, R.A.; Halling, H.; Coenen, H.H.; Zilles, K. Evaluation of ^{18}F-CPFPX, a Novel Adenosine A_1 Receptor Ligand: In Vitro Autoradiography and High-Resolution Small Animal PET. *J. Nucl. Med.* **2003**, *44*, 1682–1689.
21. Matusch, A.; Saft, C.; Elmenhorst, D.; Kraus, P.H.; Gold, R.; Hartung, H.P.; Bauer, A. Cross sectional PET study of cerebral adenosine A_1 receptors in premanifest and manifest Huntington's disease. *Eur. J. Nucl. Med. Mol. Imaging* **2014**, *41*, 1210–1220. [CrossRef] [PubMed]

22. Ishiwata, K.; Noguchi, J.; Wakabayashi, S.I.; Shimada, J.; Ogi, N.; Nariai, T.; Tanaka, A.; Endo, K.; Suzuki, F.; Senda, M. ^{11}C-Labeled KF18446: A Potential Central Nervous System Adenosine A_{2a} Receptor Ligand. *J. Nucl. Med.* **2000**, *41*, 345–354. [PubMed]
23. Ishiwata, K.; Ogi, N.; Hayakawa, N.; Oda, K.; Nagaoka, T.; Toyama, H.; Suzuki, F.; Endo, K.; Tanaka, A.; Senda, M. Adenosine A_{2A} receptor imaging with [^{11}C]KF18446 PET in the rat brain after quinolinic acid lesion: Comparison with the dopamine receptor imaging. *Ann. Nucl. Med.* **2002**, *16*, 467–475. [CrossRef] [PubMed]
24. Kendall, D.A.; Yudowski, G.A. Cannabinoid Receptors in the Central Nervous System: Their Signaling and Roles in Disease. *Front. Cell. Neurosci.* **2017**, *10*, 294. [CrossRef] [PubMed]
25. Glass, M.; Faull, R.; Dragunow, M. Loss of cannabinoid receptors in the substantia nigra in Huntington's disease. *Neuroscience* **1993**, *56*, 523–527. [CrossRef]
26. Glass, M.; Dragunow, M.; Faull, R. The pattern of neurodegeneration in Huntington's disease: a comparative study of cannabinoid, dopamine, adenosine and $GABA_A$ receptor alterations in the human basal ganglia in Huntington's disease. *Neuroscience* **2000**, *97*, 505–519. [CrossRef]
27. Van Laere, K.; Casteels, C.; Dhollander, I.; Goffin, K.; Grachev, I.; Bormans, G.; Vandenberghe, W. Widespread Decrease of Type 1 Cannabinoid Receptor Availability in Huntington Disease In Vivo. *J. Nucl. Med.* **2010**, *51*, 1413–1417. [CrossRef]
28. Thomae, D.; Morley, T.J.; Hamill, T.; Carroll, V.M.; Papin, C.; Twardy, N.M.; Lee, H.S.; Hargreaves, R.; Baldwin, R.M.; Tamagnan, G.; et al. Automated one-step radiosynthesis of the CB_1 receptor imaging agent [^{18}F]MK-9470. *J. Label. Compd. Radiopharm.* **2014**, *57*, 611–614. [CrossRef]
29. Gagnon, D.; Petryszyn, S.; Sanchez, M.G.; Bories, C.; Beaulieu, J.M.; De Koninck, Y.; Parent, A.; Parent, M. Striatal Neurons Expressing D_1 and D_2 Receptors are Morphologically Distinct and Differently Affected by Dopamine Denervation in Mice. *Sci. Rep.* **2017**, *7*, 41432. [CrossRef]
30. Bourne, J.A. SCH 23390: The First Selective Dopamine D_1-Like Receptor Antagonist. *CNS Drug Rev.* **2006**, *7*, 399–414. [CrossRef]
31. Friedman, A.M.; DeJesus, O.T.; Woolverton, W.L.; Moffaert, G.V.; Goldberg, L.I.; Prasad, A.; Barnett, A.; Dinerstein, R.J. Positron tomography of a radio-brominated analog of the D_1/DA_1 antagonist, SCH 23390. *Eur. J. Pharmacol.* **1985**, *108*, 327–328. [CrossRef]
32. Dejesus, O.; Moffaert, G.V.; Friedman, A. Evaluation of positron-emitting SCH 23390 analogs as tracers for CNS dopamine D1 receptors. *Nucl. Med. Biol.* **1989**, *16*, 47–50. [CrossRef]
33. Stenkrona, P.; Matheson, G.J.; Cervenka, S.; Sigray, P.P.; Halldin, C.; Farde, L. [^{11}C]SCH23390 binding to the D_1-dopamine receptor in the human brain—A comparison of manual and automated methods for image analysis. *EJNMMI Res.* **2018**, *8*, 74 [CrossRef] [PubMed]
34. DeJesus, O.; Moffaert, G.V.; Friedman, A. Synthesis of [^{11}C]SCH 23390 for dopamine D_1 receptor studies. *Int. J. Rad. Appl. Instr. A* **1987**, *38*, 345–348. [CrossRef]
35. Halldin, C.; Stone-Elander, S.; Farde, L.; Ehrin, E.; Fasth, K.J.; Långström, B.; Sedvall, G. Preparation of ^{11}C-labelled SCH 23390 for the in vivo study of dopamine D-1 receptors using positron emission tomography. *Appl. Radiat. Isot.* **1986**, *37*, 1039–1043. [CrossRef]
36. Sedvall, G.; Karlsson, P.; Lundin, A.; Anvret, M.; Suhara, T.; Halldin, C.; Farde, L. Dopamine D_1 receptor number—A sensitive PET marker for early brain degeneration in Huntington's disease. *Eur. Arch. Psychiatry Clin. Neurosci.* **1994**, *243*, 249–255.
37. Andrews, T.C.; Weeks, R.A.; Turjanski, N.; Gunn, R.N.; Watkins, L.H.A.; Sahakian, B.; Hodges, J.R.; Rosser, A.E.; Wood, N.W.; Brooks, D.J. Huntington's disease progression: PET and clinical observations. *Brain* **1999**, *122*, 2353–2363. [CrossRef]
38. Häggkvist, J.; Tóth, M.; Tari, L.; Varnäs, K.; Svedberg, M.; Forsberg, A.; Nag, S.; Dominguez, C.; Munoz-Sanjuan, I.; Bard, J.; et al. Longitudinal Small-Animal PET Imaging of the zQ175 Mouse Model of Huntington Disease Shows In Vivo Changes of Molecular Targets in the Striatum and Cerebral Cortex. *J. Nucl. Med.* **2017**, *58*, 617–622. [CrossRef]
39. Halldin, C.; Foged, C.; Chou, Y.H.; Karlsson, P.; Swahn, C.G.; Johan, S.; Sedvall, G.; Farde, L. Carbon-11-NNC 112: A Radioligand for PET Examination of Striatal and Neocortical D_1-Dopamine Receptors. *J. Nucl. Med.* **1998**, *39*, 2061–2068.

40. Esmaeilzadeh, M.; Farde, L.; Karlsson, P.; Varrone, A.; Halldin, C.; Waters, S.; Tedroff, J. Extrastriatal dopamine D_2 receptor binding in Huntington's disease. *Hum. Brain Mapp.* **2011**, *32*, 1626–1636. [CrossRef]
41. Ehrin, E.; Farde, L.; de Paulis, T.; Eriksson, L.; Greitz, T.; Johnström, P.; Litton, J.E.; Nilsson, J.G.; Sedvall, G.; Stone-Elander, S.; et al. Preparation of ^{11}C-labelled raclopride, a new potent dopamine receptor antagonist: Preliminary PET studies of cerebral dopamine receptors in the monkey. *Int. J. Appl. Radiat. Isot.* **1985**, *36*, 269–273. [CrossRef]
42. Langer, O.; Någren, K.; Dolle, F.; Lundkvist, C.; Sandell, J.; Swahn, C.G.; Vaufrey, F.; Crouzel, C.; Maziere, B.; Halldin, C. Precursor synthesis and radiolabelling of the dopamine D_2 receptor ligand [^{11}C]raclopride from [^{11}C]methyl triflate. *J. Label. Compd. Radiopharm.* **1999**, *42*, 1183–1193. [CrossRef]
43. Lee, Y.S.; Jeong, J.M.; Cho, Y.H.; Lee, J.H.; Lee, H.J.; Kim, J.E.; Lee, Y.S.; Kang, K.W. Evaluation of base-free ^{11}C-Raclopride synthesis with various solvents. *J. Nucl. Med.* **2015**, *56*, 2501.
44. Haroun, S.; Sanei, Z.; Jivan, S.; Schaffer, P.; Ruth, T.J.; Li, P.C. Continuous-flow synthesis of [^{11}C]raclopride, a positron emission tomography radiotracer, on a microfluidic chip. *Can. J. Chem.* **2013**, *91*, 326–332. [CrossRef]
45. Gómez-Vallejo, V.; Llop, J. Fully automated and reproducible radiosynthesis of high specific activity [^{11}C]raclopride and [^{11}C]Pittsburgh compound-B using the combination of two commercial synthesizers. *Nucl. Med. Commun.* **2011**, *32*, 1011–1017. [CrossRef]
46. Antonini, A.; Leenders, K.L.; Eidelberg, D. [^{11}C]Raclopride-PET studies of the Huntington's disease rate of progression: Relevance of the trinucleotide repeat length. *Ann. Neurol.* **1998**, *43*, 253–255. [CrossRef]
47. Pavese, N.; Politis, M.; Tai, Y.F.; Barker, R.A.; Tabrizi, S.J.; Mason, S.L.; Brooks, D.J.; Piccini, P. Cortical dopamine dysfunction in symptomatic and premanifest Huntington's disease gene carriers. *Neurobiol. Dis.* **2010**, *37*, 356–361. [CrossRef]
48. Hsu, Y.T.; Chang, Y.G.; Chern, Y. Insights into $GABA_A$ergic system alteration in Huntington's disease. *Open Biol.* **2018**, *8*, 180165. [CrossRef]
49. Garret, M.; Du, Z.; Chazalon, M.; Cho, Y.H.; Baufreton, J. Alteration of GABAergic neurotransmission in Huntington's disease. *CNS Neurosci. Ther.* **2018**, *24*, 292–300. [CrossRef]
50. Künig, G.; Leenders, K.L.; Sanchez-Pernaute, R.; Antonini, A.; Vontobel, P.; Verhagen, A.; Günther, I. Benzodiazepine receptor binding in Huntington's disease: [^{11}C]Flumazenil uptake measured using positron emission tomography. *Ann. Neurol.* **2000**, *47*, 644–648. [CrossRef]
51. Cleij, M.C.; Clark, J.C.; Baron, J.C.; Aigbirhio, F.I. Rapid preparation of [^{11}C]flumazenil: captive solvent synthesis combined with purification by analytical sized columns. *J. Label. Compd. Radiopharm.* **2007**, *50*, 19–24. [CrossRef]
52. Abd-Elrahman, K.S.; Hamilton, A.; Hutchinson, S.R.; Liu, F.; Russell, R.C.; Ferguson, S.S.G. mGluR5 antagonism increases autophagy and prevents disease progression in the zQ175 mouse model of Huntington's disease. *Sci. Signal.* **2017**, *10*, eaan6387. [CrossRef] [PubMed]
53. Ribeiro, F.M.; Hamilton, A.; Doria, J.G.; Guimaraes, I.M.; Cregan, S.P.; Ferguson, S.S. Metabotropic glutamate receptor 5 as a potential therapeutic target in Huntington's disease. *Expert Opin. Ther. Targets* **2014**, *18*, 1293–1304. [CrossRef] [PubMed]
54. Ribeiro, F.M.; Paquet, M.; Ferreira, L.T.; Cregan, T.; Swan, P.; Cregan, S.P.; Ferguson, S.S.G. Metabotropic Glutamate Receptor-Mediated Cell Signaling Pathways Are Altered in a Mouse Model of Huntington's Disease. *J. Neurosci.* **2010**, *30*, 316–324. [CrossRef]
55. Ametamey, S.M.; Kessler, L.J.; Honer, M.; Wyss, M.T.; Buck, A.; Hintermann, S.; Auberson, Y.P.; Gasparini, F.; Schubiger, P.A. Radiosynthesis and Preclinical Evaluation of ^{11}C-ABP688 as a Probe for Imaging the Metabotropic Glutamate Receptor Subtype 5. *J. Nucl. Med.* **2006**, *47*, 698–705.
56. Ametamey, S.M.; Treyer, V.; Streffer, J.; Wyss, M.T.; Schmidt, M.; Blagoev, M.; Hintermann, S.; Auberson, Y.; Gasparini, F.; Fischer, U.C.; et al. Human PET Studies of Metabotropic Glutamate Receptor Subtype 5 with ^{11}C-ABP688. *J. Nucl. Med.* **2007**, *48*, 247–252.
57. DeLorenzo, C.; Milak, M.S.; Brennan, K.G.; Kumar, J.S.D.; Mann, J.J.; Parsey, R.V. In vivo positron emission tomography imaging with [^{11}C]ABP688: binding variability and specificity for the metabotropic glutamate receptor subtype 5 in baboons. *Eur. J. Nucl. Med. Mol. Imaging* **2011**, *38*, 1083–1094. [CrossRef]

58. Bertoglio, D.; Kosten, L.; Verhaeghe, J.; Thomae, D.; Wyffels, L.; Stroobants, S.; Wityak, J.; Dominguez, C.; Mrzljak, L.; Staelens, S. Longitudinal Characterization of mGluR5 Using ^{11}C-ABP688 PET Imaging in the Q175 Mouse Model of Huntington Disease. *J. Nucl. Med.* **2018**, *59*, 1722–1727. [CrossRef]
59. Politis, M.; Pavese, N.; Tai, Y.F.; Kiferle, L.; Mason, S.L.; Brooks, D.J.; Tabrizi, S.J.; Barker, R.A.; Piccini, P. Microglial activation in regions related to cognitive function predicts disease onset in Huntington's disease: A multimodal imaging study. *Hum. Brain Mapp.* **2011**, *32*, 258–270. [CrossRef]
60. Dupont, A.C.; Largeau, B.; Santiago Ribeiro, M.; Guilloteau, D.; Tronel, C.; Arlicot, N. Translocator Protein-18 kDa (TSPO) Positron Emission Tomography (PET) Imaging and Its Clinical Impact in Neurodegenerative Diseases. *Int. J. Mol. Sci.* **2017**, *18*, 785. [CrossRef]
61. Hickman, S.; Izzy, S.; Sen, P.; Morsett, L.; El Khoury, J. Microglia in neurodegeneration. *Nat. Neurosci.* **2018**, *21*, 1359–1369. [CrossRef]
62. Toyama, H.; Hatano, K.; Suzuki, H.; Ichise, M.; Momosaki, S.; Kudo, G.; Ito, F.; Kato, T.; Yamaguchi, H.; Katada, K.; et al. In vivo imaging of microglial activation using a peripheral benzodiazepine receptor ligand: [^{11}C]PK-11195 and animal PET following ethanol injury in rat striatum. *Ann. Nucl. Med.* **2008**, *22*, 417–424. [CrossRef]
63. Alves, V.H.; Abrunhosa, A.J.; Castelo-Branco, M. Optimisation of synthesis, purification and reformulation of (R)-[N-Methyl-^{11}C]PK11195 for in vivo PET imaging studies. In Proceedings of the 2013 IEEE 3rd Portuguese Meeting in Bioengineering (ENBENG), Braga, Portugal, 20–23 February 2013; pp. 1–5.
64. Tai, Y.F.; Pavese, N.; Gerhard, A.; Tabrizi, S.J.; Barker, R.A.; Brooks, D.J.; Piccini, P. Microglial activation in presymptomatic Huntington's disease gene carriers. *Brain* **2007**, *130*, 1759–1766. [CrossRef]
65. Simmons, D.A.; James, M.L.; Belichenko, N.P.; Semaan, S.; Condon, C.; Kuan, J.; Shuhendler, A.J.; Miao, Z.; Chin, F.T.; Longo, F.M. TSPO-PET imaging using [^{18}F]PBR06 is a potential translatable biomarker for treatment response in Huntington's disease: preclinical evidence with the p75^{NTR} ligand LM11A-31. *Hum. Mol. Genet.* **2018**, *27*, 2893–2912. [CrossRef]
66. Fujimura, Y.; Kimura, Y.; Siméon, F.G.; Dickstein, L.P.; Pike, V.W.; Innis, R.B.; Fujita, M. Biodistribution and Radiation Dosimetry in Humans of a New PET Ligand, ^{18}F-PBR06, to Image Translocator Protein (18 kDa). *J. Nucl. Med.* **2010**, *51*, 145–149. [CrossRef] [PubMed]
67. Imaizumi, M.; Briard, E.; Zoghbi, S.S.; Gourley, J.P.; Hong, J.; Musachio, J.L.; Gladding, R.; Pike, V.W.; Innis, R.B.; Fujita, M. Kinetic evaluation in nonhuman primates of two new PET ligands for peripheral benzodiazepine receptors in brain. *Synapse* **2007**, *61*, 595–605. [CrossRef]
68. Wang, M.; Gao, M.; Miller, K.D.; Zheng, Q.H. Synthesis of [^{11}C]PBR06 and [^{18}F]PBR06 as agents for positron emission tomographic (PET) imaging of the translocator protein (TSPO). *Steroids* **2011**, *76*, 1331–1340. [CrossRef]
69. Briard, E.; Zoghbi, S.S.; Siméon, F.G.; Imaizumi, M.; Gourley, J.P.; Shetty, H.U.; Lu, S.; Fujita, M.; Innis, R.B.; Pike, V.W. Single-Step High-Yield Radiosynthesis and Evaluation of a Sensitive ^{18}F-Labeled Ligand for Imaging Brain Peripheral Benzodiazepine Receptors with PET. *J. Med. Chem.* **2009**, *52*, 688–699. [CrossRef]
70. Ooms, M.; Attili, B.; Celen, S.; Koole, M.; Verbruggen, A.; Van Laere, K.; Bormans, G. [^{18}F]JNJ42259152 binding to phosphodiesterase 10A, a key regulator of medium spiny neuron excitability, is altered in the presence of cyclic AMP. *J. Neurochem.* **2016**, *139*, 897–906. [CrossRef]
71. Hebb, A.; Robertson, H.; Denovan-Wright, E. Striatal phosphodiesterase mRNA and protein levels are reduced in Huntington's disease transgenic mice prior to the onset of motor symptoms. *Neuroscience* **2004**, *123*, 967–981. [CrossRef]
72. Boscutti, G.; A Rabiner, E.; Plisson, C. PET Radioligands for imaging of the PDE10A in human: Current status. *Neurosci. Lett.* **2019**, *691*, 11–17. [CrossRef] [PubMed]
73. Andrés, J.I.; De Angelis, M.; Alcázar, J.; Iturrino, L.; Langlois, X.; Dedeurwaerdere, S.; Lenaerts, I.; Vanhoof, G.; Celen, S.; Bormans, G. Synthesis, In Vivo Occupancy, and Radiolabeling of Potent Phosphodiesterase Subtype-10 Inhibitors as Candidates for Positron Emission Tomography Imaging. *J. Med. Chem.* **2011**, *54*, 5820–5835. [CrossRef] [PubMed]
74. Van Laere, K.; Ahmad, R.U.; Hudyana, H.; Dubois, K.; Schmidt, M.E.; Celen, S.; Bormans, G.; Koole, M. Quantification of ^{18}F-JNJ-42259152, a Novel Phosphodiesterase 10A PET Tracer: Kinetic Modeling and Test-Retest Study in Human Brain. *J. Nucl. Med.* **2013**, *54*, 1285–1293. [CrossRef] [PubMed]

75. Van Laere, K.; Ahmad, R.U.; Hudyana, H.; Celen, S.; Dubois, K.; Schmidt, M.E.; Bormans, G.; Koole, M. Human biodistribution and dosimetry of ^{18}F-JNJ42259152, a radioligand for phosphodiesterase 10A imaging. *Eur. J. Nucl. Med. Mol. Imaging* **2013**, *40*, 254–261. [CrossRef] [PubMed]
76. Ahmad, R.; Bourgeois, S.; Postnov, A.; Schmidt, M.E.; Bormans, G.; Van Laere, K.; Vandenberghe, W. PET imaging shows loss of striatal PDE10A in patients with Huntington disease. *Neurology* **2014**, *82*, 279–281. [CrossRef] [PubMed]
77. Barret, O.; Thomae, D.; Tavares, A.; Alagille, D.; Papin, C.; Waterhouse, R.; McCarthy, T.; Jennings, D.; Marek, K.; Russell, D.; et al. In Vivo Assessment and Dosimetry of 2 Novel PDE10A PET Radiotracers in Humans: ^{18}F-MNI-659 and ^{18}F-MNI-654. *J. Nucl. Med.* **2014**, *55*, 1297–1304. [CrossRef]
78. Russell, D.S.; Barret, O.; Jennings, D.L.; Friedman, J.H.; Tamagnan, G.D.; Thomae, D.; Alagille, D.; Morley, T.J.; Papin, C.; Papapetropoulos, S.; et al. The Phosphodiesterase 10 Positron Emission Tomography Tracer, [^{18}F]MNI-659, as a Novel Biomarker for Early Huntington Disease. *JAMA Neurol.* **2014**, *71*, 1520. [CrossRef]
79. Russell, D.S.; Jennings, D.L.; Barret, O.; Tamagnan, G.D.; Carroll, V.M.; Caillé, F.; Alagille, D.; Morley, T.J.; Papin, C.; Seibyl, J.P.; et al. Change in PDE10 across early Huntington disease assessed by [^{18}F]MNI-659 and PET imaging. *Neurology* **2016**, *86*, 748–754. [CrossRef]
80. Plisson, C.; Weinzimmer, D.; Jakobsen, S.; Natesan, S.; Salinas, C.; Lin, S.F.; Labaree, D.; Zheng, M.Q.; Nabulsi, N.; Marques, T.R.; et al. Phosphodiesterase 10A PET Radioligand Development Program: From Pig to Human. *J. Nucl. Med.* **2014**, *55*, 595–601. [CrossRef]
81. Niccolini, F.; Haider, S.; Reis Marques, T.; Muhlert, N.; Tziortzi, A.C.; Searle, G.E.; Natesan, S.; Piccini, P.; Kapur, S.; Rabiner, E.A.; et al. Altered PDE10A expression detectable early before symptomatic onset in Huntington's disease. *Brain* **2015**, *138*, 3016–3029. [CrossRef]
82. Heurling, K.; Ashton, N.J.; Leuzy, A.; Zimmer, E.R.; Blennow, K.; Zetterberg, H.; Eriksson, J.; Lubberink, M.; Schöll, M. Synaptic vesicle protein 2A as a potential biomarker in synaptopathies. *Mol. Cell. Neurosci.* **2019**, *97*, 34–42. [CrossRef] [PubMed]
83. Vogl, C.; Mochida, S.; Wolff, C.; Whalley, B.J.; Stephens, G.J. The Synaptic Vesicle Glycoprotein 2A Ligand Levetiracetam Inhibits Presynaptic Ca^{2+} Channels through an Intracellular Pathway. *Mol. Pharmacol.* **2012**, *82*, 199–208. [CrossRef] [PubMed]
84. Mercier, J.; Provins, L.; Valade, A. Discovery and development of SV2A PET tracers: Potential for imaging synaptic density and clinical applications. *Drug Discov. Today Technol.* **2017**, *25*, 45–52. [CrossRef] [PubMed]
85. Mercier, J.; Archen, L.; Bollu, V.; Carré, S.; Evrard, Y.; Jnoff, E.; Kenda, B.; Lallemand, B.; Michel, P.; Montel, F.; et al. Discovery of Heterocyclic Nonacetamide Synaptic Vesicle Protein 2A (SV2A) Ligands with Single-Digit Nanomolar Potency: Opening Avenues towards the First SV2A Positron Emission Tomography (PET) Ligands. *ChemMedChem* **2014**, *9*, 693–698. [CrossRef]
86. Bahri, M.A.; Plenevaux, A.; Aerts, J.; Bastin, C.; Becker, G.; Mercier, J.; Valade, A.; Buchanan, T.; Mestdagh, N.; Ledoux, D.; et al. Measuring brain synaptic vesicle protein 2A with positron emission tomography and [^{18}F]UCB-H. *Alzheimer's Dement. Transl. Res. Clin. Interv.* **2017**, *3*, 481–486. [CrossRef]
87. Nabulsi, N.B.; Mercier, J.; Holden, D.; Carre, S.; Najafzadeh, S.; Vandergeten, M.C.; Lin, S.f.; Deo, A.; Price, N.; Wood, M.; et al. Synthesis and Preclinical Evaluation of ^{11}C-UCB-J as a PET Tracer for Imaging the Synaptic Vesicle Glycoprotein 2A in the Brain. *J. Nucl. Med.* **2016**, *57*, 777–784. [CrossRef]
88. DiFilippo, A.; Murali, D.; Ellison, P.; Barnhart, T.; Engle, J.; Christian, B. Improved synthesis of [^{11}C]UCB-J for PET imaging of synaptic density. *J. Nucl. Med.* **2019**, *60*, 1624.
89. Bertoglio, D.; Verhaeghe, J.; Miranda, A.; Kertesz, I.; Cybulska, K.; Korat, Š.; Wyffels, L.; Stroobants, S.; Mrzljak, L.; Dominguez, C.; et al. Validation and noninvasive kinetic modeling of [^{11}C]UCB-J PET imaging in mice. *J. Cereb. Blood Flow Metab.* **2019**. [CrossRef]

© 2020 by the authors. Licensee MDPI, Basel, Switzerland. This article is an open access article distributed under the terms and conditions of the Creative Commons Attribution (CC BY) license (http://creativecommons.org/licenses/by/4.0/).

Communication

Late-Stage Copper-Catalyzed Radiofluorination of an Arylboronic Ester Derivative of Atorvastatin

Gonçalo S. Clemente [1,†], Tryfon Zarganes-Tzitzikas [2,†], Alexander Dömling [2] and Philip H. Elsinga [1,*]

1. Department of Nuclear Medicine and Molecular Imaging, University Medical Center Groningen, University of Groningen, 9713 GZ Groningen, The Netherlands; g.dos.santos.clemente@umcg.nl
2. Department of Drug Design, Groningen Research Institute of Pharmacy, University of Groningen, 9713 AV Groningen, The Netherlands; ttzitzikas@yahoo.com (T.Z.-T.); a.s.s.domling@rug.nl (A.D.)
* Correspondence: p.h.elsinga@umcg.nl; Tel.: +31-50-361-3247
† These authors contributed equally to this work.

Academic Editor: Svend Borup Jensen
Received: 9 October 2019; Accepted: 18 November 2019; Published: 20 November 2019

Abstract: There is an unmet need for late-stage ^{18}F-fluorination strategies to label molecules with a wide range of relevant functionalities to medicinal chemistry, in particular (hetero)arenes, aiming to obtain unique in vivo information on the pharmacokinetics/pharmacodynamics (PK/PD) using positron emission tomography (PET). In the last few years, Cu-mediated oxidative radiofluorination of arylboronic esters/acids arose and has been successful in small molecules containing relatively simple (hetero)aromatic groups. However, this technique is sparsely used in the radiosynthesis of clinically significant molecules containing more complex backbones with several aromatic motifs. In this work, we add a new entry to this very limited database by presenting our recent results on the ^{18}F-fluorination of an arylboronic ester derivative of atorvastatin. The moderate average conversion of [^{18}F]F$^-$ (12%), in line with what has been reported for similarly complex molecules, stressed an overview through the literature to understand the radiolabeling variables and limitations preventing consistently higher yields. Nevertheless, the current disparity of procedures reported still hampers a consensual and conclusive output.

Keywords: fluorine-18; radiochemistry; late-stage radiofluorination; drug development; copper-catalyzed; boronic pinacol ester

---

1. Introduction

Being already a clinically established molecular imaging modality, positron emission tomography (PET) increasingly broadened its application field by also becoming an essential partner of the pharmaceutical industry [1,2]. Its unique combination of spatial resolution, quantification, and detection sensitivity provides essential in vivo information at an early stage by directly measuring tissue uptake concentrations of the radiolabeled molecules of interest. Ideally, the radionuclide should be added to the desired molecular structure causing as little disturbance as possible, especially in the vicinity of the active site(s), and at the latest possible stage in the process to avoid radiation loss and exposure. Historically, radiochemistry found an unparalleled ally in nucleophilic substitution reactions with [^{18}F]F$^-$ [3,4]. However, this became more challenging when the focus fell on the labeling of (hetero)arenes that are not easily reactive to aromatic nucleophilic substitutions. The ubiquitous role of heteroaromatic pharmacophores in drug development and medicinal chemistry stressed out the need for improved radiofluorination techniques to overcome the typically far-from-ideal electrophilic fluorination with carrier-added [^{18}F]F$_2$. Recently, several methods have been published aiming for a practical, transversal, and straightforward ^{18}F-fluorination of electron-rich, -poor, and –neutral (hetero)arenes [5–15]. One of these strategies, the late-stage copper-mediated oxidative ^{18}F-fluorination of arylboronic ester and acid derivatives, has received great attention from radiochemistry research groups but is still not routinely applied in the production of clinical PET radiopharmaceuticals. Numerous basic-research proposals for improving this Cu-catalyzed reaction have been successfully reported and conceptualized with simple heteroaromatic groups [15–33], but advanced applications to more complex molecules with potential clinical value are sparse and generally reveal very fluctuating ^{18}F-fluorination efficiencies [33–38]. Following previous work from our group [39], where we applied this Cu-mediated strategy to several structurally different drug-like molecules and investigated the influence of a range of temperature, solvents, catalyst, and precursor amounts, we aimed to go up in terms of complexity, applicability, and relevance. As a proof-of-concept, we synthesized an arylboronic ester derivative of atorvastatin (6), the highest-selling drug of all time and one of the most clinically prescribed. The presence of three phenyl groups and an electron-rich pyrrole core, together with a flexible hydrophobic side-chain, entails an increasingly challenging ^{18}F-fluorination test to this Cu-catalyzed strategy when compared to our previous simple drug-like molecules or even to the majority of the molecules reported in the literature. Thus, to highlight the potentialities and drawbacks of this radiolabeling strategy, herein we present and discuss one of the most complex labeling precursor scaffolds that have been submitted to Cu-catalyzed radiofluorination. With this, we add a new and significant entry to the still very structure-limited database of bioactive molecules that have been radiolabeled via this strategy. Moreover, the existence of a radiolabeled atorvastatin analog has the potential to become a widespread research tool to aid in the understanding of the recently reported pleiotropic and off-target mechanisms of statins [40,41], enabling the study of cellular and subcellular interactions through high sensitive nuclear analytical and imaging techniques. The findings using [^{18}F]atorvastatin (8) may then be inferred to the native molecule increasing the knowledge related to its pharmacokinetics/pharmacodynamics, which represents a practical example of the synergy that can exist between PET imaging and the pharmaceutical industry. As atorvastatin is a widely characterized and registered drug, any envisaged clinical assays with this radiotracer are also facilitated by the fact that its toxicological profile is already well described.

2. Results

The introduction of a labile boronic pinacol ester (Bpin) in the position to be radiofluorinated, facilitates the intermediate transmetalation with [Cu(OTf)$_2$(py)$_4$] and further coordination to [^{18}F]F$^-$, to yield, after oxidation and reductive elimination, the desired [^{18}F]fluorobenzene derivative (Scheme 1). With this procedure, a radioactive analog of the atorvastatin intermediate (7) was synthesized since the original structure is preserved with the native fluorine being solely substituted by its β$^+$-emitting radioisotope (conserving physicochemical and biological properties).

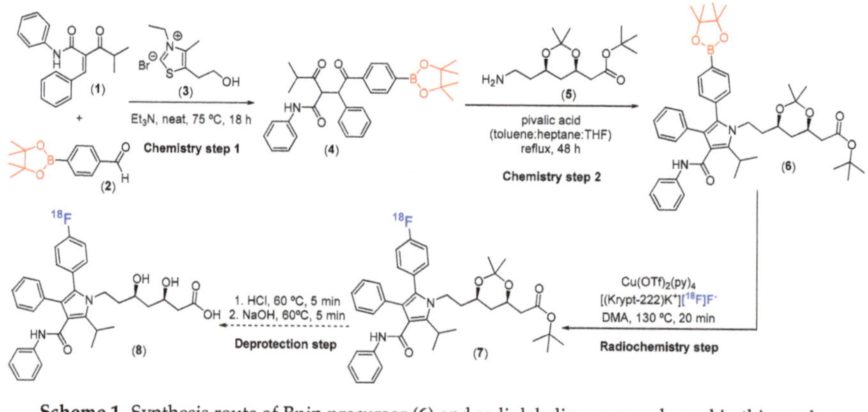

Scheme 1. Synthesis route of Bpin precursor (6) and radiolabeling approach used in this work.

For the ^{18}F-fluorination of the Bpin labeling precursor (**6**), the aqueous [^{18}F]F$^-$ produced in a biomedical cyclotron was quantitatively trapped (>95%) in an anion-exchange cartridge. The presence of an excess of basic salts and phase-transfer agents, typically used to efficiently recover the trapped [^{18}F]F$^-$ and enhance its reactivity, are known to be detrimental to the Cu catalyst stability, disturbing the essential oxidation/reduction cycle for the radiolabeling. Thus, to not significantly affect [^{18}F]F$^-$ elution, a previously optimized [39] balanced compromise was achieved using 3.15 mg of kryptofix 2.2.2 (Krypt-2.2.2), 50 µg of K$_2$CO$_3$, and 0.5 mg of K$_2$C$_2$O$_4$ in 1 mL 80% CH$_3$CN (elution efficiency: 80.3% ± 2.5%, n = 7, when performed dropwise). The recovered [(Krypt-2.2.2)K$^+$][^{18}F]F$^-$ solution was then azeotropically dried at 105 °C under gentle magnetic stirring and a light stream of argon (directly over the solution and not in the solution), without ever letting the mixture to completely dry. The softness of this drying step seems to be important to minimize the often significant losses of activity by evaporation and adsorption of the [^{18}F]F$^-$ to the borosilicate glass reaction vial walls (Table S1). After increasing the temperature to 130 °C and adding an optimized [39] solution of 60 µmol of Bpin labeling precursor (**6**) and 20 µmol of [Cu(OTf)$_2$(py)$_4$] in 0.8 mL dimethylacetamide (DMA), the reaction mixture was left to react under vigorous stirring for 20 min, as increasing the reaction up to 60 min only improved the final [^{18}F]F$^-$ conversion yield by approximately 3%. At the very beginning and after 10 min of the reaction, the sealed vial was purged with 5 mL of dried atmospheric air (passed through a P$_2$O$_5$ cartridge) to facilitate the re-oxidation of the copper complex, as the Cu(III) species seem to be responsible for the nucleophilic aromatic substitution [33,42]. However, this procedure does not appear to be relevant for the success of the reaction as the absence of it led to identical radiolabeling results.

Although achievable, the approach used only yielded an inconsistent radiofluorination of the labeling precursor (12% ± 11% determined by multiplying the radio-TLC conversion of [^{18}F]F$^-$ with radio-HPLC purity, n = 7). The absence of products of degradation and radiochemical impurities from the chromatographic spectra (Figure S3), associated with the still visible signal of the intact Bpin labeling precursor (**6**), suggests that the ^{18}F-fluorination might have been hampered by a reduction of the Cu catalyst reactivity. It is known from the literature that the atorvastatin side chain [43], the presence of the two non-functionalized mono-substituted benzene rings and a pyrrole core [44], and the basic salts in solution [16] can all influence copper oxidation states, which might explain the limited [^{18}F]F$^-$ conversion. Nevertheless, the radiofluorination yields obtained in this work are in line with what has been reported for complex heteroaromatic molecules, especially if containing several phenyl groups in its structure [33,35,38], and should still be sufficient to proceed for the development of [^{18}F]atorvastatin (**8**) preclinical screening assays after a fast and nearly quantitative deprotection of the side chain [45] (Figure S4).

3. Discussion

The Cu-mediated oxidative ^{18}F-fluorination strategy improved the radiochemistry field by supplying a practical solution for the labeling of (hetero)arenes. The proof-of-concept radiofluorination of arylboronic esters and acid derivatives, without the presence of extensive heteroaromatic functional groups, has already been proven successful. But the translation to larger scales and more complex biologically active molecules aimed for PET application/evaluation is generally associated with low to moderate ^{18}F-fluorination yields and reproducibility. The radiofluorination herein presented with an arylboronic ester derivative of atorvastatin (**6**) proved to be in line with these findings and led us to an overall review through the literature to understand the radiolabeling variables and limitations preventing consistently higher yields. This late-stage Cu-mediated radiofluorination strategy has already shown to be very dependent of the type and complexity of the labeling precursor used, and very sensitive to all the processes associated with the method–from the additives used to enable [^{18}F]F$^-$ elution, passing through the azeotropic drying harshness, the anhydrous environment level, and reaction solvents used, to the temperature, reagent amounts and Cu catalyst type. The base and phase transfer catalyst amounts used for the radiolabeling of the atorvastatin intermediate have been previously optimized in our latest work [39]. Higher amounts invariably ended up in no detecTable ^{18}F-fluorination of the Bpin atorvastatin precursor (possibly due to the formation of copper adducts [22]), and lower amounts resulted in poorer elution efficiencies without improving the final [^{18}F]F$^-$ conversion yield to the radiofluorinated product. The softness of the drying step can also be essential for the procedure to not fall in one of the drawbacks of this radiolabeling methodology—the significant reduction of [^{18}F]F$^-$ availability for the reaction due to the extensive escape of activity and adsorption to the borosilicate glass reaction vial walls. In our work, we reached the best results by preventing the [(Krypt-2.2.2)K$^+$][^{18}F]F$^-$ solution from tumultuous boiling and harsh agitation, as this avoids splashing of the complex to upper regions of the reaction vial that will not be in contact with the subsequent Bpin labeling precursor/Cu-catalyst solution. Additionally, it is also beneficial to not let the [(Krypt-2.2.2)K$^+$][^{18}F]F$^-$ solution completely evaporate (3 azeotropic drying cycles with 0.5 mL anhydrous acetonitrile, each one starting after the previous volume has almost vanished, followed by the dilution with 100 μl of anhydrous DMA immediately after the last cycle has nearly evaporated completely and the further addition of the remaining solvent with the precursor (6) and [Cu(OTf)$_2$(py)$_4$]). To circumvent the downsides of azeotropic evaporation, especially when automated where manipulation and close control of the conditions are challenging, a few solid-phase extraction (SPE) drying procedures have been rising in the literature [21,27,28,30], some even able to avoid the use of bases [29]. However, being very recent, they still lack a proper multicentre evaluation and assessment into more than just simple (hetero)arenes, as some authors claim not being able to reproduce them [38] and when attempted by us for the ^{18}F-fluorination of the arylboronic ester derivative of atorvastatin (6), invariably led to no detectable [^{18}F]F$^-$ conversion (despite shown to be successful when tested first in some of the same simple aryl boronic acid esters used in our previous work [39]).

Currently, late-stage Cu-mediated ^{18}F-fluorination of precursors containing multi (hetero)arenes in their structure is still very dependent on a range of variables and on the existing expertise in the radiochemistry lab performing it. Therefore, a case-by-case optimization still seems to be necessary, being extremely difficult to reach a standardized procedure for every labeling precursor, which might explain the reason why the exact same methodology has been very rarely repeated in the literature. An analysis through the Cu-mediated works published, and hereby referenced [15–39], shows that [Cu(OTf)$_2$(py)$_4$] is still by far the most common catalyst used (against other options such as Cu(OTf)$_2$, Cu(OTf)$_2$(associated with diverse pyridine derivatives), or Cu(CF$_3$SO$_3$)$_2$), and the typical amounts for all of them are between 5 to 30 μmol while the Bpin labeling precursor may vary from 4 to 60 μmol. The reaction temperatures are usually kept around 120 °C ± 10 °C while anhydrous DMA and dimethylformamide (DMF) are the solvents almost exclusively reported, with the first one having the propensity for better conversion efficacies [39] which can arguably be due to its higher boiling point and resistance to bases. Numerous base additives (e.g., potassium oxalate/trifluoromethanesulfonate, dimethylaminopyridine, tetraethylammonium bicarbonate/bromide, tetrabutylammonium fluoride, and trichlorophenylethenesulfonate) have been used for [^{18}F]F$^-$ elution, with potassium carbonate being preferentially chosen. Carrier-added ([^{19}F]KF) radiolabeling reactions to simulate conventional fluorinations showed no improvement in the conversion yields [24]. The reaction times reported are typically between 20 to 30 min, and an experiment prolonging the reaction with the arylboronic ester derivative of atorvastatin (6) until 60 min did not result in a significant increase in [^{18}F]F$^-$ conversion.

In summary, from the analysis of the literature, a general association can be established between a higher concentration of reactants (and typically a 5 to 12 eq. excess of simple arylBpin precursor over [Cu(OTf)$_2$(py)$_4$]), and minimizing the reaction volume and the molar ratio of the added base, with ^{18}F-fluorination efficiency. Nevertheless, the direct conversion of these conditions is not always practically (and economically) achievable for complex and clinically relevant (hetero)arene precursors, since this may result in the use of several dozens of mg of valuable precursor (as it happens with the current arylboronic ester derivative of atorvastatin (6)) instead of just a few mg of the simple arenes. This is also expected to have a negative impact on the final molar activity (GBq.mmol^{-1}) of the radiotracer. Furthermore, the extensive use of a Cu-catalyst might bring additional issues, in terms of by-product formation and the need for further refined purification techniques.

4. Materials and Methods

4.1. General Procedure for the Synthesis of the Arylboronic Pinacol Ester (Bpin) Labeling Precursor (6)

Solvents and reagents, including the atorvastatin intermediate standard and atorvastatin reference (CAS 125971 95-1 and CAS 344423-98-9 from TCI Chemicals, Zwijndrecht, Belgium), were available from commercial suppliers and used without any further purification.

A mixture of 2-benzylidene-4-methyl-3-oxo-N-phenylpentanamide (1, 5 g, 17 mmol, 1.00 equiv.), 3-ethyl-5-(2-hydroxyethyl)-4-methyl-3-thiazolium bromide (3, 1.7 g, 6.8 mmol, 0.40 equiv.), triethylamine (5 mL, 36 mmol, 2.12 equiv.), and 4-formylphenylboronic acid pinacol ester (2, 4.9 g, 21 mmol, 1.20 equiv.) was heated at 75 °C under argon atmosphere with vigorous stirring for 16 h. The reaction was monitored by thin-layer chromatography (TLC) until the consumption of the N-phenylpentanamide (1). Isopropyl alcohol (25 mL) was added, and the reaction mixture was maintained at 25 °C for 4 h under stirring. The remaining solid was vacuum filtered and washed with 25 mL of water followed by 20 mL of isopropyl alcohol. The product was dried under high vacuum for 4 h, affording 4-methyl-3-oxo-2-(2-oxo-1-phenyl-2-(4-(4,4,5,5-tetramethyl-1,3,2-dioxaborolan-2-yl)phenyl)ethyl)-N-phenylpentanamide (4) as a yellowish crystalline solid in approximately 14% yield (1.8 g, 2.4 mmol).

Pivalic acid (0.5 g, 4.9 mmol, 3.77 equiv.) was added, under nitrogen atmosphere, to a solution of the previously synthesized phenylpentanamide derivative (4, 1 g, 1.3 mmol, 1.00 equiv.) and tert-butyl 2-((4R,6R)-6-(2-aminoethyl)-2,2-dimethyl-1,3-dioxan-4-yl)acetate (5, 1 g, 3.7 mmol, 2.85 equiv.) in toluene:heptane:tetrahydrofuran (1:4:1 v/v) (20 mL). The reaction mixture was refluxed for 24 h with azeotropic removal of water, monitored by TLC, cooled to room temperature, and extracted with ethyl acetate (3 × 50 mL). The organic phase was washed with saturated aqueous sodium chloride solution (50 mL). The solvent was removed under vacuum, the desired Bpin labeling precursor (6) was obtained as a pale yellow solid in approximately 60% yield (0.6 g, 0.8 mmol) after purification by column chromatography (petroleum ether:ethyl acetate).

4.2. Characterization Data

tert-butyl 2-((4R,6R)-6-(2-(2-isopropyl-4-phenyl-3-(phenylcarbamoyl)-5-(4-(4,4,5,5-tetramethyl-1,3,2-dioxaborolan-2-yl)phenyl)-1H-pyrrol-1-yl)ethyl)-2,2-dimethyl-1,3-dioxan-4-yl)acetate (Bpin labeling precursor 6):

^{1}H NMR (500 MHz, CDCl$_3$) δ 7.72 (d, J = 7.9 Hz, 2 H), 7.20–7.15 (m, 9 H), 7.06 (d, J = 7.9 Hz, 2 H), 6.97 (t, J = 7.4 Hz, 1 H), 6.88 (s, 1 H), 4.17–4.07 (m, 2 H), 3.91–3.82 (m, 1 H), 3.67–3.57 (m, 2 H), 2.35 (dd, J = 15.2, 7.3 Hz, 1 H), 2.22 (dd, J = 15.2, 5.8 Hz, 1 H), 1.72–1.58 (m, 2 H), 1.53 (dd, J = 7.1, 3.9 Hz, 6 H), 1.43 (s, 9 H), 1.34 (d, J = 2.6 Hz, 9 H), and 1.23 (s, 9 H).

^{13}C NMR (126 MHz, CDCl$_3$) δ 184.5, 170.3, 164.9, 141.8, 138.4, 135.1, 134.7, 134.6, 130.6, 130.6, 129.9, 128.6, 128.3, 126.5, 123.5, 121.7, 119.6, 115.4, 98.7, 83.9, 80.7, 66.4, 65.9, 42.5, 40.9, 38.5, 38.0, 35.9, 29.9, 28.1, 27.0, 26.0, 24.9, 24.5, 21.7, 21.6, and 19.7.

HRMS-ESI: m/z calcd. for $C_{46}H_{60}BN_2O_7$ [M + H]$^+$ 763.452, found 763.379.

4.3. General Procedure for the Cu-mediated Radiosynthesis

All procedures involving the handling of radioactive substances were carried out in a radiochemistry laboratory with the required conditions of radiological protection and safety.

The Cu-mediated radiolabeling procedure followed our previously optimized method using several structurally different drug-like molecules functionalized with a Bpin leaving group [39]. The aqueous [^{18}F]F$^-$ used in this work was produced by the ^{18}O(p,n)^{18}F nuclear reaction in an IBA (Ottignies-Louvain-la-Neuve, Belgium) Cyclone 18/9 cyclotron and then loaded (approx. 1.5 GBq) into a polystyrene-divinylbenzene in a HCO_3^- anion exchange cartridge (Chromafix 45-PS-HCO_3^-) without the need of any preconditioning. The cartridge was then washed out to a 5 mL borosilicate glass Wheaton reaction V-vial (containing a stirring bar) with 1 mL of an 80% acetonitrile solution of 3.15 mg Krypt-2.2.2, 0.05 mg K_2CO_3, and 0.5 mg $K_2C_2O_4$. This solution was submitted to azeotropic drying with subsequent additions of anhydrous acetonitrile at 105 °C to originate moistureless [(Krypt-2.2.2)K^+][^{18}F]F^-. Then, 0.8 mL of DMA with the boronic pinacol ester derivative labeling precursor (6, 60 µmol) was added to this same vial with the previously dissolved [Cu(OTf)$_2$(py)$_4$] catalyst (20 µmol, 0.33 equiv.). This reaction mixture was left under vigorous stirring at 130 °C for 20 min to afford 7 after a total synthesis time of under 60 min (Figure S3). The conversion to the ^{18}F-product was assessed by radio-TLC (TLC-SG developed in hexane:ethyl acetate (1:1 v/v), Rf([^{18}F]F^-) = 0.0–0.2 and Rf(7) = 0.8–1.0) and radio—High performance liquid chromatography (HPLC) (SymmetryPrep^{TM} C18 7 µm 7.8 × 300 mm; A: sodium acetate 0.05 M pH 4.7, B: acetonitrile; 0–4 min.: 90% A, 4–15 min.: 90% A to 20% A, 15–25 min.: 20% A to 5% A, 25–33 min.: 5% A 33–34 min.: 5% A to 90% A, 34–35 min.: 90% A; flow: 6 mL.min^{-1}.; Rf(8) ≈ 16 min Rf(7) ≈ 23 min). As a proof-of-concept, 7 was converted to [^{18}F]atorvastatin (8) by a fast (extra 10 min of synthesis time) and nearly quantitative deprotection of the side chain [45] with HCl followed by NaOH. The final product (8) was then isolated (approx. 25 MBq) by HPLC (Figure S4).

5. Conclusions

Despite being potentially attainable with the Cu-mediated ^{18}F-fluorination strategy, our goal for an enhanced automatable approach to achieve [^{18}F]atorvastatin (8) in a larger production scale with practical and sufficient yields will continue, as the ultimate purpose is to proceed for the development of preclinical screening assays and further clinical evaluation in humans. A deeper understanding of the crucial conditions to optimize the yields obtained with the Cu-catalysed radiofluorination was attempted but due to the disparity of data, procedures, and labeling precursors reported in the literature, it is hardly possible to reach to a consensual and accurate conclusion. From a review of the literature, it seems undeniable that the nature of the (hetero)arene labeling precursor plays a major role in the efficiency of ^{18}F-fluorination. The wise approach still seems to be to perform an individual "one variable at a time" optimization for each scaffold to be radiolabeled, despite the fact that this might ignore the influence of multifactorial interactions [46]. Thus, the search for more robust late-stage radiofluorination procedures compatible with suitable heteroaromatic pharmacophores remains a very stimulating topic that, ultimately, can lead not only to refined radiopharmaceutical drug discovery but also to aid the pharmaceutical industry to evaluate pharmacokinetics/dynamics and better understand certain mechanisms of action.

Supplementary Materials: The following are available online at http://www.mdpi.com/1420-3049/24/23/4210/s1, Figure S1: ^1H NMR characterization of the Bpin labeling precursor, Figure S2: ^{13}C NMR characterization of the Bpin labeling precursor, Figure S3: Chromatographic profile of the compounds used and synthesized, Figure S4: Chromatographic profile of [^{18}F]atorvastatin (8), Table S1: Influence of azeotropic drying procedure in [^{18}F]F^- availability to the radiolabeling reaction.

Author Contributions: Conceptualization, G.S.C., T.Z.-T., A.D. and P.H.E.; methodology, G.S.C. and T.Z.-T.; synthesis and characterization, T.Z.-T. and G.S.C.; radiochemistry, G.S.C.; writing—Original draft preparation, G.S.C.; writing—Review and editing, G.S.C., T.Z.-T., A.D. and P.H.E.; supervision, A.D. and P.H.E.

Funding: The work in A.D.'s laboratory was financially supported from the NIH (NIH 2R01GM097082-05) and the European Union's Horizon 2020 research and innovation program under MSC ITN "Accelerated Early staGe drug dIScovery" (AEGIS, grant agreement No 675555) and COFUND ALERT (grant agreement No 665250) and KWF Kankerbestrijding grant (grant agreement No 10504) and the Qatar National Research Foundation (NPRP6-065-3-012). G.S.C. would like to thank the Dutch Open Technologieprogramma from NWO Toegepaste en Technische Wetenschappen (project n° 13547) for the scholarship funding.

Conflicts of Interest: The authors declare no conflict of interest.

References

1. Elsinga, P.H.; van Waarde, A.; Paans, A.M.J.; Dierckx, R.A.J.O. *Trends on the Role of PET in Drug Development*; World Scientific Pub Co Inc.: Singapore, 2012.
2. Fernandes, E.; Barbosa, Z.; Clemente, G.; Alves, F.; Abrunhosa, A.J. Positron emitting tracers in pre-clinical drug development. *Curr. Radiopharm.* **2012**, *5*, 90–98. [CrossRef] [PubMed]
3. Miller, P.W.; Long, N.J.; Vilar, R.; Gee, A.D. Synthesis of ^{11}C, ^{18}F, ^{15}O, and ^{13}N radiolabels for positron emission tomography. *Angew. Chem. Int. Ed.* **2008**, *47*, 8998–9033. [CrossRef] [PubMed]
4. Jacobson, O.; Kiesewetter, D.O.; Chen, X. Fluorine-18 Radiochemistry, Labeling Strategies and Synthetic Routes. *Bioconjugate Chem.* **2015**, *26*, 1–18. [CrossRef] [PubMed]
5. Teare, H.; Robins, E.G.; Kirjavainen, A.; Forsback, S.; Sandford, G.; Solin, O.; Luthra, S.K.; Gouverneur, V. Radiosynthesis and Evaluation of [^{18}F]Selectfluor bis(triflate). *Angew. Chem. Int. Ed.* **2010**, *49*, 6821–6824. [CrossRef]
6. Pretze, M.; Große-Gehling, P.; Mamat, C. Cross-Coupling Reactions as Valuable Tool for the Preparation of PET Radiotracers. *Molecules* **2011**, *16*, 1129–1165. [CrossRef]
7. Lee, E.; Kamlet, A.S.; Powers, D.C.; Neumann, C.N.; Boursalian, G.B.; Furuya, T.; Choi, D.C.; Hooker, J.M.; Ritter, T. A Fluoride-Derived Electrophilic Late-Stage Fluorination Reagent for PET Imaging. *Science* **2011**, *334*, 639–642. [CrossRef]
8. Ichiishi, N.; Brooks, A.F.; Topczewski, J.J.; Rodnick, M.E.; Sanford, M.S.; Scott, P.J.H. Copper-Catalyzed [^{18}F]Fluorination of (Mesityl)(aryl)iodonium Salts. *Org. Lett.* **2014**, *16*, 3224–3227. [CrossRef]
9. Neumann, C.N.; Hooker, J.M.; Ritter, T. Concerted nucleophilic aromatic substitution with ^{19}F− and ^{18}F−. *Nature* **2016**, *534*, 369–373. [CrossRef]
10. Makaravage, K.J.; Brooks, A.F.; Mossine, A.V.; Sanford, M.S.; Scott, P.J.H. Copper-Mediated Radiofluorination of Arylstannanes with [^{18}F]KF. *Org. Lett.* **2016**, *18*, 5440–5443. [CrossRef]
11. Beyzavi, M.H.; Mandal, D.; Strebl, M.G.; Neumann, C.N.; D'Amato, E.M.; Chen, J.; Hooker, J.M.; Ritter, T. ^{18}F-Deoxyfluorination of Phenols via Ru π-Complexes. *Acs Cent. Sci.* **2017**, *3*, 944–948. [CrossRef]
12. Tredwell, M.; Preshlock, S.M.; Taylor, N.J.; Gruber, S.; Huiban, M.; Passchier, J.; Mercier, J.; Génicot, C.; Gouverneur, V. A General Copper-Mediated Nucleophilic ^{18}F Fluorination of Arenes. *Angew. Chem. Int. Ed.* **2014**, *53*, 7751–7755. [CrossRef] [PubMed]
13. Coenen, H.H.; Ermert, J. ^{18}F-labelling innovations and their potential for clinical application. *Clin. Transl. Imaging* **2018**, *6*, 169–193. [CrossRef]
14. Deng, X.; Rong, J.; Wang, L.; Vasdev, N.; Zhang, L.; Josephson, L.; Liang, S.H. Chemistry for Positron Emission Tomography: Recent Advances in ^{11}C-, ^{18}F-, ^{13}N-, and ^{15}O-Labeling Reactions. *Angew. Chem. Int. Ed.* **2019**, *58*, 2580–2605. [CrossRef] [PubMed]
15. Preshlock, S.; Tredwell, M.; Gouverneur, V. ^{18}F-Labeling of Arenes and Heteroarenes for Applications in Positron Emission Tomography. *Chem. Rev.* **2016**, *116*, 719–766. [CrossRef]
16. Zlatopolskiy, B.D.; Zischler, J.; Krapf, P.; Zarrad, F.; Urusova, E.A.; Kordys, E.; Endepols, H.; Neumaier, B. Copper-Mediated Aromatic Radiofluorination Revisited: Efficient Production of PET Tracers on a Preparative Scale. *Chem. A Eur. J.* **2015**, *21*, 5972–5979. [CrossRef]
17. Mossine, A.V.; Brooks, A.F.; Makaravage, K.J.; Miller, J.M.; Ichiishi, N.; Sanford, M.S.; Scott, P.J.H. Synthesis of [^{18}F]Arenes via the Copper-Mediated [^{18}F]Fluorination of Boronic Acids. *Org. Lett.* **2015**, *17*, 5780–5783. [CrossRef]
18. Preshlock, S.; Calderwood, S.; Verhoog, S.; Tredwell, M.; Huiban, M.; Hienzsch, A.; Gruber, S.; Wilson, T.C.; Taylor, N.J.; Cailly, T.; et al. Enhanced copper-mediated ^{18}F-fluorination of aryl boronic esters provides eight radiotracers for PET applications. *Chem. Commun.* **2016**, *52*, 8361–8364. [CrossRef]
19. Schäfer, D.; Weiß, P.; Ermert, J.; Castillo Meleán, J.; Zarrad, F.; Neumaier, B. Preparation of No-Carrier-Added 6-[^{18}F]Fluoro-l-tryptophan via Cu-Mediated Radiofluorination. *Eur. J. Org. Chem.* **2016**, *2016*, 4621–4628. [CrossRef]
20. Giglio, B.C.; Fei, H.; Wang, M.; Wang, H.; He, L.; Feng, H.; Wu, Z.; Lu, H.; Li, Z. Synthesis of 5-[(18)F]Fluoro-α-methyl Tryptophan: New Trp Based PET Agents. *Theranostics* **2017**, *7*, 1524–1530. [CrossRef]
21. Zischler, J.; Kolks, N.; Modemann, D.; Neumaier, B.; Zlatopolskiy, B.D. Alcohol-Enhanced Cu-Mediated Radiofluorination. *Chem. A Eur. J.* **2017**, *23*, 3251–3256. [CrossRef]

22. Mossine, A.V.; Brooks, A.F.; Ichiishi, N.; Makaravage, K.J.; Sanford, M.S.; Scott, P.J.H. Development of Customized [^{18}F]Fluoride Elution Techniques for the Enhancement of Copper-Mediated Late-Stage Radiofluorination. *Sci. Rep.* **2017**, *7*, 233. [CrossRef] [PubMed]
23. Blevins, D.W.; Kabalka, G.W.; Osborne, D.R.; Akula, M.R. Effect of Added Cu(OTf)2 on the Cu(OTf)2(Py)4-Mediated Radiofluorination of Benzoyl and Phthaloylglycinates. *Nat. Sci.* **2018**, *10*, 125–133. [CrossRef]
24. Zhou, D.; Chu, W.; Katzenellenbogen, J. Exploration of alcohol-enhanced Cu-mediated radiofluorination towards practical labeling. *J. Nucl. Med.* **2018**, *59*, 187.
25. Mossine, A.V.; Brooks, A.F.; Bernard-Gauthier, V.; Bailey, J.J.; Ichiishi, N.; Schirrmacher, R.; Sanford, M.S.; Scott, P.J.H. Automated synthesis of PET radiotracers by copper-mediated ^{18}F-fluorination of organoborons: Importance of the order of addition and competing protodeborylation. *J. Label. Compd. Radiopharm.* **2018**, *61*, 228–236. [CrossRef]
26. Lahdenpohja, S.; Keller, T.; Rajander, J.; Kirjavainen, A.K. Radiosynthesis of the norepinephrine transporter tracer [^{18}F]NS12137 via copper-mediated ^{18}F-labelling. *J. Label. Compd. Radiopharm.* **2019**, *62*, 259–264. [CrossRef]
27. Antuganov, D.; Zykov, M.; Timofeev, V.; Timofeeva, K.; Antuganova, Y.; Orlovskaya, V.; Fedorova, O.; Krasikova, R. Copper-Mediated Radiofluorination of Aryl Pinacolboronate Esters: A Straightforward Protocol by Using Pyridinium Sulfonates. *Eur. J. Org. Chem.* **2019**, *2019*, 918–922. [CrossRef]
28. Zhang, B.; Fraser, B.H.; Klenner, M.A.; Chen, Z.; Liang, S.H.; Massi, M.; Robinson, A.J.; Pascali, G. [^{18}F]Ethenesulfonyl Fluoride as a Practical Radiofluoride Relay Reagent. *Chem. A Eur. J.* **2019**, *25*, 7613–7617. [CrossRef]
29. Zhang, X.; Basuli, F.; Swenson, R.E. An azeotropic drying-free approach for copper-mediated radiofluorination without addition of base. *J. Label. Compd. Radiopharm.* **2019**, *62*, 139–145. [CrossRef]
30. Zlatopolskiy, B.D.; Zischler, J.; Krapf, P.; Richarz, R.; Lauchner, K.; Neumaier, B. Minimalist approach meets green chemistry: Synthesis of ^{18}F- labeled (hetero)aromatics in pure ethanol. *J. Label. Compd. Radiopharm.* **2019**, *62*, 404–410. [CrossRef]
31. Mossine, A.V.; Tanzey, S.S.; Brooks, A.F.; Makaravage, K.J.; Ichiishi, N.; Miller, J.M.; Henderson, B.D.; Skaddan, M.B.; Sanford, M.S.; Scott, P.J.H. One-pot synthesis of high molar activity 6-[^{18}F]fluoro-l-DOPA by Cu-mediated fluorination of a BPin precursor. *Org. Biomol. Chem.* **2019**. [CrossRef]
32. Lahdenpohja, S.O.; Rajala, N.A.; Rajander, J.; Kirjavainen, A.K. Fast and efficient copper-mediated ^{18}F-fluorination of arylstannanes, aryl boronic acids, and aryl boronic esters without azeotropic drying. *Ejnmmi Radiopharm. Chem.* **2019**, *4*, 28. [CrossRef] [PubMed]
33. Taylor, N.J.; Emer, E.; Preshlock, S.; Schedler, M.; Tredwell, M.; Verhoog, S.; Mercier, J.; Genicot, C.; Gouverneur, V. Derisking the Cu-Mediated ^{18}F-Fluorination of Heterocyclic Positron Emission Tomography Radioligands. *J. Am. Chem. Soc.* **2017**, *139*, 8267–8276. [CrossRef] [PubMed]
34. Drerup, C.; Ermert, J.; Coenen, H.H. Synthesis of a Potent Aminopyridine-Based nNOS-Inhibitor by Two Recent No-Carrier-Added ^{18}F-Labelling Methods. *Molecules* **2016**, *21*, 1160. [CrossRef] [PubMed]
35. Lien, V.T.; Klaveness, J.; Olberg, D.E. One-step synthesis of [^{18}F]cabozantinib for use in positron emission tomography imaging of c-Met. *J. Label. Compd. Radiopharm.* **2018**, *61*, 11–17. [CrossRef]
36. Bernard-Gauthier, V.; Mossine, A.V.; Mahringer, A.; Aliaga, A.; Bailey, J.J.; Shao, X.; Stauff, J.; Arteaga, J.; Sherman, P.; Grand'Maison, M.; et al. Identification of [^{18}F]TRACK, a Fluorine-18-Labeled Tropomyosin Receptor Kinase (Trk) Inhibitor for PET Imaging. *J. Med. Chem.* **2018**, *61*, 1737–1743. [CrossRef]
37. Wilson, T.; Xavier, M.-A.; Knight, J.; Verhoog, S.; Torres, J.B.; Mosley, M.; Hopkins, S.; Wallington, S.; Allen, D.; Kersemans, V.; et al. PET imaging of PARP expression using [^{18}F]olaparib. *J. Nucl. Med.* **2018**, *60*, 504–510. [CrossRef]
38. Guibbal, F.; Meneyrol, V.; Ait-Arsa, I.; Diotel, N.; Patché, J.; Veeren, B.; Bénard, S.; Gimié, F.; Yong-Sang, J.; Khantalin, I.; et al. Synthesis and Automated Labeling of [^{18}F]Darapladib, a Lp-PLA2 Ligand, as Potential PET Imaging Tool of Atherosclerosis. *Acs. Med. Chem. Lett.* **2019**, *10*, 743–748. [CrossRef]
39. Zarganes-Tzitzikas, T.; Clemente, G.S.; Elsinga, P.H.; Dömling, A. MCR Scaffolds Get Hotter with ^{18}F-Labeling. *Molecules* **2019**, *24*, 1327. [CrossRef]
40. Mohammad, S.; Nguyen, H.; Nguyen, M.; Abdel-Rasoul, M.; Nguyen, V.; Nguyen, C.D.; Nguyen, K.T.; Li, L.; Kitzmiller, J.P. Pleiotropic Effects of Statins: Untapped Potential for Statin Pharmacotherapy. *Curr. Vasc. Pharmacol.* **2019**, *17*, 239–261. [CrossRef]

41. Fatehi Hassanabad, A.; McBride, S.A. Statins as Potential Therapeutics for Lung Cancer: Molecular Mechanisms and Clinical Outcomes. *Am. J. Clin. Oncol.* **2019**, *42*, 732–736. [CrossRef]
42. Fier, P.S.; Luo, J.; Hartwig, J.F. Copper-Mediated Fluorination of Arylboronate Esters. Identification of a Copper(III) Fluoride Complex. *J. Am. Chem. Soc.* **2013**, *135*, 2552–2559. [CrossRef] [PubMed]
43. Refat, M.S.; Al-Saif, F.A. Synthesis, spectral, thermal, and antimicrobial studies of transition metal complexes of atorvastatin calcium as a lipid-lowering agent. *J. Therm. Anal. Calorim.* **2015**, *120*, 863–878. [CrossRef]
44. Hatcher, L.Q.; Karlin, K.D. Ligand Influences in Copper-Dioxygen Complex-Formation and Substrate Oxidations. In *Advances in Inorganic Chemistry*; Academic Press: Cambridge, MA, USA, 2006; Volume 58, pp. 131–184.
45. Novozhilov, Y.V.; Dorogov, M.V.; Blumina, M.V.; Smirnov, A.V.; Krasavin, M. An improved kilogram-scale preparation of atorvastatin calcium. *Chem. Cent. J.* **2015**, *9*, 7. [CrossRef] [PubMed]
46. Bowden, G.D.; Pichler, B.J.; Maurer, A. A Design of Experiments (DoE) Approach Accelerates the Optimization of Copper-Mediated ^{18}F-Fluorination Reactions of Arylstannanes. *Sci. Rep.* **2019**, *9*, 11370. [CrossRef] [PubMed]

Sample Availability: Samples of the compounds are available from the authors.

 © 2019 by the authors. Licensee MDPI, Basel, Switzerland. This article is an open access article distributed under the terms and conditions of the Creative Commons Attribution (CC BY) license (http://creativecommons.org/licenses/by/4.0/).

Review

Recent Advances in Bioorthogonal Click Chemistry for Efficient Synthesis of Radiotracers and Radiopharmaceuticals

Sajid Mushtaq [1,†], Seong-Jae Yun [2,†] and Jongho Jeon [3,*]

1. Department of Nuclear Engineering, Pakistan Institute of Engineering and Applied Sciences, Islamabad 45650, Pakistan; sajidmushtaq@pieas.edu.pk
2. IT Convergence Materials Group, Korea Institute of Industrial Technology, Cheonan 31056, Korea; ysj8570@gmail.com
3. Department of Applied Chemistry, School of Applied Chemical Engineering, Kyungpook National University, Daegu 41566, Korea
* Correspondence: jeonj@knu.ac.kr; Tel.: +82-53-950-5584
† These authors contributed equally to this paper.

Academic Editor: Svend Borup Jensen
Received: 4 September 2019; Accepted: 27 September 2019; Published: 2 October 2019

Abstract: In recent years, several catalyst-free site-specific reactions have been investigated for the efficient conjugation of biomolecules, nanomaterials, and living cells. Representative functional group pairs for these reactions include the following: (1) azide and cyclooctyne for strain-promoted cycloaddition reaction, (2) tetrazine and trans-alkene for inverse-electron-demand-Diels–Alder reaction, and (3) electrophilic heterocycles and cysteine for rapid condensation/addition reaction. Due to their excellent specificities and high reaction rates, these conjugation methods have been utilized for the labeling of radioisotopes (e.g., radiohalogens, radiometals) to various target molecules. The radiolabeled products prepared by these methods have been applied to preclinical research, such as in vivo molecular imaging, pharmacokinetic studies, and radiation therapy of cancer cells. In this review, we explain the basics of these chemical reactions and introduce their recent applications in the field of radiopharmacy and chemical biology. In addition, we discuss the significance, current challenges, and prospects of using bioorthogonal conjugation reactions.

Keywords: radiolabeling; bioorthogonal reaction; click chemistry; site-specific reaction; radiopharmaceuticals; radioisotopes; molecular imaging

---

1. Introduction

The term 'click chemistry' has been introduced to describe specific chemical reactions, which are fast, reliable and can be selectively applied to the synthesis of functional materials and biomolecule conjugates [1–6]. Click chemistry can be broadly defined as a ligation reaction in which two reactants are joined under ambient conditions to provide the desired product in high chemical yield and short time [7–10]. Over the last two decades, tremendous development and progress has been achieved in these conjugation reactions to encompass wide substrate scopes in the click reaction. Additionally, in several cases, these ligations proceed in aqueous media without significant decrease of the selectivity and reaction rate. Furthermore, click chemistries enable the facile isolation of the desired products from the reaction mixtures and facilitate the removal of the non-reacted substrates and byproducts, without the need for sophisticated separation methods [11–16]. Therefore, click chemistry-based conjugation methods have been applied to several avenues of research, including biochemical sciences, material sciences [17–24], drug discovery [25–28], pharmaceutical sciences [29–34], and synthesis of radiolabeled products [35–41]. Several typically used ligation reactions which are closely related to click chemistry include the thiol-Michael addition reaction [42], ring-opening reactions of aziridinium ions and epoxides [43], hydrazone and oxime formation from an aldehyde group [44] and so on. However, these reactions showed certain disadvantages such as poor specificity and stability under aqueous conditions, because of the reactivity of these functional groups with biomolecule residues and water. In 2003, K. B. Sharpless and M. G. Finn et al. reported that copper(I)-catalyzed azide-alkyne [3+2] cycloaddition reaction (CuAAC) can be employed as a new class of click reactions for rapid and reliable bioconjugation [45]. As both azide and alkyne groups are unreactive toward protein residues or other biomolecules, this ligation brought about a great impact and has been utilized as an efficient site-specific ligation methodology. Later, some researchers reported that the exogenous metals used to catalyze the click reaction (e.g., copper) could cause mild to severe cytotoxic effects and thus the use of metal catalyst-free chemical reaction has been recommended for several applications [46]. Therefore, catalyst-free, rapid, biocompatible, and bioorthogonal reactions such as strain-promoted azide-alkyne cycloaddition reaction (SPAAC) [47] and inverse-electron-demand Diels–Alder reaction (IEDDA) [48] have been developed as useful alternatives, and have been extensively used in various research fields (Figure 1).

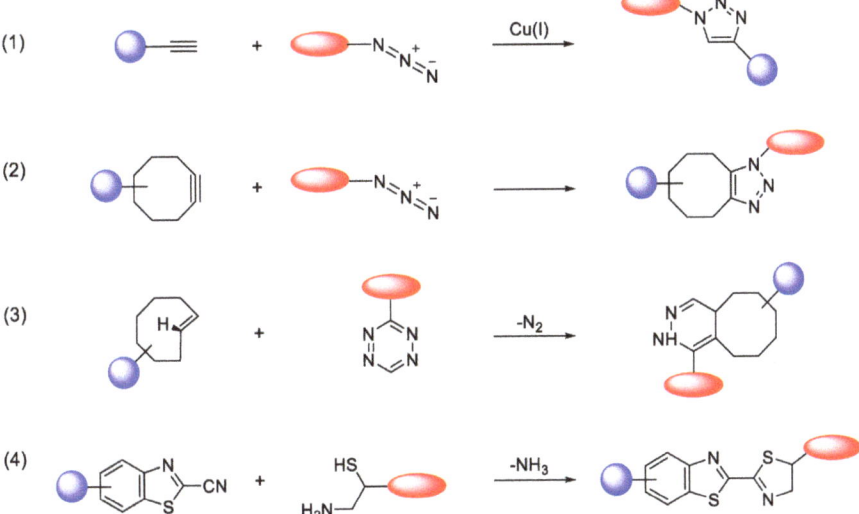

Figure 1. Selected bioorthogonal conjugation reactions. (**1**) Copper-catalyzed azide-alkyne cycloaddition reaction (CuAAC); (**2**) strain-promoted azide-alkyne cycloaddition reaction (SPAAC); (**3**) tetrazine and trans-alkene substrates for inverse electron-demand-Diels–Alder reaction (IEDDA); (**4**) condensation reaction between 2-cyanobenzothiazole (CBT) and 1,2-aminothiol (*N*-terminal cysteine).

In recent years, these conjugation reactions have also been applied to the synthesis of radioisotope-labeled molecules, which have been used for nuclear imaging using positron emission tomography (PET) and single-photon emission computed tomography (SPECT) as well as for therapeutic applications. Particularly, several important diagnostic radioisotopes including ^{11}C ($t_{1/2}$ = 20 min), ^{18}F ($t_{1/2}$ = 110 min), ^{99m}Tc ($t_{1/2}$ = 360 min), and ^{68}Ga ($t_{1/2}$ = 68 min) have short half-lives, and thus their radiolabeling procedures require rapid and efficient reactions which can provide reliable radiochemical results, such as high radiochemical yield (RCY) and purity, and minimal undesired by-product formation [49]. In this regard, the catalyst-free click reactions can be highly useful tools for radiolabeling complex small molecules and biomacromolecules, which are sensitive to harsh reaction conditions such as elevated temperatures, extreme pH, and the presence of metal catalysts [50]. In addition to in vitro radiolabeling applications, these ligation methods have also been investigated for in vivo pre-targeted strategies for specific imaging and cancer therapy in animal xenograft models [51].

This review aims to highlight the recent and noteworthy results for the synthesis of radiolabeled molecules using site-specific click reactions. In detail, this review will mainly focus on the following bioconjugation reactions: (1) strain-promoted azide-alkyne cycloaddition (SPAAC); (2) inverse-electron-demand Diels–Alder cycloaddition reaction (IEDDA); (3) rapid condensation/cycloaddition reactions based on electrophilic heterocycles. The review will also showcase the advantages of these reactions, which have empowered radiochemists in the production of radiolabeled products and radiopharmaceuticals for imaging and therapeutic purposes. Finally, future directions and emerging trends of these ligation methods will be discussed.

2. Strained Promoted Copper-Free Click Reaction for Synthesis of Radiolabeled Molecules

In Vitro Radiolabeling of Biomolecules

In SPAAC, the ring strain of cyclic alkynes such as dibenzocyclooctyne (DBCO) is used to drive the reaction with azide groups in the absence of copper(I) catalysis [52,53]. Generally, two strategies have been employed for SPAAC-based radiolabeling. The first is the synthesis of radiolabeled cyclooctyne precursors, which can be used for the labeling of azide containing biomolecules, and the other is the preparation of radioisotope-tagged azide tracers which are reacted with cyclooctyne modified biomolecules. In 2011, Feringa group investigated SPAAC reaction for the efficient ^{18}F-labeling of biomolecules [54]. In this study, three ^{18}F-labeled azides were synthesized, and the prepared tracers were conjugated with DBCO modified bombesin peptide derivatives. Notably, the reaction proceeded with high efficiency to provide ^{18}F-labeled cancer-targeting peptides in 15 min with good radiochemical yields (RCYs) (Figure 2). Particularly, radiolabeling studies using these reactions were also explored in human plasma to determine their reactivity and specificity in biological media.

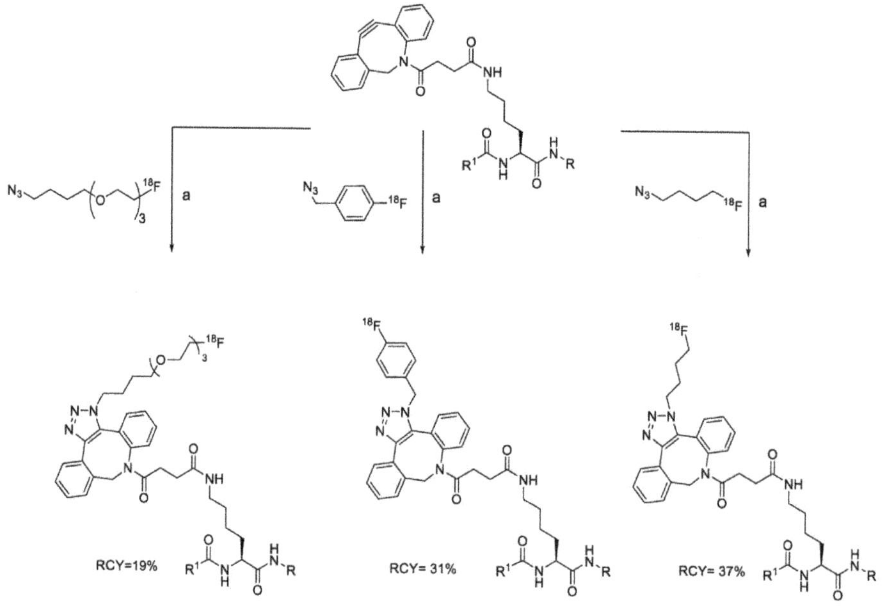

Figure 2. ^{18}F-radiolabeling of bombesin derivative using SPAAC: **a)** human plasma or dimethyl sulfoxide (DMSO), room temperature, 15 min. R = Pyr-Gln, Pyr = pyroglutamic acid, R_1 = Leu-Gly-Asn-Gln-Trp-Ala-Val-Gly-His-Leu-Met-NH$_2$, RCY = radiochemical yield.

Wuest et al. reported the synthesis of an ^{18}F-labeled DBCO analog for efficient preparation of a diagnostic probe. (Table 1, entry 1) This tracer was reacted with several azide group-bearing geldanamycine moieties and carbohydrates to furnish the corresponding ^{18}F-labeled products. Importantly, these radiolabeling reactions were performed in various media, including in methanol, DMSO/water (1:1), and bovine serum albumin, wherein the observed RCYs did not decrease significantly [55]. Along similar lines, Carpenter et al. used a modified ^{18}F-labeled DBCO analog for the radiolabeling of azide conjugated substrates (Table 1, entry 2). The radiolabeling was performed at room temperature in N,N-dimethylformamide (DMF) to afford the desired radiolabeled products [56]. The peptide A20FMDV2 (Table 1, entry 3), which has a strong binding affinity with integrin $\alpha_v\beta_6$-receptor, was successfully labeled with ^{18}F, using a SPAAC-based ligation. The radiolabeling of the azide group bearing A20FMDV2 was performed at ambient temperature to give the product in 11% of isolated RCY. The radiolabeled peptide was highly stable in rat serum, and its binding affinity towards the target receptor was not affected. However, in vivo studies revealed its decreased targeting ability due to the structural differences and increased lipophilicity compared to the parent structure [57]. Several other ^{18}F-DBCO analogs have shown good RCY for the preparation of radiolabeled peptides for targeting cancer [58,59]. To improve the efficiency of ^{18}F radiolabeling, Roche et al. explored a new ^{18}F-labeled azide prosthetic group, ^{18}F-FPyZIDE (Table 1, entry 6). In their study, the radiolabeled tracers were evaluated in both CuAAC- and SPAAC-based ligations and the labeling results showed that both radiolabeling methods provided high RCYs under mild reaction conditions (room temperature or 40 °C) [60]. Evans et al. investigated the radiosynthesis of ^{68}Ga-labeled peptide using azide group-bearing 1,4,7,10-tetraazacyclododecane-tetraacetic acid (DOTA) chelator and DBCO group conjugated cRGD peptide (Table 1, entry 7) [61]. The developed novel bioorthogonal click reaction has been used in the design and preparation of multimodal imaging tracers. Ghosh et al. studied a dual-modal scaffold in which the precursor was first labeled with ^{68}Ga using a DOTA chelator, and then, a near-infrared (NIR)-absorbing fluorescent dye, IR Dye 800CW, was incorporated into the tracer using SPAAC ligation. The dual-labeled tracer was then applied to the targeted imaging of a somatostatin receptor and the quantification of its biological uptake in vivo (Table 1, entry 8) [62].

Table 1. Examples of SPAAC in labeling reactions using short half-life radioisotopes.

Entry	DBCO Precursor	Azide Precursor	Product [a]	RCY(%)	Ref
1		N$_3$–R R = 4-azidoaniline, 11-azido-3,6,9-trioxaundane-1-amine, 6-azido-6-deoxyglucose, 2-azido-deoxyglucose, azido-geldanamycin		69–98	[55]
2		N$_3$–R R = Ph, PEGylated acid		64–75	[56]
3		R = A20FMDV2 peptide		12	[57]

Table 1. Cont.

Entry	DBCO Precursor	Azide Precursor	Product [a]	RCY(%)	Ref
4	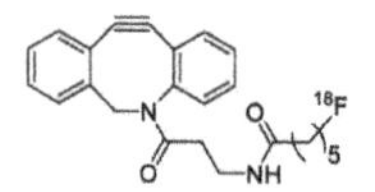	R = Tyr^3-octreotate peptide	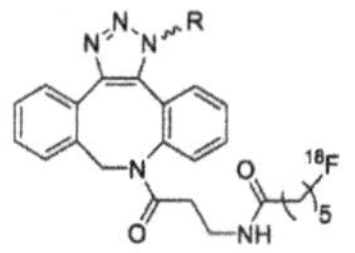	95	[58]
5	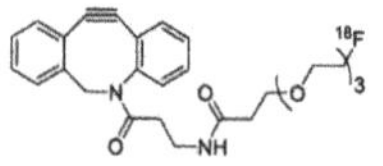	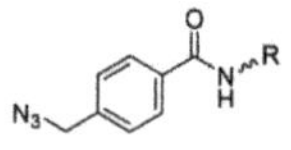 R = cRGD peptide	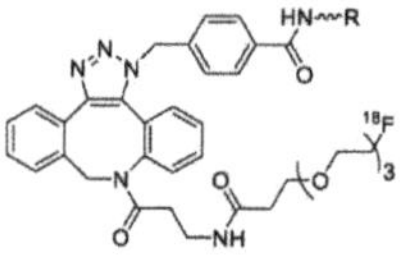	93	[59]
6	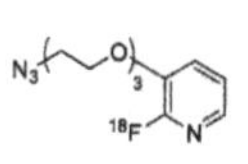	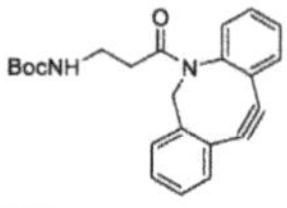	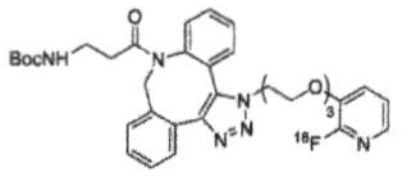	>95	[60]
7	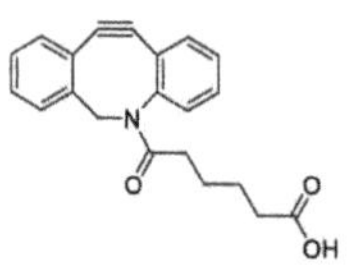	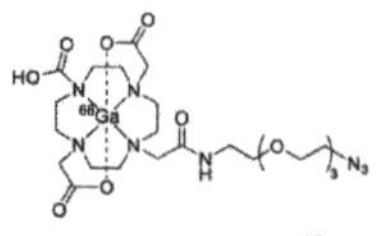 R = (PEG)$_3$-DOTA-^{68}Ga	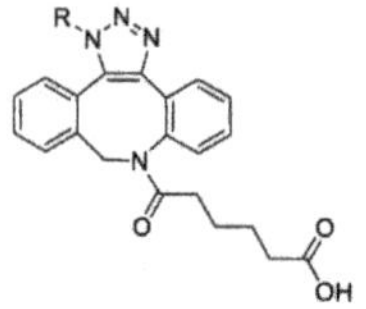	94–100	[61]
8	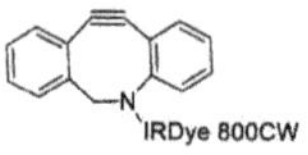	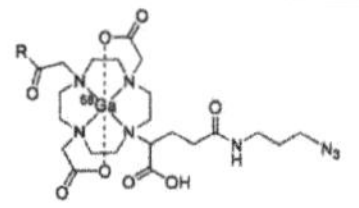 R = Tyr^3-octreotate peptide	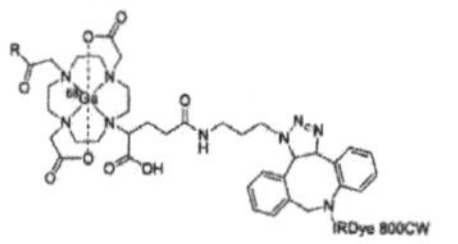	80	[62]

[a] Products were obtained as isomeric mixtures.

Generally, most SPAAC ligations based on DBCO derivatives display second-order rate constants in the 1–2 M^{-1} s^{-1} range with azide groups [63] due to which, the observed RCY is not satisfactory when using low substrate concentrations. To improve reaction kinetics, ^{18}F-labeled oxa-dibenzocyclooctyne (ODIBO), which has a k$_2$ value of 45 M^{-1} s^{-1}, was synthesized to label azide containing biomolecules with high efficiency (Figure 3) [64,65]. This new prosthetic group enabled site-specific radiolabeling using much smaller amounts (about one-tenth) of azide bearing molecules than those of DBCO-based reactions.

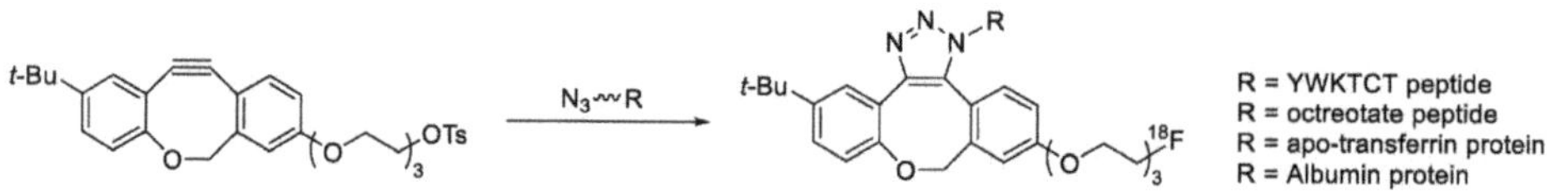

Figure 3. Radiolabeling of peptides or proteins using ^{18}F-labeled ODIBO.

R = YWKTCT peptide
R = octreotate peptide
R = apo-transferrin protein
R = Albumin protein

The study reported by Kim et al. used SPAAC ligation in both radiolabeling reaction and purification steps [66]. For this application, ^{18}F-labeled azide tracer was first reacted with DBCO-modified cancer-targeting peptide (cRGD) and then the desired product was separated from unreacted peptide substrates using an azide modified resin as a scavenger for the DBCO group. The remarkable two-steps process provided the radiolabeled peptide in high decay-corrected RCY (92%) and radiochemical purity (98%) (Figure 4). Notably, PET imaging and biodistribution data confirmed the high tumor uptake value of the ^{18}F-labeled peptides in U87MG xenograft along with significant enhancement of the tumor to background ratio [67].

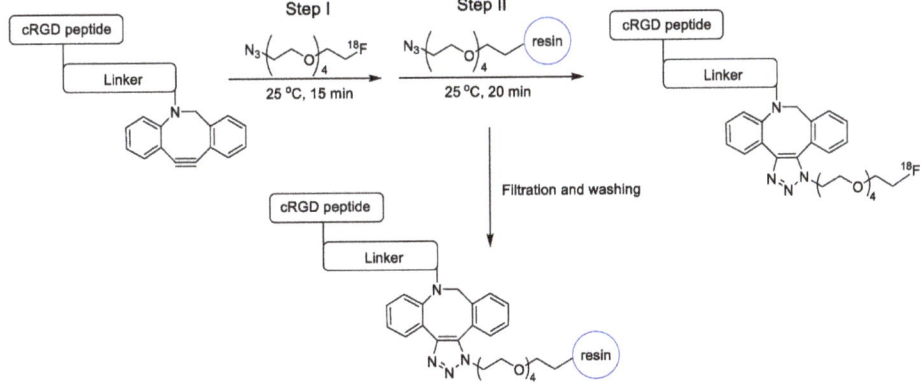

Figure 4. ^{18}F-radiolabeling of DBCO-modified cRGD dimer using ^{18}F-labeled azide precursor and polystyrene-supported azide-modified resin for purification of unreacted substrate.

SPAAC reaction has also been applied to the labeling of radioisotopes with longer half-lives, such as radioactive metals and radioactive iodine. We reported the use of ^{125}I-labeled azide prosthetic groups for synthesizing radiolabeled biomolecules and nanomaterials. In this process, DBCO group modified cRGD peptides were efficiently conjugated with ^{125}I-labeled azides in high RCY and radiochemical purity after HPLC purification (Figure 5) [68,69]. It was reported that ^{64}Cu could be labeled with cross-bridged cyclam chelator CB-TE1K1P under mild conditions. To employ this chelator for radiolabeling biomolecules, Anderson et al. synthesized a DBCO modified chelator (DBCO-PEG$_4$-CB-TE1K1P) and reacted it with an azide-bearing Cetuximab by SPAAC ligation (Figure 6). The ^{64}Cu labeling proceeded in high RCY (>95%) at 37 °C, and the radiolabeled antibody showed enhanced serum stability when compared with those of previously reported ^{64}Cu chelators [70].

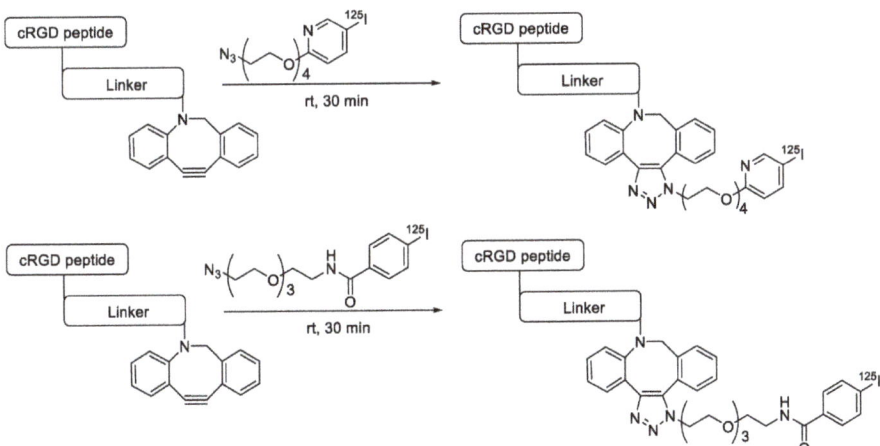

Figure 5. Radiolabeling of DBCO-conjugated cRGD peptide using ^{125}I-labeled azide tracers.

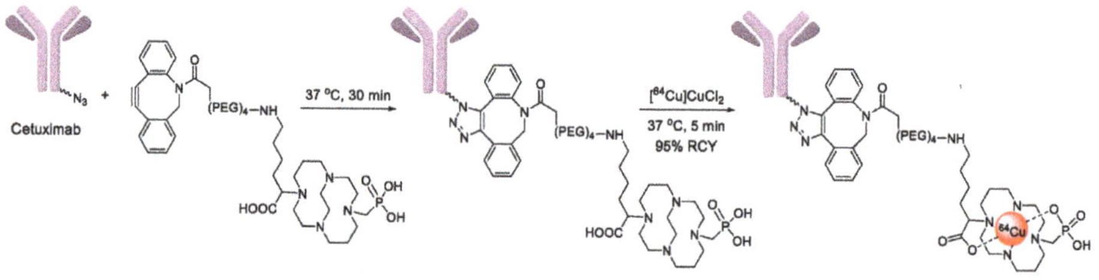

Figure 6. Reaction of azide-conjugated Cetuximab antibody with DBCO-conjugated crossed bridged macrocyclic CB-TE1K1P chelator for ^{64}Cu radiolabeling.

Yuan et al. explored the synthesis of ^{89}Zr-labeled PET imaging agents using SPAAC ligation on the surface of the superparamagnetic feraheme (FH). For this study, azide-functionalized FH nanoparticles were prepared and were mixed with ^{89}Zr under elevated temperature to deliver the ^{89}Zr-labeled azide-FH. In the next step, DBCO-conjugated RGD peptide, or Cy5.5 tagged protamine was reacted with ^{89}Zr-azide-FH to give the desired radiolabeled products with good radiochemical results and specific radioactivity (Figure 7) [71]. This strategy provided an efficient approach for the preparation of multimodal/multifunctional nanoprobes, which are suitable for a wide range of diagnostic and therapeutic applications.

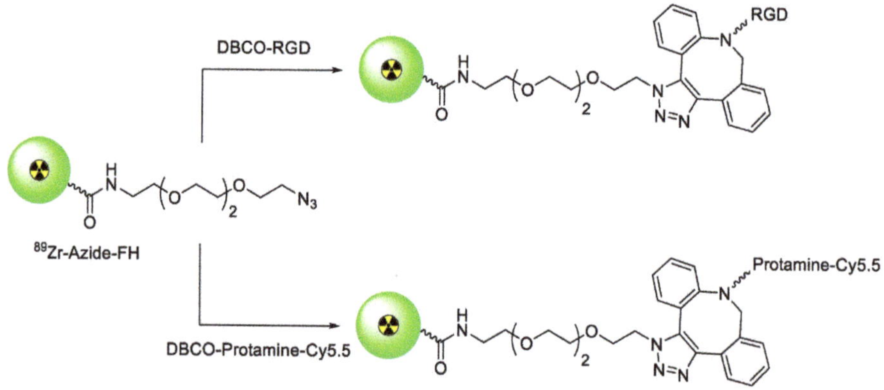

Figure 7. Preparation of ^{89}Zr-labeled multifunctional nanoprobes using SPAAC ligation.

Several kinds of liposomes are known to be useful vehicles for targeted delivery in biomedical research as well as clinically approved platforms [72]. Hood and co-workers used SPAAC ligation for efficient conjugation between ^{111}In-labeled liposomes and single-chain variable fragments (scFv) or monoclonal antibodies. The radiolabeled tracer, ^{111}In-liposomes-mAb/scFv, was used in the targeted imaging of the platelet-endothelial cell adhesion molecule (PECAM-1) and intracellular adhesion molecule (ICAM-1). The uptake value of ^{111}In-liposomes/scFv into the target cells was much higher than that of ^{111}In-liposomes/mAb [73]. Recently, thermosensitive hydrogels comprising polyisocyanopeptide (PIC) were labeled with ^{111}In via a SPAAC method. In this research, azide-modified PIC hydrogel was first conjugated with DBCO-modified diethylenetriaminepentaacetic acid (DTPA) chelator to afford the PIC-DTPA conjugate. Next, PIC-DTPA was reacted with ^{111}InCl$_3$ to give the ^{111}In-labeled PIC in high RCY. The radiolabeled PIC was applied in a SPECT imaging study for evaluating the efficacy of PIC gels in wound mouse models [74]. Figure 8 shows the ^{99m}Tc labeling of human serum albumin (HSA) via a SPAAC reaction. After labeling ^{99m}Tc(CO)$_3$ with an azide group-modified dipyridine chelator, it was then reacted with ADIBO bearing HSA under mild condition to give the radiolabeled protein in high RCY (76–99%). The ^{99m}Tc-labeled HSA prepared by this procedure showed better stability in vivo, as compared with those previously reported ^{99m}Tc-labeled HSA, which were obtained by direct ^{99m}Tc labeling [75]. The radiolabeled HSA thus prepared, was used in blood pool imaging using SPECT.

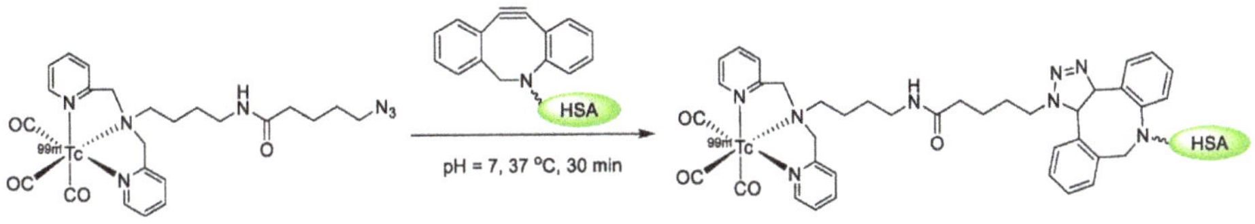

Figure 8. SPAAC for ^{99m}Tc-based radiolabeling of human serum protein.

3. Inverse-Electron-Demand Diels–Alder Reaction for Synthesis of Radiolabeled Molecules

3.1. In Vitro Radiolabeling of Biomolecules

The inverse electron demand Diels–Alder (IEDDA) between 1,2,4,5-tetrazine and strained alkene (such as trans-cyclooctene, TCO) is a well-established bioorthogonal reaction, which is typically regarded as the fastest click reaction with first-order rate constants ranging up to 10^5 M^{-1} S^{-1} [76–79]. Since the first report on IEDDA reaction, several kinds of strained alkenes/alkynes and tetrazine analogs have been synthesized, and these functional group pairs have been applied to the radiolabeling of various small molecules, biomolecules, and nanomaterials [80,81]. Due to the extremely rapid reaction rate of IEDDA under mild conditions such as room temperature, neutral pH, and in aqueous media, this reaction has been a highly useful ligation approach for labeling radioisotopes with short half-lives. In 2010, Fox et al. reported the IEDDA-mediated ^{18}F-labeling of small molecules. The radiolabeled TCO (Table 2, entry 1) could be synthesized by a nucleophilic substitution reaction of the tosylated precursor in 71% RCY. Remarkably, the IEDDA reaction between a model tetrazine substrate and ^{18}F-labeled TCO provided the desired product in more than 98% RCY in 10 seconds [82]. Conti et al. applied IEEDA to the synthesis of an ^{18}F-labeled cancer-targeting peptide [83]. The labeling reaction of a tetrazine conjugated cRGD peptide was carried out using an ^{18}F-labeled TCO analog, which was prepared using a similar protocol, and delivered the radiolabeled peptide in excellent RCY (Figure 9). The ^{18}F-labeled cRGD thus prepared, was evaluated in the U87MG xenograft model and exhibited clear visualization of tumor cells by PET imaging.

Table 2. IEDDA-mediated in vitro radiolabeling.

Entry	Tetrazine	Dienophile	Product [a]	RCY (%)	Ref
1	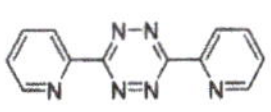	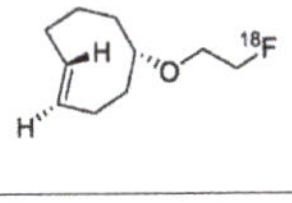	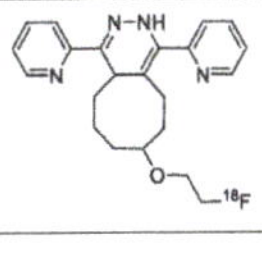	>98	[82]
2	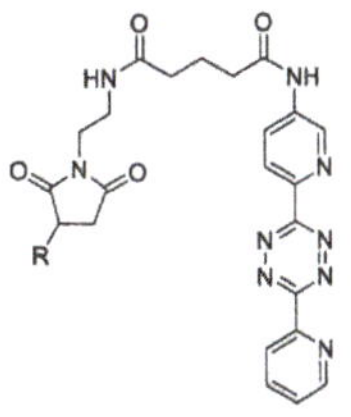 R = c(RGDyC) or VEGF protein	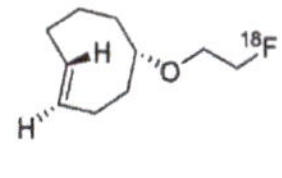	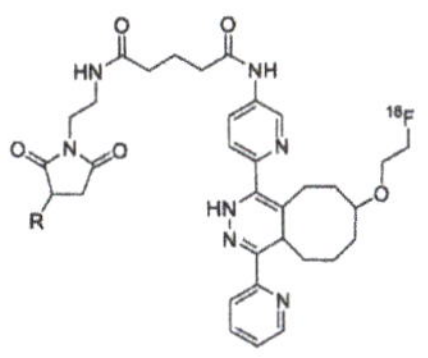	95 c(RGDyC), 75 (VEGF)	[84]

Table 2. *Cont.*

Entry	Tetrazine	Dienophile	Product [a]	RCY (%)	Ref
3	R= AZD2281			92	[85]
4	R= Cys40-exendin-4			>80	[86]
5				>99	[87]
6	R = Bombesin			46	[88]
7				>98	[89]
8	Tz-polymer	anti-A33 antibody		-	[90]
9		anti-VEGFR2 antibody		69	[91]
10		R = cRGD peptide, HSA protein		>99 (cRGD), 93 (HSA)	[92]

[a] Products were obtained as isomeric mixtures.

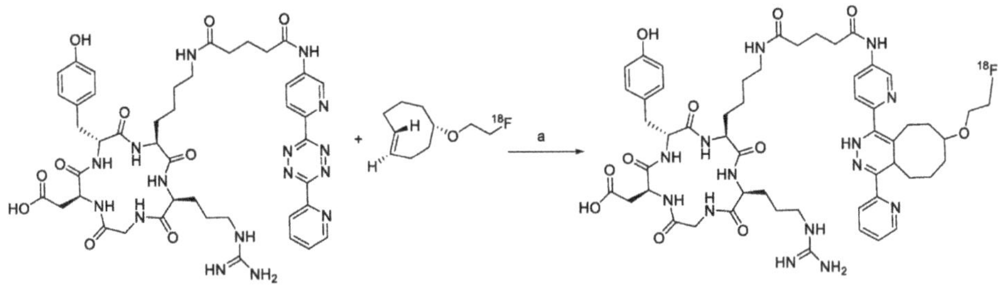

Figure 9. Radiolabeling of tetrazine conjugated cRGD peptide using ^{18}F-labeled TCO; **a)** DMSO, 10 s, room temperature.

Later, the same group reported a maleimide-conjugated tetrazine analog, which was used to incorporate the tetrazine group onto biomolecules comprising free cysteine moieties. The tetrazine bearing biomolecules (cRGD peptide and VEGF protein) prepared by the above method was then reacted with ^{18}F-labeled TCO to give PET-imaging tracers for diagnosis of cancer cells in vivo (Table 2, entry 2) [84]. Weissleder and coworkers synthesized ^{18}F-AZD2281, a poly-ADP-ribose-polymerase 1, as a PET imaging tracer (Table 2, entry 3). In this report, ^{18}F-labeled TCO and a tetrazine group-bearing AZD2281 were incubated for 3 minutes, and the crude product was purified using a magnetic TCO-scavenger resin for removing the unreacted substrate, without the need for carrying out the traditional HPLC purification. The process delivered the ^{18}F-labeled AZD2281 in 92% RCY using the scavenger-assisted method [85]. The prepared radiolabeled tracer was then evaluated in xenograft models to visualize MDA-MB-436 tumors. Wu and coworkers extended the application of IEDDA ligation to the radiolabeling of the exendin-4 peptide and applied the ^{18}F-labeled exendin-4 to the targeted imaging of GLP-1R receptor in an animal model [86]. In 2015, the Schirrmacher group reported the novel silicon-fluoride acceptor (SiFA) labeling method, which is based on an isotopic exchange reaction (Table 2, entry 5). This simple labeling step (^{19}F →^{18}F), which is based on a silicon-fluorine scaffold, provided the ^{18}F-labeled tetrazine in much higher RCY (78%) than those realized with other ^{18}F chemistries [87].

Norbornene analogs are known to be reactive toward tetrazines. Although the reaction rate was much slower than those of TCO analogs, the preparation of a norbornene substrate is straightforward. Importantly, norbornene analogs are known to be more stable than TCO analogs, which are prone to isomerization to their *cis*-isomers under physiological conditions. Knight and coworkers reported the reaction of the tetrazine group-conjugated bombesin peptide with an ^{18}F-labeled norbornene prosthetic group, to provide the radiolabeled product with high efficiency and radiochemical purity (Table 2, entry 6) [88]. In addition to the radioactive fluoride, ^{11}C is another important cyclotron-produced radioisotope for preclinical and clinical PET imaging. Particularly, the ^{11}C-labeled methyl triflate and methyl iodide are the most prominent synthons for nucleophilic methylation of alcohols, amines, and thiols, which are commonly used for the production of various radiotracers and radiopharmaceuticals. Herth and coworkers reported the first synthesis of an ^{11}C-labeled tetrazine and its reaction with a strained cyclooctene (Table 2, entry 7) [89]. The radioactive precursor [^{11}C]CH$_3$I was reacted with a tetrazine-conjugated phenol group to give the desired radiolabeled tetrazine in 33% RCY, which underwent a click reaction with a trans-cyclooctenol in 20 seconds, suggesting the suitability of this conjugation method for preparation of radiolabeled molecules with short-lived isotopes such as ^{11}C. Devaraj et al. reported the development of a ^{68}Ga-labeled tetrazine modified dextran polymer for increasing the half-life and in vivo stability of the tracer in blood (Table 2, entry 8), and evaluated its use in human colon cancer cells (LS174T) and xenograft models [90].

Radioactive iodines have been used for the preparation of various radiotracers for in vivo imaging and biodistribution studies. The traditional radioiodination method via an electrophilic substitution reaction typically provides high RCY in a short time. However, the radiolabeled tracer synthesized using the above reaction generally exhibited considerable deiodination in the living subjects, and the liberated radioactive iodines rapidly accumulated in the thyroid and stomach which and resulted in high background signals in the images. Moreover, the use of a strong oxidant that requires radioiodination often resulted in decreased biological activity of the molecules. To address these problems, the radioactive iodine-labeled tetrazine can be used as an alternative method for the efficient radiolabeling of biomolecules. Valliant et al. reported the rapid radiolabeling of antibody based on IEDDA. In this study, the ^{125}I-labeled tetrazine analog was incubated with the TCO-modified anti-VEGFR2 for 5 minutes to afford the desired product in 69% RCY. Interestingly, the radiolabeled antibody, which was prepared using this procedure, displayed a 10-fold increase in stability to in vivo deiodination, then the same antibody prepared by direct radioiodination using iodogen (Table 2, entry 9) [91]. Along similar lines, we investigated a modified ^{125}I-labeled tetrazine tracer via oxidative halo-destannylation of the corresponding precursor (Table 2, entry 10) [92]. The prepared radiolabeled tetrazine was then applied to the labeling of TCO derived cRGD peptide and human serum albumin (HSA) and delivered excellent RCYs (>99%). The biodistribution study of the ^{125}I-labeled HSA in normal ICR mice demonstrated enhanced in vivo stability toward deiodination than the radiolabeled HSA obtained using the conventional iodination method. Valliant group also synthesized $^{123/125}$I-labeled carborane-tetrazine and employed it for the radiolabeling of TCO-bound H520 cells [93].

Several radioactive metal-labeled tracers have also been prepared by IEDDA ligation for diagnostic purposes. Lewis et al. reported tetrazine conjugated metal-chelating agents such as DOTA and deferoxamine (DFO) for the radiolabeling of norbornene bearing trastuzumab using ^{64}Cu or ^{89}Zr (Figure 10) [94]. By using this procedure, radiolabeled trastuzumab was obtained in high RCY (>80%) and high specific radioactivity (>2.9 mCi/mg). Furthermore, PET imaging studies demonstrated that radiolabeled antibodies were quite stable in vivo conditions and showed specific uptake in HER2-positive BT-474 tumor cells. In 2018, IEDDA ligation was employed for the preparation of therapeutic radioisotope-labeled human antibodies 5B1 and huA33 (Figure 11) [95]. In this study, a tetrazine conjugated DOTA chelator was synthesized, and labeled with ^{225}Ac, a useful therapeutic radioisotope. The radiolabeled tetrazine tracer was then reacted with TCO-modified antibodies to give the desired products within 5 min. This two-step method provided superior RCYs compared to the conventional approaches used in clinical applications. In addition, the biodistribution results demonstrated that the ^{225}Ac-labeled antibody showed high tumor uptake values and relatively low non-specific accumulation in normal organs.

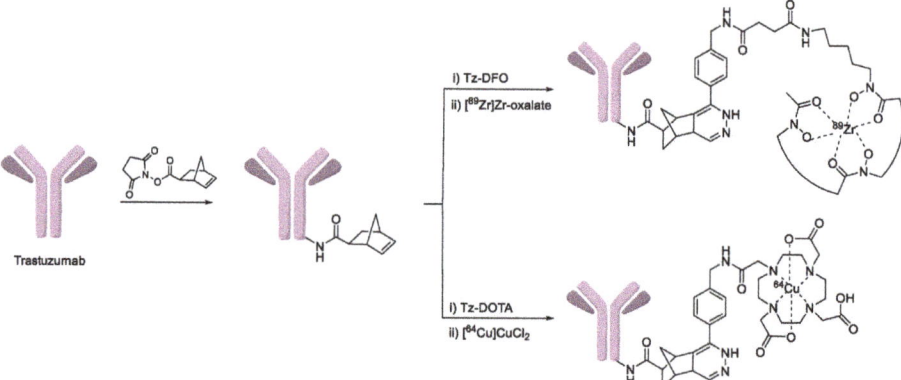

Figure 10. IEDDA-mediated radiolabeling of trastuzumab.

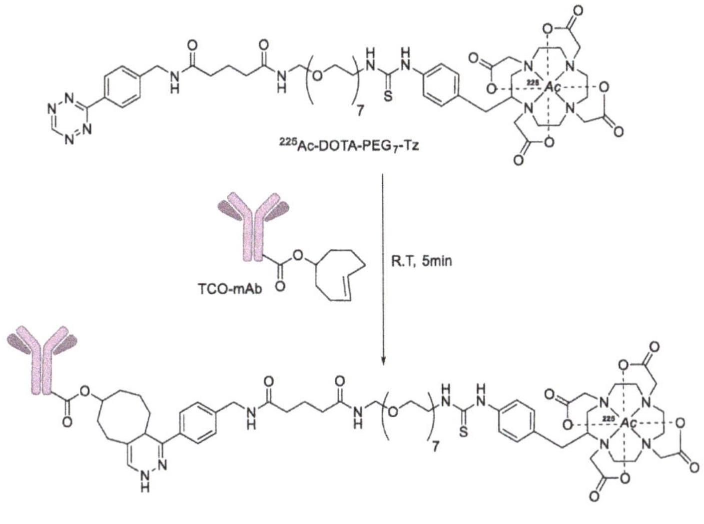

Figure 11. IEDDA-mediated synthesis of [225]Ac-labeled monoclonal antibody.

Recently, Long et al. reported the radiolabeling process for microbubble, which is a contrast agent used in ultrasound imaging and relies on an IEDDA reaction for its operation. First, a tetrazine-bearing metal chelator (HBED-CC) was labeled with [68]Ga. The TCO-modified phospholipids were then treated with ^{68}Ga-HBED-CC-tetrazine under mild conditions to give the ^{68}Ga-labeled lipid molecule (^{68}Ga-PE). Next, the prepared ^{68}Ga-PE was combined with other types of lipids, and the resulting formulation was activated to form gas-filled microbubbles (Figure 12). This strategy enabled the PET-based real-time monitoring and pharmacokinetic study of newly developed contrast agents for ultrasound analysis [96].

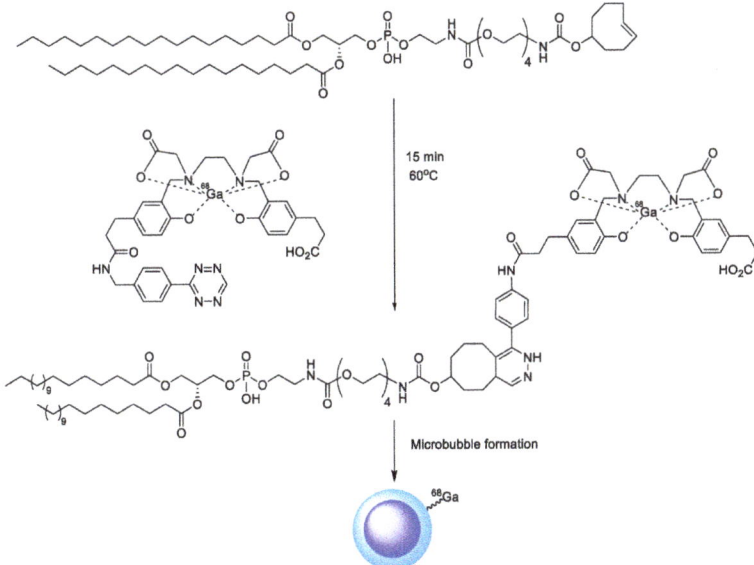

Figure 12. Synthesis of ^{68}Ga-labeled microbubble using IEDDA.

3.2. In Vivo Pre-Targeted Imaging and Therapy

The tetrazine and TCO groups are not reactive towards amine or thiol nucleophiles and show high reaction specificity in biological media. In addition, IEDDA can proceed with fast kinetics even at very low reactant concentrations. Due to these reasons, IEDDA-based ligation is one of the most potent tools for pre-targeting applications among the existing click reaction approaches. In the pre-targeted approach, a cancer-targeting ligand and a radiolabeled small molecule are administered separately into a living subject. Generally, the TCO (or tetrazine) conjugated tumor-targeting antibody is injected first into the tumor xenograft model and is allowed to accumulate in the tumor cells for a certain period (Figure 13). Next, the radiolabeled tetrazine (or TCO) group is administered after the excess amount of antibody in healthy tissues is excreted from the body. The in vivo click reaction through the above procedure decreases the circulation time of the radioligand and results in reduced non-specific uptake of radioactivity in healthy tissues. Furthermore, this approach also facilitates the delivery of radioisotopes with short half-lives, which would not be feasible with antibody-based imaging studies [97]. Table 3 shows in vivo pre-targeted studies that were conducted using IEDDA-based ligation in animal models.

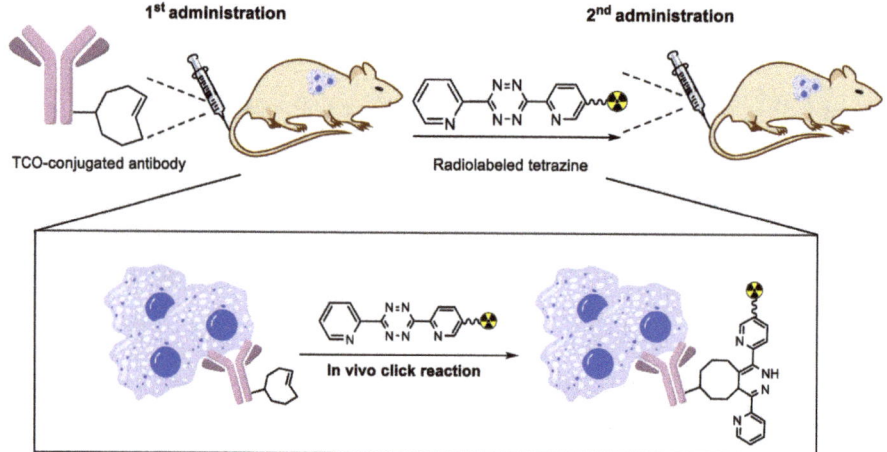

Figure 13. General strategy for pre-targeted imaging and therapy using IEDDA.

Table 3. IEDDA-based in vivo pre-targeted approach.

Entry	Biomolecule	Radiotracer	Animal Model	Ref
1	CC49-TCO antibody	^{111}In-labeled tetrazine	LS174T cells (Balb/C mouse)	[98]
2	CC49-TCO antibody	^{111}In-labeled tetrazine	LS174T cells (Balb/C mouse)	[99]
3	CC49-TCO antibody	^{177}Lu-labeled tetrazine	LS174T cells (Balb/C mouse)	[100]
4	CC49-TCO antibody	^{177}Lu-labeled tetrazine	LS174T cells (Balb/C mouse)	[101]
5	AVP04-07-TCO diabody	^{177}Lu-labeled tetrazine	LS174T cells (Balb/C mouse)	[102]
6	Z_{2395}-TCO affibody	^{111}In-labeled tetrazine ^{177}Lu-labeled tetrazine	SKOV-3 cells (Balb/C mouse)	[103]

Table 3. Cont.

Entry	Biomolecule	Radiotracer	Animal Model	Ref
7	PEGylated-TCO	^{18}F-labeled tetrazine	Healthy Balb/C mouse	[104]
8	5B1-TCO antibody	^{18}F-labeled tetrazine	BxPC3 cells (athymic nude mice)	[105]
9	Cetuximab-TCO antibody Trastuzumab-TCO antibody	^{18}F-labeled tetrazine	A431 cells (nu/nu mouse) BT-474 cells (nu/nu mouse)	[106]
10	Porous silicon-TCO nanoparticle	^{18}F-labeled tetrazine	Healthy (Balb/C mouse)	[107]
11	PSMA antagonist-tetrazine conjugate	^{18}F-labeled TCO	LNCaP cells (Balb/C mouse)	[108]
12	Trastuzumab-tetrazine antibody	^{18}F-labeled TCO	SKOV-3 cells (Balb/C mouse)	[109]
13	A33-TCO antibody	^{18}F-labeled tetrazine	LS174T cells (Balb/C mouse) A431 cells (Balb/C mouse)	[110]
14	Mesoporous silica-TCO nanoparticle	^{11}C-labeled tetrazine	Healthy Balb/C mouse	[111]
15	Polyglutamic acid-TCO	^{11}C-labeled tetrazine	CT26 cell (Balb/C mouse)	[112]
16	A33-TCO antibody	^{64}Cu-labeled tetrazine	SW1222 cell (mouse)	[113]
17	HuA33-TCO antibody	^{64}Cu-labeled tetrazine	SW1222 cell (mouse)	[114]
18	5B1-TCO antibody	^{64}Cu-labeled tetrazine	BxPC3 and Capan-2 cells (athymic nude mice)	[115]
19	HuA33-dye-800-TCO	^{64}Cu-labeled tetrazine	SW1222 cell (mouse)	[116]
20	C225-TCO antibody	^{68}Ga-labeled tetrazine	A431 cells (Balb/C mouse)	[117]
21	HuA33-TCO antibody	^{68}Ga-labeled tetrazine	SW1222 cell (CrTac:NCr-Foxn1^{nu} mouse)	[118]
22	Bisphosphonate-TCO conjugate	^{177}Lu-labeled tetrazine ^{99m}Tc-labeled tetrazine	Healthy Balb/C mouse	[119]
23	Bevacizumab-TCO antibody	^{99m}Tc-labeled tetrazine	B16-F10 cell (C57 Bl/6J mouse)	[120]
24	CC49-TCO antibody	^{212}Pb-labeled tetrazine	LS174T cells (Balb/C mouse)	[121]

In 2010, the Robillard group reported their pioneering work on in vivo pre-targeted imaging of cancers using IEDDA ligation [98]. In the first step, TCO group bearing CC49 antibody was injected to target colon cancer cells in a mouse model. Post administration of the antibody (24 h), only a small excess (3.4 equivalent) amount of ^{111}In-labeled tetrazine tracer was injected into the same mouse model. The obtained SPECT images showed the efficient delivery of the radioisotope into the tumor and indicated a high tumor-to-normal tissue ratio (Table 3, entry 1). In the next study, the same research group revealed that TCO could be converted to its (Z)-isomer, which is unreactive to tetrazine in the presence of copper-containing proteins [99]. Thus, a shorter linker was introduced in the tetrazine tracer to impede interactions with the copper-containing proteins in albumin. By this structural modification, the reactivity and isomerization half-life of TCO was increased compared to that of the previously used TCO analog. Later, Robillard and coworkers reported the use of tetrazine-functionalized clearing agents as a modified pre-targeting system (Table 3, entry 3) [100]. While a portion of the administered antibody accumulated in the tumor tissue in this approach, a significant portion of it still remained in the blood. This accumulated antibody could cause a reduced target-to-background ratio because IEDDA reaction is also feasible at non-specific areas in the body. To address this problem, the group added one more step in the animal study (Figure 14). After the TCO-modified antibody was injected into the xenograft model to target tumor cells in vivo, the tetrazine bearing galactose-albumin conjugate was injected as a TCO clearance agent to mask the unbound TCO-modified antibody in the blood. The radiolabeled tetrazine was then injected to enable the IEDDA reaction at the surface of tumor site. This approach demonstrated that the use of a clearing agent could lead to the doubling of the tetrazine tumor uptake and a greater than 100-fold improvement of the tumor-to-blood ratio at 3 h could be realized after injection of the radiolabeled tetrazine.

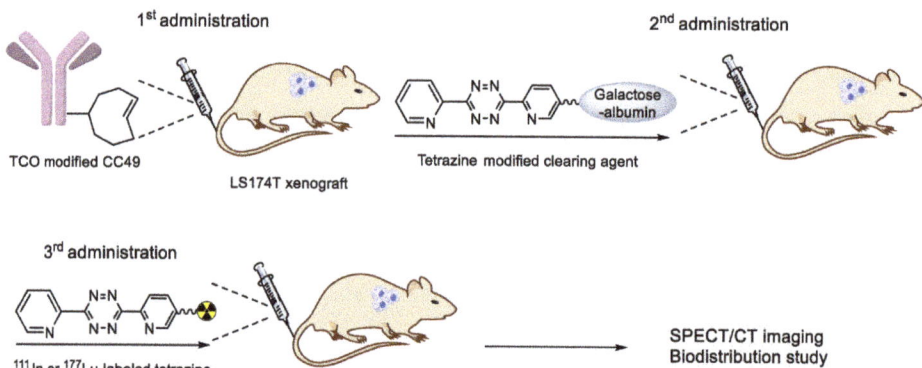

Figure 14. A modified strategy using tetrazine-bearing clearing agent.

The same group reported a pre-targeted radioimmunotherapy study using a similar strategy. To achieve high tumor uptake and improved tumor-to-blood ratio, the group employed a linker with higher hydrophilicity to prepare the TCO-tagged CC49 antibody [101]. In 2015, TCO-functionalized diabody, AVP04-07 was evaluated in the pre-targeted strategy [102]. In this study, the TAG72-targeting dimers of single-chain Fv fragments and ^{177}Lu-labeled tetrazine tracers were evaluated in the LS174T tumor xenograft. As the diabody showed rapid renal clearance kinetics, this strategy could provide high tumor-to-blood ratio and low non-specific retention in the kidneys. In a related study, the authors successfully performed an IEDDA-based pre-targeted study by employing HER2 affibody molecules and ^{111}In/^{177}Lu-labeled tetrazine tracers (Table 3, entry 6) [103].

In addition to these results, several research groups have investigated a variety of pre-targeted approaches using short half-life radioisotope-labeled TCO or tetrazine derivatives. Denk et al. developed a novel ^{18}F-labeled tetrazine by the direct ^{18}F-fluorination of the tosylated precursor, which proceeded in an RCY up to 18% (Table 3, entry 7) [104]. The PET imaging study exhibited fast homogeneous biodistribution of the ^{18}F-labeled tetrazine, which can also cross the blood–brain barrier. The high reactivity of this tracer towards TCO-bearing molecules and favorable pharmacokinetic properties indicated that ^{18}F-labeled tetrazine can be a useful tracer for bioorthogonal PET imaging. Lewis et al., reported ^{18}F-based pre-targeted PET imaging studies using TCO-modified anti-CA19.9 antibody 5B1 and a 1,4,7-triazacyclononane-1,4,7-triacetic acid (NOTA)-conjugated tetrazine analog. The complexation reaction using AlCl$_3$ and [^{18}F]F$^-$ provided the desired radioligand in 54–65% decay-corrected RCY [105]. The in vivo pre-targeted images displayed its effective targeting ability with radioactivity up to 6.4% ID/g in the tumors at 4 h post administration. Sarparanta and co-workers investigated in vivo IEDDA reaction between TCO conjugated monoclonal antibodies and ^{18}F-labeled tetrazine molecule [106]. For this study, TCO conjugated antibody (trastuzumab and cetuximab) was injected into tumor-bearing (BT-474 cells and A431 cells) mice and the ^{18}F-labeled tetrazine-containing hydrophilic linker was injected into the same xenograft models after given time points (1, 2, or 3 days). The highest tumor-to-background ratio was observed when the radioisotope was injected after 3 days post the administration of the TCO-modified antibody. In addition, the ^{18}F-labeled tetrazine was applied to the pre-targeted in vivo imaging of TCO-modified porous silicon nanoparticles (Table 3, entry 10) [107]. Bormans et al. developed a new ^{18}F-labeled TCO tracer for in vivo IEDDA. To prepare radiolabeled TCO, the authors synthesized a dioxolane-fused TCO analog from its cis isomer by using a micro-flow photochemistry process (Figure 15) [108]. The nucleophilic substitution of mesylated precursor using dry K[^{18}F]F, K$_{222}$ complex provided the desired ^{18}F-labeled tracer in 12% RCY and >99% radiochemical purity. This product showed excellent reactivity and stability toward a tetrazine and thus it was applied to pre-targeted PET Imaging. In this approach, a tetrazine-modified trastuzumab monoclonal antibody was injected initially into SKOV-3 xenograft models (Figure 16). After 2 or 3 days, the ^{18}F-labeled TCO was injected, and then the PET images were obtained after 2 h post the administration of the radioligand. The obtained results showed that the pre-targeted imaging strategy provided better tumor-to-muscle ratio when compared to that of control groups which did not use the pre-targeting approach (Table 3, entry 12) [109].

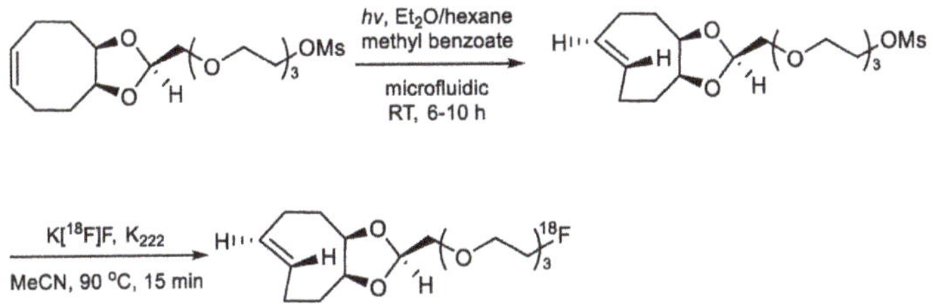

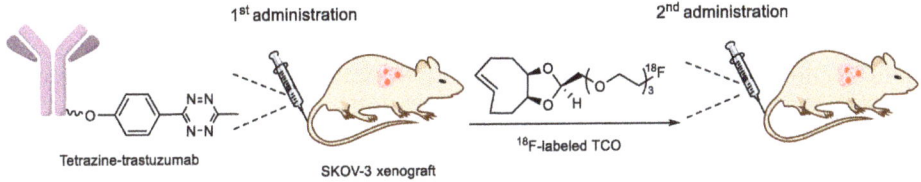

Figure 15. Radiosynthesis of ^{18}F-labeled TCO.

Figure 16. Two-step pre-targeting strategy using ^{18}F-labeled TCO.

The radiolabeled small molecule tracers often underwent rapid renal or hepatobiliary clearance, and therefore, the efficiency of in vivo click reactions is reduced. To increase the blood circulation time of the functional group, the Weissleder group designed the tetrazine group-bearing polymers comprising dextran scaffolds [110]. An ^{18}F-labeled polymer-modified tetrazine and TCO-bearing CD45 monoclonal antibodies were investigated in a living mouse, and the PET imaging study revealed excellent conversion of reactants and high tumor uptake in the tumor xenograft, which suggested that the radiolabeled polymer will be a promising candidate for pre-targeted imaging. The use of IEDDA for pre-targeted PET imaging has also been investigated with ^{11}C. In 2016, Mikula et al. reported the use of ^{11}C-labeled tetrazine for in vivo click reaction. An amino tetrazine analog was reacted with [^{11}C]CH$_3$OTf to provide ^{11}C-labeled tetrazine in 52% of RCY (Figure 17) [111], and the resulting product exhibited high reaction rate with TCO derivatives. Furthermore, the product also showed good stability under physiological conditions and demonstrated rapid clearance kinetics in mice. This ^{11}C-labeled tracer was then applied to animal imaging studies with TCO-modified mesoporous silica nanoparticles in normal mice. Herth et al. reported the improved radiosynthesis of ^{11}C-labeled tetrazine for pre-targeted PET imaging (Table 3, entry 15) [112]. In this study, the radioligand was evaluated with TCO-functionalized polyglutamic acid and indicated potential use for brain imaging.

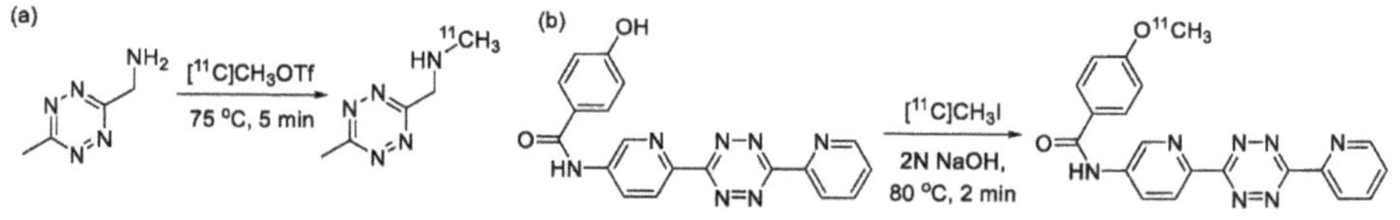

Figure 17. ^{11}C-labeled tetrazine tracers for in vivo IEDDA reaction, (a) from ref [111], (b) from ref [112].

The use of the TCO-tetrazine ligation in living subjects was also extended to several metal radioisotopes. Lewis et al. reported a pre-targeted strategy using a TCO-bearing huA33 antibody and ^{64}Cu-labeled NOTA-tetrazine conjugate (Table 3, entry 16). In this study, a tumor-targeted antibody was administered to SW1222 colorectal cancer xenografts, and the tetrazine tracer was then injected one-day post administration of the antibody. This approach exhibited enhanced tumor-to-background ratio and reduced non-specific radiation dose in normal tissues [113]. In the following study, the authors reported a site-specific conjugation method to construct the huA33-TCO immunoconjugate by using enzymatic transformations and a bifunctional linker [114]. A similar bioconjugation strategy was also applied to the preparation of a TCO and fluorescent dye-bearing antibody (huA33-Dye800-TCO) for bimodal PET/optical pre-targeted imaging of colorectal cancer cells (Table 3, entry 19) [116] using a ^{64}Cu-sarcophagine-based tetrazine tracer. This strategy demonstrated the non-invasive visualization of tumors and the image-guided excision of malignant tumor tissue. Aboagye et al., synthesized ^{68}Ga-labeled tetrazine to study its use in pre-targeted PET imaging of EGFR-expressing A431 tumor. After administration of the TCO-bearing cetuximab, the ^{68}Ga-labeled tracer was injected to the mouse model, and PET imaging showed a significant improvement in the tumor-to-background ratio compared to that with the traditional direct radiolabeling method [117]. Recently, Lewis et al. used a modified pre-targeted PET imaging strategy for obtaining a better tumor-to-blood ratio. The authors employed a tetrazine-modified dextran polymer to reduce injected TCO-bearing antibody, which remained in blood circulation. After the TCO-modified antibody was injected into the xenograft model to target tumor cells, the TCO scavenger was administrated to mask unbounded TCO modified antibody in the blood. Next, ^{68}Ga-labeled tetrazine radioligand was injected to allow the IEDDA reaction at the surface of tumor cells. Further, the use of the TCO masking agent in this study showed a significant improvement in the PET image quality and tumor-to-background ratio (Table 3, entry 21) [118].

The Valliant group demonstrated a pre-targeted strategy for bone imaging and radiotherapy based on the IEDDA between the TCO-conjugated bisphosphonate and radiolabeled tetrazines (Figure 18) [119]. In this experiment, TCO-bisphosphonate conjugate was first injected into an animal model for accumulation of the dienophile in the skeleton. After 12 h post administration, ^{99m}Tc-labeled tetrazine was administered intravenously, and the acquired SPECT/CT imaging revealed high radioactivity in the knees and shoulder, which suggested that the TCO-bisphosphonate can be a useful probe for targeting functionalized tetrazine in the bone tissue. A therapeutic radioisotope (^{177}Lu)-labeled radioligand was also investigated in the same study.

In 2018, Garcia et al. investigated an antibody pre-targeting approach using TCO-bearing bevacizumab and ^{99m}Tc-labeled tetrazine tracer. To increase renal clearance kinetics of the radioisotope, a hydrophilic peptide linker was introduced between tetrazine and the 6-hydrazinonicotinyl group, which is a well-known chelator of ^{99m}Tc. The pre-targeted bevacizumab SPECT imaging was then investigated in B16-F10 melanoma cell's xenograft [120]. In addition to various diagnostic research, the alpha-particle emitting radioisotope (^{212}Pb) was applied to the pre-targeted radioimmunotherapy by Quinn et al. (Table 4, entry 24). In this study, the LS174T tumor-bearing mice were injected with CC49-TCO monoclonal antibody. Two doses of the tetrazine bearing Galactose-albumin as a TCO clearing agent were injected after 30 and 48 h to remove the unbound antibodies in blood and normal organs. Then, ^{212}Pb-labeled tetrazine was injected for targeted tumor therapy. This pre-targeted alpha-particle therapy successfully reduced the tumor growth and improved the survival of model mice [121].

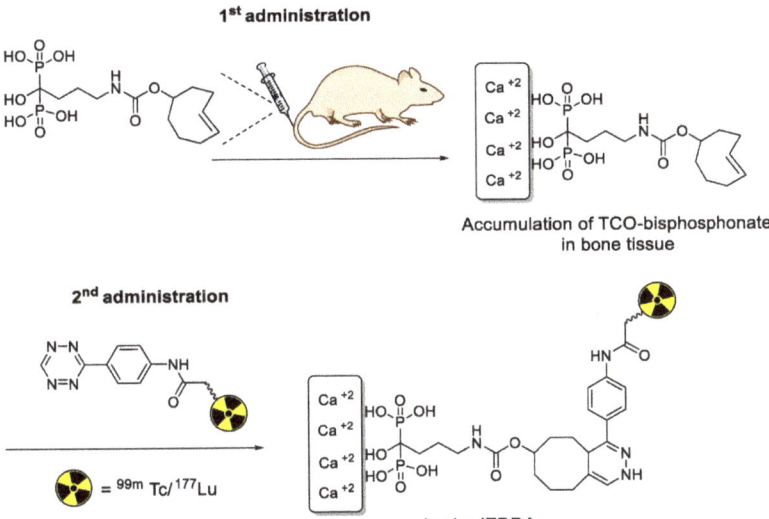

Figure 18. Pre-targeted IEDDA ligation between TCO-bisphosphonate and radiolabeled tetrazine in bone tissue.

4. Other Click Reactions Based on Aromatic Prosthetic Groups

4.1. Condensation/Addition Reactions Using Aromatic Compounds

As shown in previous sections, SPAAC and IEDDA have been two of the most frequently used radiolabeling methods for several years. To apply these reactions in the labeling procedure, the target molecule (e.g., peptide, antibody) needs to be modified to incorporate an artificial functional group, which is then reacted with the radiolabeled prosthetic group. For example, a TCO analog needs to be conjugated with the target molecule, to facilitate its reaction with a radioisotope-containing tetrazine. Such modification of biomolecule requires additional synthetic, and purification steps. Furthermore, the presence of excess amounts of randomly conjugated functional groups can cause decreased biological activities of the molecules. Therefore, several labeling procedures, which do not involve a modification of the biomolecules, have been developed. In many cases, these methods utilized electrophilic aromatic prosthetic groups that displayed rapid reaction rates and high selectivities toward a specific nucleophile such as thiol or 1,2-amino thiol. Table 4 summarizes recent studies on the applications of aromatic prosthetic groups for radiolabeling reactions.

In 2012, Jeon et al. investigated the rapid condensation reaction between ^{18}F-labeled 2-cyanobenzothiazole (^{18}F-CBT) and N-terminal cysteine-bearing biomolecules (Table 4, entry 1) [122]. The ^{18}F-CBT was synthesized from the corresponding tosylated precursor using K[^{18}F]F and 18-crown-6 as the phase transfer catalyst. This radiolabeled CBT (^{18}F-CBT) can be reacted with N-terminal cysteine with a second-order reaction rate of ca. 9 M^{-1} s^{-1}. The rapid condensation reaction between the N-terminal cysteine-bearing dimeric cRGD peptide and the ^{18}F-CBT provided the ^{18}F-labeled peptide (^{18}F-CBT-RGD$_2$) in a high (>80%) RCY under mild conditions, and the prepared ^{18}F-CBT-RGD$_2$ was investigated for its use in specific tumor imaging in U87MG xenograft models. Later, ^{18}F-CBT was also applied for the efficient radiolabeling of EGFR-targeting affibody molecules (Z$_{EFGR:1907}$), and the radiolabeled affibody provided clear visualization of the A431 tumors in animal models [123]. As the heterocyclic adducts, which result from the condensation reaction between CBT and N-terminal cysteine are hydrophobic, the injected tracers prepared by the above method showed high non-specific uptake in normal organs. Therefore, the Seimbille group synthesized a more hydrophilic ^{18}F-labeled CBT tracer containing a diethylene glycol linker and 2-fluoropyridine moiety (Table 4, entry 2) The optimized radiolabeling condition provided ^{18}F-labeled cancer-targeting peptide, which is more hydrophilic than the ones reported in the previous studies [124]. The same research group also reported the synthesis of the metal-chelating agent-conjugated CBT prosthetic groups for ^{68}Ga-labeled tracers for PET imaging of tumor hypoxia [125]. The rapid condensation for radiolabeling procedure provided the desired radiotracers in high RCY under mild conditions (Table 4, entry 3). In another study, the same group synthesized the two bifunctional chelators, the desferrioxamine B-bearing CBT (DFO-CBT) and the cysteine-bearing CBT (DFO-Cys) for efficient radiolabeling. These chelators were employed in the labeling with the [^{89}Zr]Zr-oxalate and rapid conjugation with cRGD peptide. The two-step radiochemical process exhibited high RCY under mild reaction conditions [126]. As CBT structure contained a hydroxy group, it can be a good substrate for facile labeling of radioactive iodines [127]. Thus, we synthesized a ^{125}I-labeled CBT (^{125}I-CBT) via electrophilic iodination reaction under mild reaction conditions. The ^{125}I-CBT was then applied to the rapid radiolabeling of N-terminal cysteine-bearing cRGD peptide in high RCY (Table 4, entry 4).

In 2013, Barbas III group reported the chemoselective ligation of thiol-containing proteins using methylsulfonyl derivatives [128]. They showed that phenyloxadiazole methylsulfone and phenyltetrazole methylsulfone react rapidly and selectively with the sulfhydryl group of cysteine residues in aqueous media under mild conditions (at neutral pH and room temperature) In addition, the structures resulting from these ligation reactions were more stable under physiological conditions in comparison to the corresponding products obtained by maleimide-thiol chemistry. These advantages lead to the development of new prosthetic groups for site-specific radiolabeling reactions. Mindt et al. reported a ^{18}F-labeled phenyloxadiazole methylsulfone analog([^{18}F]FPOS) for the rapid and chemoselective radiolabeling of thiol-bearing biomolecules under mild conditions (Table 4, entry 5) [129]. In this study, [^{18}F]FPOS was applied to efficient radiolabeling of free thiol group-bearing biomolecules. The radiolabeled affibody ($Z_{HER2:2395}$) could be successfully applied to the PET imaging of HER2-positive tumor cells in animal models. Recently, we have reported a radioiodinated phenyltetrazole methylsulfone derivative as a new thiol-reactive prosthetic group (Table 4, entry 6) [130]. The ^{125}I-labeled (4-(5-methane-sulfonyl-[1,2,3,4]tetrazole-1-yl)-phenol) (^{125}I-MSTP) can be prepared by using a simple iodination reaction from the phenolic precursor in high RCY (73% isolated yield) and purity (>99%). The ^{125}I-MSTP was used for site-specific radiolabeling of a single free-thiol-bearing peptide and protein by using radioiodinated labeling of thiol-containing biomolecules. The radiolabeled HSA prepared by this method exhibited enhanced in vivo stability upon deiodination compared with radioiodinated HSA prepared by a direct iodination reaction. In 2018, Park et al. reported a novel condensation reaction using an aryl diamine linker and ^{125}I-labeled aldehyde prosthetic group (Table 4, entry 7) [131]. This method was applied to rapid and efficient radiolabeling of bioactive molecules and the labeled products showed high in vitro and in vivo stability. Samnick et al. proposed a new phenol-reactive prosthetic group for site-specific radiolabeling reaction of tyrosine-containing biomolecules (Table 4, entry 8) [132]. The ^{18}F-labeled 1,2,4-triazoline-3,5-dione([^{18}F]FS-PTAD) was reacted with the model compounds such as phenol, L-tyrosine and N-acetyl-L-tyrosine methyl amide to evaluate the efficacy of the labeling reaction, which proceeded rapidly under mild aqueous conditions to furnish the corresponding radiolabeled compounds in good RCY (45–58%) within 5 min.

Table 4. Aromatic prosthetic groups for radiolabeling reactions.

Entry	Radiotracer	Target Molecule	Product	RCY (%)	Ref
1		R = cRGD$_2$ peptide, RLuc8 protein, Z$_{EGFR:1907}$ affibody		80 (cRGD), 12 (RLuc8), 41 (Z$_{EGFR:1907}$)	[122,123]
2		R = cRGD peptide		7	[124]
3				99	[125]
4		R = cRGD peptide		>99	[127]
5		R= BBN peptide, Z$_{HER2:2395}$ affibody		>99 (BBN), 40 (Z$_{HER2:2395}$)	[129]
6		HSA protein, GCQRPPR peptide	R = HSA protein, GCQRPPR peptide	65 (HSA), 99 (GCQRPPR)	[130]
7		R = cRGD peptide, HSA protein		99 (cRGD), 94 (HSA)	[131]
8				45	[132]

4.2. Miscellaneous

Neumaier and coworkers demonstrated the ^{18}F-radiolabeling of biomolecules using [3+2] cycloaddition reactions between ^{18}F-labeled nitrone and maleimide-bearing molecules [133]. This reaction can provide high efficiency for the synthesis of radiolabeled small molecules. However, the cycloaddition reaction required elevated temperatures in organic solvents, and thus, this method was not suitable for radiolabeling of proteins or antibodies. Continuing this theme, the same group explored more efficient [3+2] cycloaddition reactions using ^{18}F-labeled nitriloxides and N-hydroxyimidoyl chloride (Figure 19). Interestingly, these radiolabeled tracers showed high reactivity with a strained alkene and norbornene analogs under ambient temperature, suggesting that this method can be a useful alternative to the copper-free azide–alkyne click reactions for the radiolabeling of biomolecules [134].

Recently, Wuest et al. demonstrated the first application of the sulfo-click chemistry in the ^{18}F-labeling reaction (Figure 20). In this study, ^{18}F-labeled thiol acids were synthesized and treated with sulfonyl azide-modified small molecules and peptide substrates to afford the corresponding radiolabeled products in moderate to good RCYs [135]. Furthermore, this labeling reaction can be selectively performed in aqueous solvents with a high degree of functional group compatibility.

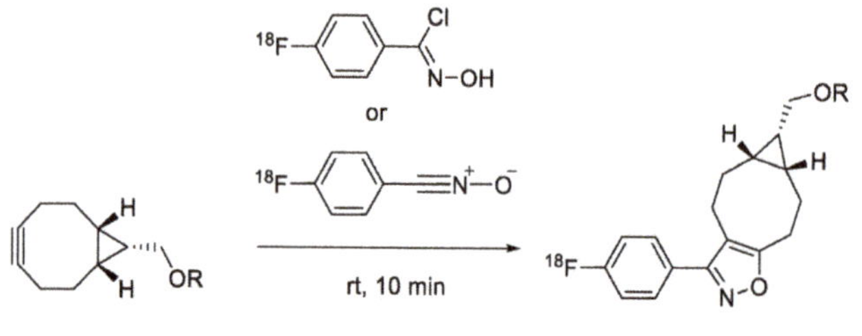

Figure 19. [3+2] cycloaddition reaction using ^{18}F-labeled nitriloxides or N-hydroxyimidoyl chloride.

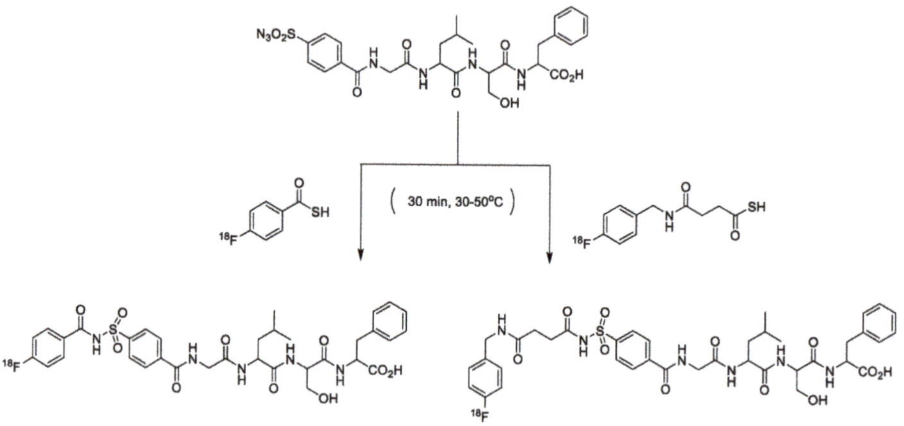

Figure 20. Radiolabeling of biomolecules using sulfo-click chemistry.

Recently, a novel photochemical conjugation reaction was developed for one-pot radiolabeling of antibodies by the Holland group [136,137]. The group synthesized the ^{68}Ga-labeled photoactivatable ligand, which contained an aryl azide group ([^{68}Ga]GaNODAGA-PEG$_3$-ArN$_3$) (Figure 21). The prepared radiotracer underwent a facile reaction with an amino group of the antibodies, including GMP-grade Herceptin^{TM} upon light irradiation (λ_{max} ~ 365 nm) within 5 min. The radiolabeled product was also utilized for the specific tumor imaging in SK-OV-3 tumor xenograft. A similar method was also applied to the radiosynthesis of ^{89}Zr-labeled antibody by using a desferrioxamine B conjugated aryl azide group [138]. As the radiolabeling of trastuzumab has been carried out over a short time with high efficiency and purity, this approach will be applicable for the efficient radiolabeling of various biologically active molecules.

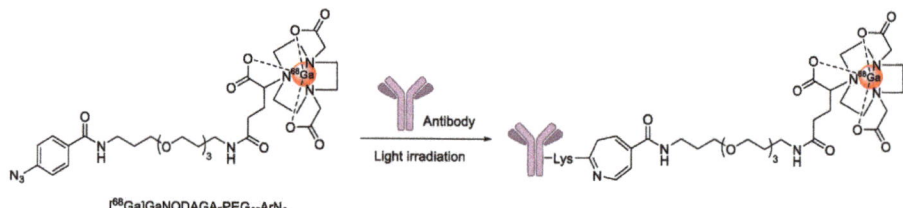

Figure 21. Radiolabeling of antibodies using photochemical conjugation reaction.

5. Conclusions and Future Perspectives

In this review, we focused on recent examples that highlight the application of bioorthogonal click chemistries for the preparation of radiolabeled products. For many years, rapid and selective conjugation reactions including SPAAC and IEDDA have been successfully employed for the straightforward, site-specific, and efficient labeling of various radioisotopes to the small molecules, biomacromolecules, functional nanomaterials, and living cells. In addition, electrophilic aromatic prosthetic groups which display fast reaction kinetics and high selectivity towards the amine or thiol groups could also be the preferred methods for the radiolabeling procedure, because these reactions do not need the introduction of an artificial functional group to the target the biomolecule. Regarding future perspectives, it is anticipated that the relevance of bioorthogonal strategies will continue to be applicable beyond the rapid labeling of a radioisotope to a target molecule of interest. For example, the development of in vivo ligation based on IEDDA enabled the investigation of various approaches for specific tumor imaging with decreased non-specific accumulation of radioligand in normal tissues. Particularly, the introduction of clearing agents before administration of radiotracers demonstrated improved tumor-to-background ratio with enhanced uptake values in the target sites. Although some recent advancements can provide potent tools in nuclear medicine, several key challenges need to be addressed for their further development. The functional groups and resulting adducts obtained by bioorthogonal ligations are normally hydrophobic, which may result in non-specific uptake and retarded excretion kinetics in a living subject. Moreover, conjugation of a relatively large functional group to the small molecule probes or short peptides will affect their pharmacokinetic profiles and induce undesired accumulation or retention of radioactive signals in healthy tissues. For instance, we have synthesized ^{18}F-labeled dimeric cRGD peptide by using the condensation reaction between CBT and N-terminal cysteine (Table 4, entry 1). This method provided an efficient radiochemical result. However, the hydrophobic adduct produced by the radiolabeling reaction afforded high uptake values in normal organs, including in liver and kidneys compared with [^{18}F]FPPRGD$_2$, which is a clinically approved radiopharmaceutical [122]. Such undesired biodistribution results would hamper further investigation of new radiotracers. Therefore, development of fine-tuned functional group pairs, which are smaller and less lipophilic, and at the same time possess high reactivity and selectivity must be investigated to maximize specific targeting ability of the radioligand with minimal background signal. Consequently, bioorthogonal click reactions have exhibited enormous potential for development of radiopharmaceuticals and applications in the field of nuclear medicine. The optimization of these ligation methods will enable the exploration of advanced theranostic strategies as well as the investigation of sophisticated biological phenomena. We expect that these tools will continue to be used as a key technology for the development of various radiolabeled molecules and radiopharmaceuticals, which can offer benefits across preclinical studies and ultimately in clinical applications in the future.

Author Contributions: S.M., and S.-J.Y. wrote the review paper. J.J. conceived the main concept and edited the review paper.

Funding: This work was supported by the research grant from the National Research Foundation of Korea (grant number: 2017M2A2A6A01070858).

Acknowledgments: We would like to thank Lubna Ghani for formatting the reference list.

Conflicts of Interest: The authors declare no conflicts of interest.

Abbreviations

CBT	2-cyanobenzothiazole
CuAAC	copper(I)-catalyzed azide-alkyne [3+2] cycloaddition reaction
DBCO	dibenzocyclooctyne
DFO	deferoxamine
DMF	N,N-dimethylformamide
DMSO	dimethyl sulfoxide
DOTA	1,4,7,10-tetraazacyclododecane-tetraacetic acid

DTPA	diethylenetriaminepentaacetic acid
FH	feraheme
HPLC	high-performance liquid chromatography
HAS	human serum albumin
ICAM-1	intercellular adhesion molecule
IEDDA	inverse-electron-demand Diels–Alder reaction
MSTP	(4-(5-methane-sulfonyl-[1,2,3,4]tetrazole-1-yl)-phenol)
NOTA	1,4,7-triazacyclononane-1,4,7-triacetic acid
ODIBO	oxa-dibenzocyclooctyne
PBS	phosphate-buffered saline
PECAM-1	platelet-endothelial cell adhesion molecule
PET	positron emission tomography
PIC	polyisocyanopeptide
RCY	radiochemical yield
SPAAC	strain-promoted azide-alkyne cycloaddition reaction
SPECT	single-photon emission computed tomography
TCO	trans-cyclooctene

References

1. Schulze, B.; Schubert, U.S. Beyond click chemistry–supramolecular interactions of 1,2,3-triazoles. *Chem. Soc. Rev.* **2014**, *43*, 2522–2571. [CrossRef] [PubMed]
2. Jewett, J.C.; Bertozzi, C.R. Synthesis of a fluorogenic cyclooctyne activated by Cu-free click chemistry. *Org. Lett.* **2011**, *13*, 5937–5939. [CrossRef] [PubMed]
3. Xu, L.; Li, Y.; Li, Y. Application of "click" chemistry to the construction of supramolecular functional systems. *Chem. Asian J.* **2014**, *3*, 582–602. [CrossRef]
4. McKay, C.S.; Finn, M.G. Click chemistry in complex mixtures: Bioorthogonal bioconjugation. *Chem. Biol.* **2014**, *21*, 1075–1101. [CrossRef] [PubMed]
5. Hua, Y.; Flood, A.H. Click chemistry generates privileged CH hydrogen-bonding triazoles: The latest addition to anion supramolecular chemistry. *Chem. Soc. Rev.* **2010**, *39*, 1262–1271. [CrossRef] [PubMed]
6. Aucagne, V.; Leigh, D.A. Chemoselective formation of successive triazole linkages in One Pot: "Click– Click" chemistry. *Org. Lett.* **2006**, *8*, 4505–4507. [CrossRef]
7. Ganesh, V.; Sudhir, V.S.; Kundu, T.; Chandrasekaran, S. 10 Years of Click Chemistry: Synthesis and Applications of Ferrocene-Derived Triazoles. *Chem. Asian J.* **2011**, *6*, 2670–2694. [CrossRef]
8. Becer, C.R.; Hoogenboom, R.; Schubert, U.S. Click chemistry beyond metal-catalyzed cycloaddition. *Angew. Chem. Int. Ed.* **2009**, *48*, 4900–4908. [CrossRef]
9. Agalave, S.G.; Maujan, S.R.; Pore, V.S. Click chemistry: 1, 2, 3-triazoles as pharmacophores. *Chem. Asian J.* **2011**, *6*, 2696–2718. [CrossRef]
10. Li, L.; Zhang, Z. Development and Applications of the Copper-Catalyzed Azide-Alkyne Cycloaddition (CuAAC) as a Bioorthogonal Reaction. *Molecules* **2016**, *21*, 1393. [CrossRef]
11. Kappe, C.O.; Van der Eycken, E. Click chemistry under non-classical reaction conditions. *Chem. Soc. Rev.* **2010**, *39*, 1280–1290. [CrossRef] [PubMed]
12. Mamidyala, S.K.; Finn, M.G. In situ click chemistry: Probing the binding landscapes of biological molecules. *Chem. Soc. Rev.* **2010**, *39*, 1252–1261. [CrossRef] [PubMed]
13. Van Steenis, D.J.V.; David, O.R.; van Strijdonck, G.P.; van Maarseveen, J.H.; Reek, J.N. Click-chemistry as an efficient synthetic tool for the preparation of novel conjugated polymers. *Chem. Commun.* **2005**, *34*, 4333–4335. [CrossRef] [PubMed]
14. Wong, C.H.; Zimmerman, S.C. Orthogonality in organic, polymer, and supramolecular chemistry: From Merrifield to click chemistry. *Chem. Commun.* **2013**, *49*, 1679–1695. [CrossRef]
15. Van Dijk, M.; Rijkers, D.T.; Liskamp, R.M.; van Nostrum, C.F.; Hennink, W.E. Synthesis and applications of biomedical and pharmaceutical polymers via click chemistry methodologies. *Bioconjugate Chem.* **2009**, *20*, 2001–2016. [CrossRef] [PubMed]
16. Moses, J.E.; Moorhouse, A.D. The growing applications of click chemistry. *Chem. Soc. Rev.* **2007**, *36*, 1249–1262. [CrossRef]

17. Sanchez-Sanchez, A.; Pérez-Baena, I.; Pomposo, J. Advances in click chemistry for single-chain nanoparticle construction. *Molecules* **2013**, *18*, 3339–3355. [CrossRef] [PubMed]
18. Lummerstorfer, T.; Hoffmann, H. Click chemistry on surfaces: 1, 3-dipolar cycloaddition reactions of azide-terminated monolayers on silica. *J. Phys. Chem. B* **2004**, *108*, 3963–3966. [CrossRef]
19. Devadoss, A.; Chidsey, C.E. Azide-modified graphitic surfaces for covalent attachment of alkyne-terminated molecules by "click" chemistry. *J. Am. Chem. Soc.* **2007**, *129*, 5370–5371. [CrossRef]
20. Schlossbauer, A.; Schaffert, D.; Kecht, J.; Wagner, E.; Bein, T. Click chemistry for high-density biofunctionalization of mesoporous silica. *J. Am. Chem. Soc.* **2008**, *130*, 12558–12559. [CrossRef]
21. Li, N.; Binder, W.H. Click-chemistry for nanoparticle-modification. *J. Mater. Chem.* **2011**, *21*, 16717–16734. [CrossRef]
22. Xi, W.; Scott, T.F.; Kloxin, C.J.; Bowman, C.N. Click chemistry in materials science. *Adv. Funct. Mater.* **2014**, *24*, 2572–2590. [CrossRef]
23. Tasdelen, M.A. Diels–Alder "click" reactions: Recent applications in polymer and material science. *Polym. Chem.* **2011**, *2*, 2133–2145. [CrossRef]
24. Nandivada, H.; Jiang, X.; Lahann, J. Click chemistry: Versatility and control in the hands of materials scientists. *Adv. Mater.* **2007**, *19*, 2197–2208. [CrossRef]
25. Thirumurugan, P.; Matosiuk, D.; Jozwiak, K. Click chemistry for drug development and diverse chemical–biology applications. *Chem. Rev.* **2013**, *113*, 4905–4979. [CrossRef]
26. Kolb, H.C.; Sharpless, K.B. The growing impact of click chemistry on drug discovery. *Drug Discov. Today* **2003**, *8*, 1128–1137. [CrossRef]
27. Whiting, M.; Muldoon, J.; Lin, Y.C.; Silverman, S.M.; Lindstrom, W.; Olson, A.J.; Kolb, H.C.; Finn, M.G.; Sharpless, K.B.; Elder, J.H.; et al. Inhibitors of HIV-1 protease by using in situ click chemistry. *Angew. Chem. Int. Ed.* **2006**, *45*, 1435–1439. [CrossRef]
28. Pagliai, F.; Pirali, T.; Del Grosso, E.; Di Brisco, R.; Tron, G.C.; Sorba, G.; Genazzani, A.A. Rapid synthesis of triazole-modified resveratrol analogues via click chemistry. *J. Med. Chem.* **2006**, *49*, 467–470. [CrossRef]
29. Kamal, A.; Shankaraiah, N.; Reddy, C.R.; Prabhakar, S.; Markandeya, N.; Srivastava, H.K.; Sastry, G.N. Synthesis of bis-1,2,3-triazolo-bridged unsymmetrical pyrrolobenzodiazepine trimers via 'click' chemistry and their DNA-binding studies. *Tetrahedron* **2010**, *66*, 5498–5506. [CrossRef]
30. De Geest, B.G.; Van Camp, W.; Du Prez, F.E.; De Smedt, S.C.; Demeester, J.; Hennink, W.E. Degradable multilayer films and hollow capsules via a 'click' strategy. *Macromol. Rapid Commun.* **2008**, *29*, 1111–1118. [CrossRef]
31. Xu, X.D.; Chen, C.S.; Lu, B.; Wang, Z.C.; Cheng, S.X.; Zhang, X.Z.; Zhuo, R.X. Modular synthesis of thermosensitive P (NIPAAm-co-HEMA)/β-CD based hydrogels via click chemistry. *Macromol. Rapid Commun.* **2009**, *30*, 157–164. [CrossRef] [PubMed]
32. Moorhouse, A.D.; Santos, A.M.; Gunaratnam, M.; Moore, M.; Neidle, S.; Moses, J.E. Stabilization of G-quadruplex DNA by highly selective ligands via click chemistry. *J. Am. Chem. Soc.* **2006**, *128*, 15972–15973. [CrossRef] [PubMed]
33. De Geest, B.G.; Van Camp, W.; Du Prez, F.E.; De Smedt, S.C.; Demeester, J.; Hennink, W.E. Biodegradable microcapsules designed via 'click' chemistry. *Chem. Commun.* **2008**, 190–192. [CrossRef] [PubMed]
34. Lallana, E.; Fernandez-Megia, E.; Riguera, R. Surpassing the use of copper in the click functionalization of polymeric nanostructures: A strain-promoted approach. *J. Am. Chem. Soc.* **2009**, *131*, 5748–5750. [CrossRef]
35. Schieferstein, H.; Betzel, T.; Fischer, C.R.; Ross, T.L. ^{18}F-click labeling and preclinical evaluation of a new 18 F-folate for PET imaging. *EJNMMI Res.* **2013**, *3*, 68. [CrossRef] [PubMed]
36. Notni, J.; Šimeček, J.; Hermann, P.; Wester, H.J. TRAP, a Powerful and Versatile Framework for Gallium-68 Radiopharmaceuticals. *Chem. Asian J.* **2011**, *17*, 14718–14722. [CrossRef]
37. Pretze, M.; Mamat, C. Automated preparation of [^{18}F]AFP and [^{18}F]BFP: Two novel bifunctional ^{18}F-labeling building blocks for Huisgen-click. *J. Fluorine Chem.* **2013**, *150*, 25–35. [CrossRef]
38. Roberts, M.P.; Pham, T.Q.; Doan, J.; Jiang, C.D.; Hambley, T.W.; Greguric, I.; Fraser, B.H. Radiosynthesis and 'click' conjugation of ethynyl-4-[^{18}F]fluorobenzene - an improved [^{18}F]synthon for indirect radiolabeling. *J. Label. Compd. Radiopharm.* **2015**, *58*, 473–478. [CrossRef]
39. Mindt, T.L.; Müller, C.; Melis, M.; de Jong, M.; Schibli, R. "Click-to-chelate": In vitro and in vivo comparison of a ^{99m}Tc(CO)$_3$-labeled N(τ)-histidine folate derivative with its isostructural, clicked 1,2,3-triazole analogue. *Bioconjugate Chem.* **2008**, *19*, 1689–1695. [CrossRef]

40. Zeng, D.; Lee, N.S.; Liu, Y.; Zhou, D.; Dence, C.S.; Wooley, K.L.; Katzenellenbogen, J.A.; Welch, M.J. [64]Cu Core-labeled nanoparticles with high specific activity via metal-free click chemistry. *ACS Nano* **2012**, *6*, 5209–5219. [CrossRef]
41. Liu, Z.; Li, Y.; Lozada, J.; Schaffer, P.; Adam, M.J.; Ruth, T.J.; Perrin, D.M. Stoichiometric Leverage: Rapid 18F-Aryltrifluoroborate Radiosynthesis at High Specific Activity for Click Conjugation. *Angew. Chem. Int. Ed.* **2013**, *52*, 2303–2307. [CrossRef] [PubMed]
42. Nair, D.P.; Podgorski, M.; Chatani, S.; Gong, T.; Xi, W.; Fenoli, C.R.; Bowman, C.N. The thiol-Michael addition click reaction: A powerful and widely used tool in materials chemistry. *Chem. Mater.* **2013**, *26*, 724–744. [CrossRef]
43. Sweeney, J.B. Aziridines: Epoxides' ugly cousins? *Chem. Soc. Rev.* **2002**, *31*, 247–258. [CrossRef] [PubMed]
44. Crisalli, P.; Kool, E.T. Water-soluble organocatalysts for hydrazone and oxime formation. *J. Org. Chem.* **2013**, *78*, 1184–1189. [CrossRef]
45. Wang, Q.; Chan, T.R.; Hilgraf, R.; Fokin, V.V.; Sharpless, K.B.; Finn, M.G. Bioconjugation by copper (I)-catalyzed azide-alkyne [3+2] cycloaddition. *J. Am. Chem. Soc.* **2003**, *125*, 3192–3193. [CrossRef] [PubMed]
46. Dai, L.; Xue, Y.; Qu, L.; Choi, H.J.; Baek, J.B. Metal-free catalysts for oxygen reduction reaction. *Chem. Rev.* **2015**, *115*, 4823–4892. [CrossRef]
47. Jewett, J.C.; Bertozzi, C.R. Cu-free click cycloaddition reactions in chemical biology. *Chem. Soc. Rev.* **2010**, *39*, 1272–1279. [CrossRef]
48. Juhl, K.; Jørgensen, K.A. The First Organocatalytic Enantioselective Inverse-Electron-Demand Hetero-Diels–Alder Reaction. *Angew. Chem. Int. Ed.* **2003**, *42*, 1498–1501. [CrossRef]
49. Li, Z.; Conti, P.S. Radiopharmaceutical chemistry for positron emission tomography. *Adv. Drug Deliv. Rev.* **2010**, *62*, 1031–1051. [CrossRef]
50. Walsh, J.C.; Kolb, H.C. Applications of click chemistry in radiopharmaceutical development. *CHIMIA Int. J. Chem.* **2010**, *64*, 29–33. [CrossRef]
51. Šečkutė, J.; Devaraj, N.K. Expanding room for tetrazine ligations in the in vivo chemistry toolbox. *Curr. Opin. Chem. Biol.* **2013**, *17*, 761–767. [CrossRef] [PubMed]
52. Agard, N.J.; Prescher, J.A.; Bertozzi, C.R. A strain-promoted [3+2] azide–alkyne cycloaddition for covalent modification of biomolecules in living systems. *J. Am. Chem. Soc.* **2004**, *126*, 15046–15047. [CrossRef] [PubMed]
53. Mbua, N.E.; Guo, J.; Wolfert, M.A.; Steet, R.; Boons, G.J. Strain-promoted alkyne–azide cycloadditions (SPAAC) reveal new features of glycoconjugate biosynthesis. *ChemBioChem* **2011**, *12*, 1912–1921. [CrossRef] [PubMed]
54. Campbell-Verduyn, L.S.; Mirfeizi, L.; Schoonen, A.K.; Dierckx, R.A.; Elsinga, P.H.; Feringa, B.L. Strain-promoted copper-free "click" chemistry for ^{18}F radiolabeling of bombesin. *Angew. Chem. Int. Ed.* **2011**, *50*, 11117–11120. [CrossRef] [PubMed]
55. Bouvet, V.; Wuest, M.; Wuest, F. Copper-free click chemistry with the short-lived positron emitter fluorine-18. *Org. Biomol. Chem.* **2011**, *9*, 7393–7399. [CrossRef]
56. Carpenter, R.D.; Hausner, S.H.; Sutcliffe, J.L. Copper-free click for PET: Rapid 1,3-dipolar cycloadditions with a fluorine-18 cyclooctyne. *ACS Med. Chem. Lett.* **2011**, *2*, 885–889. [CrossRef] [PubMed]
57. Hausner, S.H.; Carpenter, R.D.; Bauer, N.; Sutcliffe, J.L. Evaluation of an integrin $\alpha_v\beta_6$-specific peptide labeled with [^{18}F] fluorine by copper-free, strain-promoted click chemistry. *Nucl. Med. Biol.* **2013**, *40*, 233–239. [CrossRef]
58. Arumugam, S.; Chin, J.; Schirrmacher, R.; Popik, V.V.; Kostikov, A.P. [^{18}F]Azadibenzocyclooctyne ([^{18}F]ADIBO): A biocompatible radioactive labeling synthon for peptides using catalyst free [3+2] cycloaddition. *Bioorg. Med. Chem. Lett.* **2011**, *21*, 6987–6991. [CrossRef]
59. Kettenbach, K.; Ross, T.L. A ^{18}F-labeled dibenzocyclooctyne (DBCO) derivative for copper-free click labeling of biomolecules. *Med. Chem. Comm.* **2016**, *7*, 654–657. [CrossRef]
60. Roche, M.; Specklin, S.; Richard, M.; Hinnen, F.; Génermont, K.; Kuhnast, B. [^{18}F]FPyZIDE: A versatile prosthetic reagent for the fluorine-18 radiolabeling of biologics via copper-catalyzed or strain-promoted alkyne-azide cycloadditions. *J. Label. Compd. Radiopharm.* **2019**, *62*, 95–108. [CrossRef]
61. Evans, H.L.; Carroll, L.; Aboagye, E.O.; Spivey, A.C. Bioorthogonal chemistry for ^{68}Ga radiolabelling of DOTA-containing compounds. *J. Label. Compd. Radiopharm.* **2014**, *57*, 291–297. [CrossRef] [PubMed]

62. Ghosh, S.C.; Hernandez-Vargas, S.; Rodriguez, M.; Kossatz, S.; Voss, J.; Carmon, K.S.; Reiner, T.; Schonbrunn, A.; Azhdarinia, A. Synthesis of a fluorescently labeled ^{68}Ga-DOTA-TOC analog for somatostatin receptor targeting. *ACS Med. Chem. Let.* **2017**, *8*, 720–725. [CrossRef] [PubMed]
63. Gordon, C.G.; Mackey, J.L.; Jewett, J.C.; Sletten, E.M.; Houk, K.N.; Bertozzi, C.R. Reactivity of Biarylazacyclooctynones in Copper-Free Click Chemistry. *J. Am. Chem. Soc.* **2012**, *134*, 9199–9208. [CrossRef]
64. McNitt, C.D.; Popik, V.V. Photochemical generation of oxa-dibenzocyclooctyne (ODIBO) for metal-free click ligations. *Org. Biomol. Chem.* **2012**, *10*, 8200–8202. [CrossRef] [PubMed]
65. Boudjemeline, M.; McNitt, C.D.; Singleton, T.A.; Popik, V.V.; Kostikov, A.P. [^{18}F]ODIBO: A prosthetic group for bioorthogonal radiolabeling of macromolecules via strain-promoted alkyne–azide cycloaddition. *Org. Biomol. Chem.* **2018**, *16*, 363–366. [CrossRef]
66. Sachin, K.; Jadhav, V.H.; Kim, E.M.; Kim, H.L.; Lee, S.B.; Jeong, H.J.; Lim, S.T.; Sohn, M.H.; Kim, D.W. F-18 Labeling protocol of peptides based on chemically orthogonal strain-promoted cycloaddition under physiologically friendly reaction condition. *Bioconjugate Chem.* **2012**, *23*, 1680–1686. [CrossRef] [PubMed]
67. Kim, H.L.; Sachin, K.; Jeong, H.J.; Choi, W.; Lee, H.S.; Kim, D.W. F-18 labeled RGD probes based on bioorthogonal strain-promoted click reaction for PET imaging. *ACS Med. Chem. Lett.* **2015**, *6*, 402–407. [CrossRef]
68. Jeon, J.; Kang, J.A.; Shim, H.E.; Nam, Y.R.; Yoon, S.; Kim, H.R.; Lee, D.E.; Park, S.H. Efficient method for iodine radioisotope labeling of cyclooctyne-containing molecules using strain-promoted copper-free click reaction. *Bioorg. Med. Chem.* **2015**, *23*, 3303–3308. [CrossRef]
69. Choi, M.H.; Shim, H.E.; Nam, Y.R.; Kim, H.R.; Kang, J.A.; Lee, D.E.; Park, S.H.; Choi, D.S.; Jang, B.S.; Jeon, J. Synthesis and evaluation of an ^{125}I-labeled azide prosthetic group for efficient and bioorthogonal radiolabeling of cyclooctyne-group containing molecules using copper-free click reaction. *Bioorg. Med. Chem. Lett.* **2016**, *26*, 875–878. [CrossRef]
70. Zeng, D.; Ouyang, Q.; Cai, Z.; Xie, X.Q.; Anderson, C.J. New cross-bridged cyclam derivative CB-TE1K1P, an improved bifunctional chelator for copper radionuclides. *Chem. Commun.* **2014**, *50*, 43–45. [CrossRef]
71. Yuan, H.; Wilks, M.Q.; El Fakhri, G.; Normandin, M.D.; Kaittanis, C.; Josephson, L. Heat-induced-radiolabeling and click chemistry: A powerful combination for generating multifunctional nanomaterials. *PloS ONE* **2017**, *12*, 0172722. [CrossRef] [PubMed]
72. Öztürk Öncel, M.Ö.; Garipcan, B.; Inci, F. Biomedical Applications: Liposomes and Supported Lipid Bilayers for Diagnostics, Theranostics, Imaging, Vaccine Formulation, and Tissue Engineering. In *Biomimetic Lipid Membranes: Fundamentals, Applications, and Commercialization*; Kök, F.N., Arslan Yildiz, A., Inci, F., Eds.; Springer: Cham, Switzerland, 2019; pp. 193–212.
73. Hood, E.D.; Greineder, C.F.; Shuvaeva, T.; Walsh, L.; Villa, C.H.; Muzykantov, V.R. Vascular targeting of radiolabeled liposomes with bio-orthogonally conjugated ligands: Single chain fragments provide higher specificity than antibodies. *Bioconjugate Chem.* **2018**, *29*, 3626–3637. [CrossRef] [PubMed]
74. Op't Veld, R.C.; Joosten, L.; van den Boomen, O.; Boerman, O.; Kouwer, P.H.; Middelkoop, E.; Rowan, A.; Jansen, J.A.; Walboomers, F.; Wagener, F. Monitoring ^{111}In-labelled polyisocyanopeptide (PIC) hydrogel wound dressings in full-thickness wounds. *Biomater. Sci.* **2019**, *7*, 3041–3050. [CrossRef] [PubMed]
75. Lodhi, N.A.; Park, J.Y.; Kim, K.; Kim, Y.J.; Shin, J.H.; Lee, Y.S.; Im, H.J.; Jeong, J.M.; Khalid, M.; Cheon, G.J.; et al. Development of ^{99m}Tc-Labeled Human Serum Albumin with Prolonged Circulation by Chelate-then-Click Approach: A Potential Blood Pool Imaging Agent. *Mol. Pharm.* **2019**, *16*, 1586–1595. [CrossRef] [PubMed]
76. Knall, A.C.; Slugovc, C. Inverse electron demand Diels–Alder (iEDDA)-initiated conjugation: A (high) potential click chemistry scheme. *Chem. Soc. Rev.* **2013**, *42*, 5131–5142. [CrossRef]
77. Liu, F.; Liang, Y.; Houk, K.N. Theoretical elucidation of the origins of substituent and strain effects on the rates of Diels–Alder reactions of 1,2,4,5-tetrazines. *J. Am. Chem. Soc.* **2014**, *136*, 11483–11493. [CrossRef] [PubMed]
78. Selvaraj, R.; Fox, J.M. Trans-Cyclooctene - a stable, voracious dienophile for bioorthogonal labeling. *Curr. Opin. Chem. Biol.* **2013**, *17*, 753–760. [CrossRef]
79. Versteegen, R.M.; Rossin, R.; Ten Hoeve, W.; Janssen, H.M.; Robillard, M.S. Click to release: Instantaneous doxorubicin elimination upon tetrazine ligation. *Angew. Chem. Int. Ed.* **2013**, *52*, 14112–14116. [CrossRef]
80. Mushtaq, S.; Jeon, J. Synthesis of PET and SPECT radiotracers using inverse electron-demand Diels–Alder reaction. *Appl. Chem. Eng.* **2017**, *28*, 141–152.

81. Oliveira, B.L.; Guo, Z.; Bernardes, G.J.L. Inverse electron demand Diels–Alder reactions in chemical biology. *Chem. Soc. Rev.* **2017**, *46*, 4895–4950. [CrossRef]
82. Li, Z.; Cai, H.; Hassink, M.; Blackman, M.L.; Brown, R.C.; Conti, P.S.; Fox, J.M. Tetrazine-trans-cyclooctene ligation for the rapid construction of ^{18}F labeled probes. *Chem. Commun.* **2010**, *46*, 8043–8045. [CrossRef] [PubMed]
83. Selvaraj, R.; Liu, S.; Hassink, M.; Huang, C.W.; Yap, L.P.; Park, R.; Fox, J.M.; Li, Z.; Conti, P.S. Tetrazine-trans-cyclooctene ligation for the rapid construction of integrin $\alpha_v\beta_3$ targeted PET tracer based on a cyclic RGD peptide. *Bioorg. Med. Chem. Lett.* **2011**, *21*, 5011–5014. [CrossRef] [PubMed]
84. Liu, S.; Hassink, M.; Selvaraj, R.; Yap, L.P.; Park, R.; Wang, H.; Chen, X.; Fox, J.M.; Li, Z.; Conti, P.S. Efficient ^{18}F labeling of cysteine-containing peptides and proteins using tetrazine–trans-cyclooctene ligation. *Mol. Imaging* **2013**, *12*, 121–128. [CrossRef]
85. Reiner, T.; Keliher, E.J.; Earley, S.; Marinelli, B.; Weissleder, R. Synthesis and in vivo imaging of a ^{18}F-labeled PARP1 inhibitor using a chemically orthogonal scavenger-assisted high-performance method. *Angew. Chem. Int. Ed.* **2011**, *50*, 1922–1925. [CrossRef] [PubMed]
86. Wu, Z.; Liu, S.; Hassink, M.; Nair, I.; Park, R.; Li, L.; Todorov, I.; Fox, J.M.; Li, Z.; Shively, J.E.; et al. Development and evaluation of ^{18}F-TTCO-Cys^{40}-Exendin-4: A PET Probe for Imaging Transplanted Islets. *J. Nucl. Med.* **2013**, *54*, 244–251. [CrossRef] [PubMed]
87. Zhu, J.; Li, S.; Wängler, C.; Wängler, B.; Lennox, R.B.; Schirrmacher, R. Synthesis of 3-chloro-6-((4-(di-tert-butyl[^{18}F]-fluorosilyl)-benzyl)oxy)-1,2,4,5-tetrazine ([^{18}F]SiFA-OTz) for rapid tetrazine-based ^{18}F-radiolabeling. *Chem. Commun.* **2015**, *51*, 12415–12418. [CrossRef] [PubMed]
88. Knight, J.C.; Richter, S.; Wuest, M.; Way, J.D.; Wuest, F. Synthesis and evaluation of an ^{18}F-labelled norbornene derivative for copper-free click chemistry reactions. *Org. Biomol. Chem.* **2013**, *11*, 3817–3825. [CrossRef]
89. Herth, M.M.; Andersen, V.L.; Lehel, S.; Madsen, J.; Knudsen, G.M.; Kristensen, J.L. Development of a ^{11}C-labeled tetrazine for rapid tetrazine–trans-cyclooctene ligation. *Chem. Commun.* **2013**, *49*, 3805–3807. [CrossRef]
90. Nichols, B.; Qin, Z.; Yang, J.; Vera, D.R.; Devaraj, N.K. ^{68}Ga chelating bioorthogonal tetrazine polymers for the multistep labeling of cancer biomarkers. *Chem. Commun.* **2014**, *50*, 5215–5217. [CrossRef]
91. Albu, S.A.; Al-Karmi, S.A.; Vito, A.; Dzandzi, J.P.; Zlitni, A.; Beckford-Vera, D.; Blacker, M.; Janzen, N.; Patel, R.M.; Capretta, A.; et al. ^{125}I-Tetrazines and inverse-electron-demand Diels–Alder chemistry: A convenient radioiodination strategy for biomolecule labeling, screening, and biodistribution studies. *Bioconjugate Chem.* **2016**, *27*, 207–216. [CrossRef]
92. Choi, M.H.; Shim, H.E.; Yun, S.J.; Kim, H.R.; Mushtaq, S.; Lee, C.H.; Park, S.H.; Choi, D.S.; Lee, D.E.; Byun, E.B.; et al. Highly efficient method for ^{125}I-radiolabeling of biomolecules using inverse-electron-demand Diels–Alder reaction. *Bioorg. Med. Chem.* **2016**, *24*, 2589–2594. [CrossRef] [PubMed]
93. Genady, A.R.; Tan, J.; Mohamed, E.; Zlitni, A.; Janzen, N.; Valliant, J.F. Synthesis, characterization and radiolabeling of carborane-functionalized tetrazines for use in inverse electron demand Diels–Alder ligation reactions. *J. Organomet. Chem.* **2015**, *798*, 278–288. [CrossRef]
94. Zeglis, B.M.; Mohindra, P.; Weissmann, G.I.; Divilov, V.; Hilderbrand, S.A.; Weissleder, R.; Lewis, J.S. Modular strategy for the construction of radiometalated antibodies for positron emission tomography based on inverse electron demand Diels–Alder click chemistry. *Bioconjugate Chem.* **2011**, *22*, 2048–2059. [CrossRef] [PubMed]
95. Poty, S.; Membreno, R.; Glaser, J.M.; Ragupathi, A.; Scholz, W.W.; Zeglis, B.M.; Lewis, J.S. The inverse electron-demand Diels–Alder reaction as a new methodology for the synthesis of ^{225}Ac-labelled radioimmunoconjugates. *Chem. Commun.* **2018**, *54*, 2599–2602. [CrossRef]
96. Hernández-Gil, J.; Braga, M.; Harriss, B.I.; Carroll, L.S.; Leow, C.H.; Tang, M.X.; Aboagye, E.O.; Long, N.J. Development of ^{68}Ga-labelled ultrasound microbubbles for whole-body PET imaging. *Chem. Sci.* **2019**, *10*, 5603–5615. [CrossRef] [PubMed]
97. Altai, M.; Membreno, R.; Cook, B.; Tolmachev, V.; Zeglis, B.M. Pretargeted imaging and therapy. *J. Nucl. Med.* **2017**, *58*, 1553–1559. [CrossRef]
98. Rossin, R.; Renart Verkerk, P.; Van Den Bosch, S.M.; Vulders, R.C.; Verel, I.; Lub, J.; Robillard, M.S. In vivo chemistry for pretargeted tumor imaging in live mice. *Angew. Chem. Int. Ed.* **2010**, *49*, 3375–3378. [CrossRef]

99. Rossin, R.; van den Bosch, S.M.; ten Hoeve, W.; Carvelli, M.; Versteegen, R.M.; Lub, J.; Robillard, M.S. Highly Reactive trans-cyclooctene tags with improved stability for Diels–Alder chemistry in living systems. *Bioconjugate Chem.* **2013**, *24*, 1210–1217. [CrossRef]
100. Rossin, R.; Läppchen, T.; van den Bosch, S.M.; Laforest, R.; Robillard, M.S. Diels–Alder reaction for tumor pretargeting: In vivo chemistry can boost tumor radiation dose compared with directly antibody. *J. Nucl. Med.* **2013**, *54*, 1989–1995. [CrossRef]
101. Rossin, R.; van Duijnhoven, S.M.; Läppchen, T.; van den Bosch, S.M.; Robillard, M.S. Trans-Cyclooctene tag with improved properties for tumor pretargeting with the Diels–Alder reaction. *Mol. Pharm.* **2014**, *11*, 3090–3096. [CrossRef]
102. Van Duijnhoven, S.M.; Rossin, R.; van den Bosch, S.M.; Wheatcroft, M.P.; Hudson, P.J.; Robillard, M.S. Diabody pretargeted with click chemistry in vivo. *J. Nucl. Med.* **2015**, *56*, 1422–1428. [CrossRef] [PubMed]
103. Altai, M.; Perols, A.; Tsourma, M.; Mitran, B.; Honarvar, H.; Robillard, M.; Rossin, R.; ten Hoeve, W.; Lubberink, M.; Orlova, A.; et al. Feasibility of Affibody-Based Biorthogonal Chemistry–Mediated Radionuclide Pretargeting. *J. Nucl. Med.* **2016**, *57*, 431–436. [CrossRef] [PubMed]
104. Denk, C.; Svatunek, D.; Filip, T.; Wanek, T.; Lumpi, D.; Fröhlich, J.; Kuntner, C.; Mikula, H. Development of a ^{18}F-Labeled Tetrazine with Favorable Pharmacokinetics for Bioorthogonal PET Imaging. *Angew. Chem. Int. Ed.* **2014**, *53*, 9655–9659. [CrossRef] [PubMed]
105. Meyer, J.P.; Houghton, J.L.; Kozlowski, P.; Abdel-Atti, D.; Reiner, T.; Pillarsetty, N.V.K.; Scholz, W.W.; Zeglis, B.M.; Lewis, J.S. ^{18}F-Based Pretargeted PET Imaging Based on Bioorthogonal Diels–Alder Click Chemistry. *Bioconjugate Chem.* **2016**, *27*, 298–301. [CrossRef]
106. Keinänen, O.; Fung, K.; Pourat, J.; Jallinoja, V.; Vivier, D.; Pillarsetty, N.K.; Airaksinen, A.J.; Lewis, J.S.; Zeglis, B.M.; Sarparanta, M. Pretargeting of internalizing trastuzumab and cetuximab with a ^{18}F-tetrazine tracer in xenograft models. *EJNMMI Res.* **2017**, *7*, 95. [CrossRef]
107. Keinänen, O.; Mäkilä, E.M.; Lindgren, R.; Virtanen, H.; Liljenbäck, H.; Oikonen, V.; Sarparanta, M.; Molthoff, C.; Windhorst, A.D.; Roivainen, A.; et al. Pretargeted PET Imaging of trans-Cyclooctene-Modified Porous Silicon Nanoparticles. *ACS Omega* **2017**, *2*, 62–69.
108. Billaud, E.M.F.; Shahbazali, E.; Ahamed, M.; Cleeren, F.; Noël, T.; Koole, M.; Verbruggen, A.; Hessel, V.; Bormans, G. Micro-flow photosynthesis of new dienophiles for inverse-electron-demand Diels–Alder reactions. Potential applications for pretargeted in vivo PET imaging. *Chem. Sci.* **2017**, *8*, 1251–1258.
109. Billaud, E.M.; Belderbos, S.; Cleeren, F.; Maes, W.; Van de Wouwer, M.; Koole, M.; Verbruggen, A.; Himmelreich, U.; Geukens, N.; Bormans, G. Pretargeted PET imaging using a bioorthogonal ^{18}F-Labeled trans-cyclooctene in an ovarian carcinoma model. *Bioconjugate Chem.* **2017**, *28*, 2915–2920. [CrossRef]
110. Devaraj, N.K.; Thurber, G.M.; Keliher, E.J.; Marinelli, B.; Weissleder, R. Reactive polymer enables efficient in vivo bioorthogonal chemistry. *Proc. Natl. Acad. Sci. USA* **2012**, *109*, 4762–4767. [CrossRef]
111. Denk, C.; Svatunek, D.; Mairinger, S.; Stanek, J.; Filip, T.; Matscheko, D.; Kuntner, C.; Wanek, T.; Mikula, H. Design, Synthesis, and Evaluation of a Low-Molecular-Weight ^{11}C-Labeled Tetrazine for Pretargeted PET Imaging Applying Bioorthogonal in vivo Click Chemistry. *Bioconjugate Chem.* **2016**, *27*, 1707–1712. [CrossRef]
112. Stéen, E.J.L.; Jørgensen, J.T.; Petersen, I.N.; Nørregaard, K.; Lehel, S.; Shalgunov, V.; Birke, A.; Edem, P.E.; L'Estrade, E.T.; Hansen, H.D.; et al. Improved radiosynthesis and preliminary in vivo evaluation of the ^{11}C-labeled tetrazine [^{11}C]AE-1 for pretargeted PET imaging. *Bioorg. Med. Chem. Lett.* **2019**, *20*, 986–990. [CrossRef] [PubMed]
113. Zeglis, B.M.; Sevak, K.K.; Reiner, T.; Mohindra, P.; Carlin, S.D.; Zanzonico, P.; Weissleder, R.; Lewis, J.S. A pretargeted PET imaging strategy based on bioorthogonal Diels–Alder click chemistry. *J. Nucl. Med.* **2013**, *54*, 1389–1396. [CrossRef] [PubMed]
114. Cook, B.E.; Adumeau, P.; Membreno, R.; Carnazza, K.E.; Brand, C.; Reiner, T.; Agnew, B.J.; Lewis, J.S.; Zeglis, B.M. Pretargeted PET imaging Using a Site-Specifically Labeled Immunoconjugate. *Bioconjugate Chem.* **2016**, *27*, 1789–1795. [CrossRef] [PubMed]
115. Houghton, J.L.; Zeglis, B.M.; Abdel-Atti, D.; Sawada, R.; Scholz, W.W.; Lewis, J.S. Pretargeted Immuno-PET of Pancreatic Cancer: Overcoming Circulating Antigen and Internalized Antibody to Reduce Radiation Doses. *J. Nucl. Med.* **2016**, *57*, 453–459. [CrossRef] [PubMed]
116. Adumeau, P.; Carnazza, K.E.; Brand, C.; Carlin, S.D.; Reiner, T.; Agnew, B.J.; Lewis, J.S.; Zeglis, B.M. A Pretargeted Approach for the Multimodal PET/NIRF Imaging of Colorectal Cancer. *Theranostics* **2016**, *6*, 2267–2277. [CrossRef] [PubMed]

117. Evans, H.L.; Nguyen, Q.D.; Carroll, L.S.; Kaliszczak, M.; Twyman, F.J.; Spivey, A.C.; Aboagye, E.O. A bioorthogonal ^{68}Ga-labelling strategy for rapid in vivo imaging. *Chem. Commun.* **2014**, *50*, 9557–9560. [CrossRef]
118. Meyer, J.P.; Tully, K.M.; Jackson, J.; Dilling, T.R.; Reiner, T.; Lewis, J.S. Bioorthogonal Masking of Circulating Antibody–TCO Groups Using Tetrazine-Functionalized Dextran Polymers. *Bioconjugate Chem.* **2018**, *29*, 538–545. [CrossRef] [PubMed]
119. Yazdani, A.; Bilton, H.; Vito, A.; Genady, A.R.; Rathmann, S.M.; Ahmad, Z.; Janzen, N.; Czorny, S.; Zeglis, B.M.; Francesconi, L.C.; et al. A bone-seeking trans-cyclooctene for pretargeting and bioorthogonal chemistry: A proof of concept study using ^{99m}Tc- and ^{177}Lu-labeled tetrazines. *J. Med. Chem.* **2016**, *59*, 9381–9389. [CrossRef]
120. García, M.F.; Gallazzi, F.; de Souza Junqueira, M.; Fernández, M.; Camacho, X.; da Silva Mororó, J.; Faria, D.; de Godoi Carneiro, C.; Couto, M.; Carrión, F.; et al. Synthesis of hydrophilic HYNIC-[1,2,4,5] tetrazine conjugates and their use in antibody pretargeting with ^{99m}Tc. *Org. Biomol. Chem.* **2018**, *16*, 5275–5285. [CrossRef] [PubMed]
121. Shah, M.A.; Zhang, X.; Rossin, R.; Robillard, M.S.; Fisher, D.R.; Bueltmann, T.; Hoeben, F.J.; Quinn, T.P. Metal-free cycloaddition chemistry driven pretargeted radioimmunotherapy using α-particle radiation. *Bioconjugate chem.* **2017**, *28*, 3007–3015. [CrossRef]
122. Jeon, J.; Shen, B.; Xiong, L.; Miao, Z.; Lee, K.H.; Rao, J.; Chin, F.T. Efficient method for site-specific ^{18}F-labeling of biomolecules using the rapid condensation reaction between 2-cyanobenzothiazole and cysteine. *Bioconjugate Chem.* **2012**, *23*, 1902–1908. [CrossRef] [PubMed]
123. Su, X.; Cheng, K.; Jeon, J.; Shen, B.; Venturin, G.T.; Hu, X.; Rao, J.; Chin, F.T.; Wu, H.; Cheng, Z. Comparison of Two Site-Specifically ^{18}F-Labeled Affibodies for PET Imaging of EGFR Positive Tumors. *Mol. Pharm.* **2014**, *11*, 3947–3956. [CrossRef] [PubMed]
124. Inkster, J.A.; Colin, D.J.; Seimbille, Y. A novel 2-cyanobenzothiazole-based ^{18}F prosthetic group for conjugation to 1,2-aminothiol-bearing targeting vectors. *Org. Biomol. Chem.* **2015**, *13*, 3667–3676. [CrossRef] [PubMed]
125. Chen, K.T.; Nguyen, K.; Ieritano, C.; Gao, F.; Seimbille, Y. A Flexible Synthesis of ^{68}Ga-Labeled Carbonic Anhydrase IX (CAIX)-Targeted Molecules via CBT/1,2-Aminothiol Click Reaction. *Molecules* **2019**, *24*, 23. [CrossRef] [PubMed]
126. Gao, F.; Ieritano, C.; Chen, K.-T.; Dias, G.M.; Rousseau, J.; Bénard, F.; Seimbille, Y. Two bifunctional desferrioxamine chelators for bioorthogonal labeling of biovectors with zirconium-89. *Org. Biomol. Chem.* **2018**, *16*, 5102–5106. [CrossRef]
127. Mushtaq, S.; Choi, D.S.; Jeon, J. Radiosynthesis of ^{125}I-labeled 2-cyanobenzothiazole: A new prosthetic group for efficient radioiodination reaction. *J. Radiopharm. Mol. Probes* **2017**, *3*, 44–51.
128. Toda, N.; Asano, S.; Barbas, C.F., III. Rapid, Stable, Chemoselective Labeling of Thiols with Julia Kocieński like Reagents: A Serum-Stable Alternative to Maleimide-Based Protein Conjugation. *Angew. Chem. Int. Ed.* **2013**, *52*, 12592–12596. [CrossRef]
129. Chiotellis, A.; Sladojevich, F.; Mu, L.; Herde, A.M.; Valverde, I.E.; Tolmachev, V.; Schibli, R.; Ametamey, S.M.; Mindt, T.L. Novel chemoselective ^{18}F-radiolabeling of thiol-containing biomolecules under mild aqueous conditions. *Chem. Commun.* **2016**, *52*, 6083–6086. [CrossRef]
130. Shim, H.E.; Mushtaq, S.; Song, L.; Lee, C.H.; Lee, H.; Jeon, J. Development of a new thiol-reactive prosthetic group for site-specific labeling of biomolecules with radioactive iodine. *Bioorg. Med. Chem. Lett.* **2018**, *28*, 2875–2878. [CrossRef]
131. Mushtaq, S.; Nam, Y.R.; Kang, J.A.; Choi, D.S.; Park, S.H. Efficient and Site-Specific ^{125}I Radioiodination of Bioactive Molecules Using Oxidative Condensation Reaction. *ACS Omega* **2018**, *3*, 6903–6911. [CrossRef]
132. Al-Momani, E.; Israel, I.; Buck, A.K.; Samnick, S. Improved synthesis of [^{18}F]FS-PTAD as a new tyrosine-specific prosthetic group for radiofluorination of biomolecules. *Appl. Radiat. Isot.* **2015**, *104*, 136–142. [CrossRef] [PubMed]
133. Zlatopolskiy, B.D.; Kandler, R.; Mottaghy, F.M.; Neumaier, B. C-(4-[^{18}F]fluorophenyl)-N-phenyl nitrone: A novel ^{18}F-labeled building block for metal free [3+2] cycloaddition. *Appl. Radiat. Isot.* **2012**, *70*, 184–192. [CrossRef]
134. Zlatopolskiy, B.D.; Kandler, R.; Kobus, D.; Mottaghy, F.M.; Neumaier, B. Beyond azide–alkyne click reaction: Easy access to ^{18}F-labelled compounds via nitrile oxide cycloadditions. *Chem. Commun.* **2012**, *48*, 7134–7136. [CrossRef] [PubMed]

135. Urkow, J.; Bergman, C.; Wuest, F. Sulfo-click chemistry with ^{18}F-labeled thio acids. *Chem. Commun.* **2019**, *55*, 1310–1313. [CrossRef] [PubMed]
136. Patra, M.; Eichenberger, L.S.; Fischer, G.; Holland, J.P. Photochemical Conjugation and One-Pot Radiolabelling of Antibodies for Immune-PET. *Angew. Chem. Int. Ed.* **2019**, *58*, 1928–1933. [CrossRef] [PubMed]
137. Eichenberger, L.S.; Patra, M.; Holland, J.P. Photoactive chelates for radiolabelling proteins. *Chem. Commun.* **2019**, *55*, 2257–2260. [CrossRef] [PubMed]
138. Patra, M.; Klingler, S.; Eichenberger, L.S.; Holland, J.P. Simultaneous Photoradiochemical Labeling of Antibodies for Immuno-Positron Emission Tomography. *IScience* **2019**, *13*, 416–431. [CrossRef] [PubMed]

© 2019 by the authors. Licensee MDPI, Basel, Switzerland. This article is an open access article distributed under the terms and conditions of the Creative Commons Attribution (CC BY) license (http://creativecommons.org/licenses/by/4.0/).

Article

Fluorine-18 Labeled Fluorofuranylnorprogesterone ([^{18}F]FFNP) and Dihydrotestosterone ([^{18}F]FDHT) Prepared by "Fluorination on Sep-Pak" Method

Falguni Basuli [1,*], Xiang Zhang [1], Burchelle Blackman [1], Margaret E. White [2,3], Elaine M. Jagoda [2], Peter L. Choyke [2] and Rolf E. Swenson [1]

[1] Imaging Probe Development Center, National Heart, Lung, and Blood Institute, National Institutes of Health, Rockville, MD 20850, USA
[2] Molecular Imaging Program, National Cancer Institute, National Institutes of Health, Bethesda, MD 20892, USA
[3] Laboratory of Genitourinary Cancer Pathogenesis, National Cancer Institute, National Institutes of Health, Bethesda, MD 20892, USA
* Correspondence: bhattacharyyaf@mail.nih.gov

Academic Editor: Svend Borup Jensen
Received: 14 June 2019; Accepted: 25 June 2019; Published: 28 June 2019

Abstract: To further explore the scope of our recently developed "fluorination on Sep-Pak" method, we prepared two well-known positron emission tomography (PET) tracers 21-[^{18}F]fluoro-16α,17α-[(R)-(1'-α-furylmethylidene)dioxy]-19-norpregn-4-ene-3,20-dione furanyl norprogesterone ([^{18}F]FFNP) and 16β-[^{18}F]fluoro-5α-dihydrotestosterone ([^{18}F]FDHT). Following the "fluorination on Sep-Pak" method, over 70% elution efficiency was observed with 3 mg of triflate precursor of [^{18}F]FFNP. The overall yield of [^{18}F]FFNP was 64–72% (decay corrected) in 40 min synthesis time with a molar activity of 37–81 GBq/μmol (1000–2200 Ci/mmol). Slightly lower elution efficiency (~55%) was observed with the triflate precursor of [^{18}F]FDHT. Fluorine-18 labeling, reduction, and deprotection to prepare [^{18}F]FDHT were performed on Sep-Pak cartridges (PS-HCO$_3$ and Sep-Pak plus C-18). The overall yield of [^{18}F]FDHT was 25–32% (decay corrected) in 70 min. The molar activity determined by using mass spectrometry was 63–148 GBq/μmol (1700–4000 Ci/mmol). Applying this quantitative measure of molar activity to in vitro assays [^{18}F]FDHT exhibited high-affinity binding to androgen receptors (K$_d$~2.5 nM) providing biological validation of this method.

Keywords: fluorine-18; fluorination on Sep-Pak; [^{18}F]FFNP; [^{18}F]FDHT; mass spectrometry; molar activity

---

1. Introduction

Recently we have developed a novel method, "fluorination on Sep-Pak", to prepare prosthetic groups for radiofluorination of biomolecules which requires neither azeotropic drying of fluorine-18 nor addition of base [1–3]. Hence, it is less time consuming and more suitable for base and temperature sensitive starting materials. Fluorine-18 was trapped on an anion exchange cartridge (PS-HCO$_3$) and dried by flowing anhydrous acetonitrile. The direct incorporation of fluorine-18 was achieved by passing the precursor solution (0.5 mL/min) through this cartridge. Recently this method has been expanded to prepare a variety of fluorine-18 labeled compounds via a well-established copper-mediated radiofluorination [4]. Therefore we envisioned that "fluorination on Sep-Pak" method might be applicable to other room temperature fluorine-18 labeling reactions. To test the hypothesis, this radiolabeling method have been considered to prepare two well-established PET tracers (Figure 1), 21-[^{18}F]fluoro-16α,17α-[(R)-(1'-α-furylmethylidene)dioxy]-19-norpregn-4-ene-3,20-dione furanyl norprogesterone ([^{18}F]FFNP) and 16β-[^{18}F]fluoro-5α-dihydrotestosterone ([^{18}F]FDHT) which have been shown to have clinical relevance.

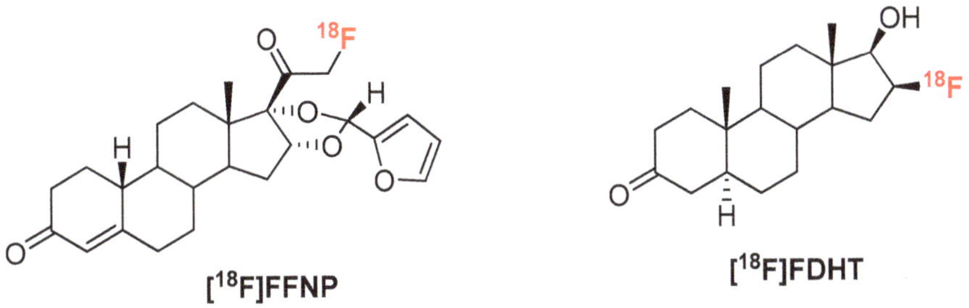

Figure 1. Structures of [^{18}F]FFNP and [^{18}F]FDHT.

Breast cancer (BC) is the most common type of cancer in women worldwide [5–7]. Estrogen receptor alpha (ERα) and progesterone receptor (PR) are important steroid hormone receptor biomarkers used to determine prognosis and to predict benefit from endocrine therapies for breast cancer patients [6,7]. Advancement of early detection and treatment of breast cancer is the key parameter for the steady decrease of breast cancer mortality. Molecular imaging is a useful technique for early detection and quantitative measurement of disease both in primary and across metastatic sites of disease [8]. The radioligand [^{18}F]FFNP demonstrates high relative binding affinity to PR and low nonspecific binding. It has been used for quantitative assessment of PR expression in vivo using PET [9–11]. Human safety and dosimetry have been assessed for identifying PR-positive breast cancer [12]. The literature reported procedure of radiolabeling to prepare [^{18}F]FFNP is by a conventional radiolabeling method using azeotropically dried [^{18}F]KF/K$_{222}$/K$_2$CO$_3$ and a triflate precursor (1) in acetonitrile either at room temperature or at elevated temperature [13,14]. With this radiosynthetic requirement for a triflate precursor and the potential clinical applications, [^{18}F]FFNP represented an ideal tracer for testing the "fluorination on Sep-Pak" method.

Prostate cancer (PC) is the most common cancer in men all over the world. It is the second leading cause of death from cancer [5]. Androgen receptor (AR) is a member of the nuclear receptor superfamily and it has a central role in prostate cancer progression [15,16]. Therefore, AR is the focus of detection and therapeutic treatment of PC [11,17,18]. PET imaging of AR can be an important tool to quantify expression in both primary tumors and metastatic sites simultaneously. It can also be a useful tool for evaluation of AR-targeted therapies. Fluorine-18 labeled 16β-fluoro-5α-dihydrotestosterone ([^{18}F]FDHT, Figure 1) has been clinically proven useful for detection and relative quantification of AR expression [19–22]. Synthesis of [^{18}F]FDHT was originally reported by reacting its triflate precursor with n-[^{18}F]Bu$_4$NF/n-Bu$_4$NOH in THF at 55 °C [23]. In this procedure, LiAlH$_4$ was used as reducing agent. Later the mild reducing agent, NaBH$_4$, was used for better compatibility in an automated synthesis module [13]. Further improvements have been reported using a fully automated radiolabeling procedure [24]. Herein we report the radiosynthesis of [^{18}F]FFNP and [^{18}F]FDHT via the "fluorination on Sep-Pak" strategy. The purity was assessed by high-performance liquid chromatography (HPLC). The quality of the [^{18}F]FDHT was further tested in vitro using saturation binding assays from which the affinity (K_d) for AR was determined.

2. Results and Discussion

2.1. [^{18}F]FFNP

Synthesis of [^{18}F]FFNP was first reported from its triflate precursor (1) with n-Bu$_4$N[^{18}F]F/n-Bu$_4$NOH in THF at room temperature with a decay-corrected radiochemical yield (RCY) of 2–13% [9].

D. Zhou et al. observed a better RCY with the triflate precursor which was purified without aqueous workup [13]. Fluorination was achieved by using K[^{18}F]F/K$_{222}$/K$_2$CO$_3$ as a fluorinating agent for the same triflate precursor (1) in acetonitrile. Up to 77% RCY (isolated after HPLC, decay-corrected) of [^{18}F]FFNP had been reported starting with 1.85 GBq (50 mCi) of [^{18}F]fluoride in 90 min [13]. In these literature reported methods, fluorination was achieved at room temperature, therefore we envisioned that our recently developed "fluorination on Sep-Pak" method might be suitable to prepare [^{18}F]FFNP. To test this hypothesis, an acetonitrile (0.5 mL) solution of the triflate precursor (1.5 mg) was passed through a PS-HCO$_3$ cartridge (0.5 mL/min) containing [^{18}F]fluoride followed by washing the cartridge with 1 mL acetonitrile. Over 40% of trapped fluoride was eluted from the cartridge (Scheme 1, Table 1).

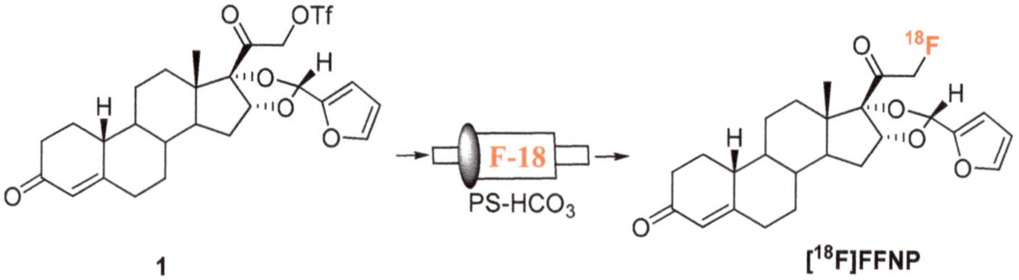

Scheme 1. Synthesis of [^{18}F]FFNP.

Table 1. Elution conditions from the Sep-Pak to prepare [^{18}F]FFNP.

Tracer	Amount of Precursor 1 or 2 (mg)	Solvent (0.5 mL)	Eluted from the Sep-Pak (%) [a]
[^{18}F]FFNP	1.5	Acetonitrile	42
	3	Acetonitrile	71–79 [b]
	5	Acetonitrile	76
		DMF	40
		DMSO	49
		THF	36

[a] Radiolabeling was performed with 0.37–0.74 GBq (10–20 mCi) of fluorine-18; [b] $n \geq 6$.

A representative HPLC of the crude product is shown in Figure 2a. The identity of the product was confirmed by co-injecting the crude reaction mixture with authentic nonradioactive standard, FFNP, (Figure 2b). Inspired by this result, the fluoride elution efficiencies were tested with a higher amount of precursor (Table 1) and over 70% elution efficiency was observed with 3 mg of the precursor. No significant improvement of the elution efficiency was noticed by further increasing the amount of precursor (5 mg). Literature methods used 2–4 mg of the precursor [13,14]. Fluoride incorporation efficiencies were also tested using different solvents (Table 1). Better elution of the product was observed with acetonitrile. After standardizing the elution efficiencies with a low amount of activity (0.37–0.74 GBq, 10–20 mCi) the full-scale reactions were performed with high amount (6.7–8.5 GBq, 180–230 mCi) of activities. The mixture was purified by HPLC using semi-preparative column (method B) to produce > 98% radiochemically pure (Figure 3a) product with a molar activity of 37–81 GBq/μmol (1000–2200 Ci/mmol, $n = 6$). The overall RCY was 64–72% (decay corrected, $n = 6$) in 40 min. The identity of the product was confirmed by comparing its HPLC retention time with co-injected authentic non-radioactive standard (Figure 3b). Using the same amount of precursor (3 mg) as reported in the literature, comparable RCY (64–72% vs. 77%) was obtained in a shorter synthesis time (40 min vs. 70 min). For a direct comparison of the RCY, synthesis of [^{18}F]FFNP was performed using [^{18}F]KF/K$_2$CO$_3$/K$_{222}$ with 3 mg of triflate precursor (1) following the literature method [13]. In a typical reaction starting with 7.2 GBq (195 mCi) of [^{18}F]fluoride, 2.0 GBq (55 mCi) of the product was obtained in 70 min (43% RCY, decay corrected). The lower RCY could be due to the aqueous workup of the triflate salt precursor as reported in the literature [13]. The workup procedure for the triflate salt in this experiment is unknown as this is from a commercial source.

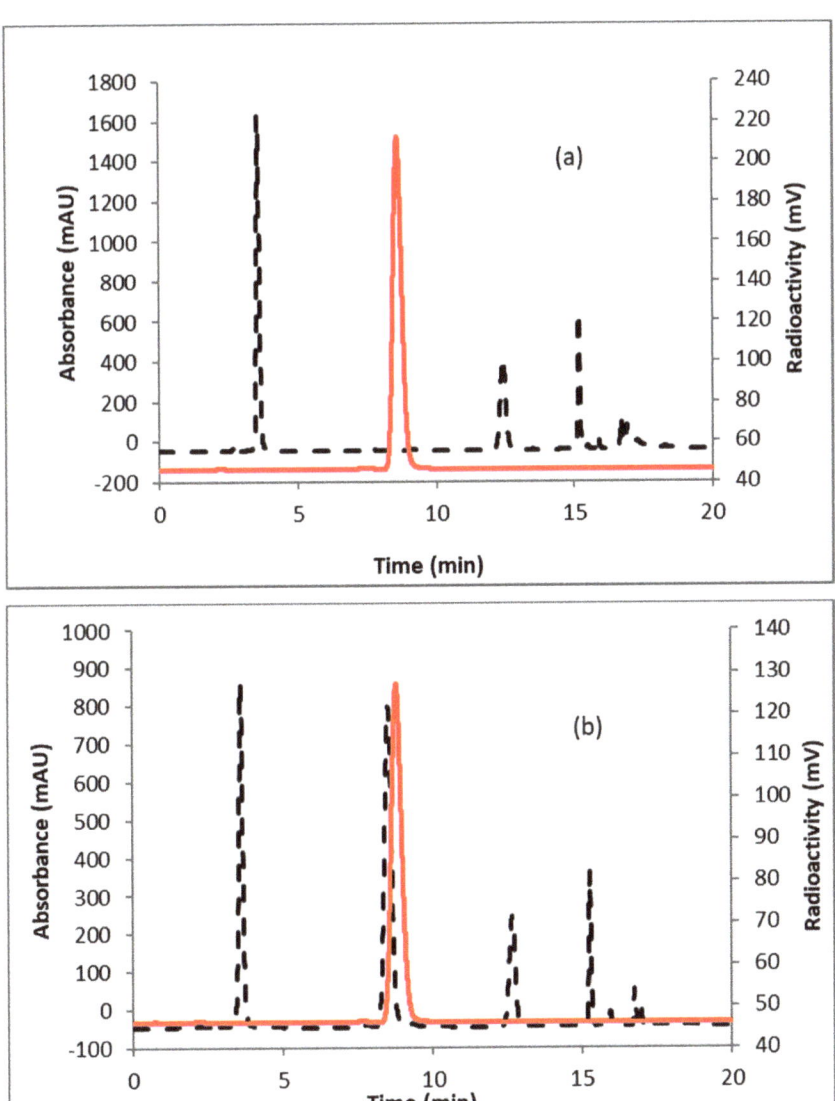

Figure 2. HPLC analysis (method A) of crude reaction mixture of (**a**) [^{18}F]FFNP; (**b**) [^{18}F]FFNP, co-injected with the non-radioactive standard. Solid line, in-line radio detector; dotted line, UV detector at 254 nm.

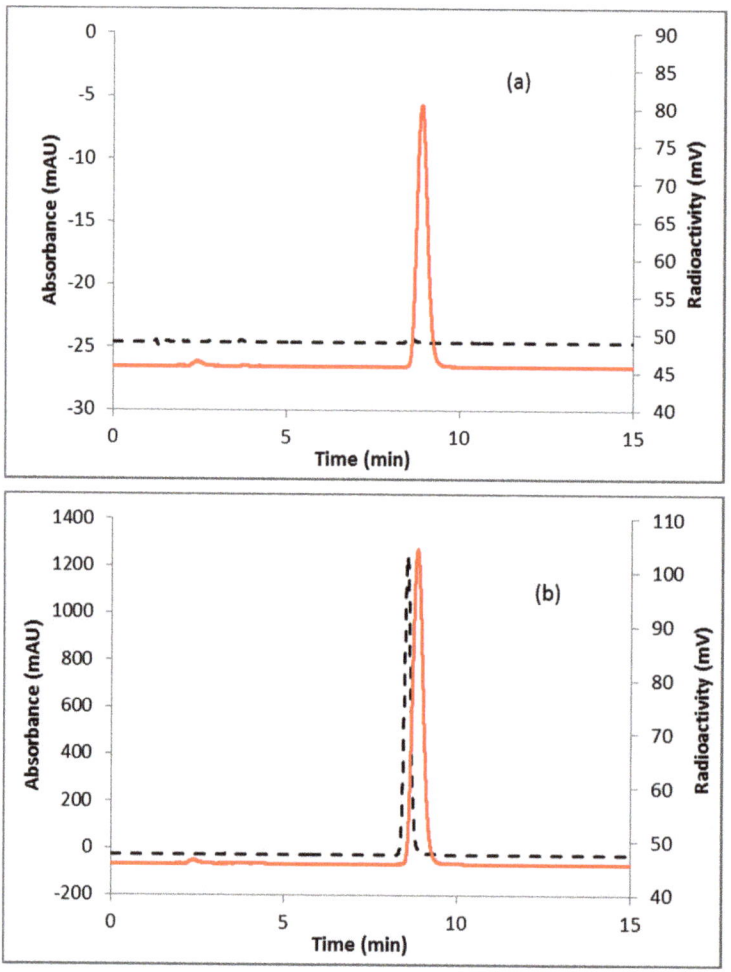

Figure 3. HPLC analysis (method A) of pure (**a**) [^{18}F]FFNP; (**b**) [^{18}F]FFNP, co-injected with the non-radioactive standard. Solid line, in-line radiodetector; dotted line, UV detector at 254 nm.

2.2. [^{18}F]FDHT

Fluorination efficiency to prepare intermediate **3** (Scheme 2) was similarly tested using different conditions (Table 2). The highest fluoride incorporation efficiency was obtained using 5 mg of precursor in 0.5 mL anhydrous tetrahydrofuran (THF) followed by washing the Sep-Pak with 1 mL acetonitrile. The next step was to reduce the intermediate **3**. The originally reported literature method used a strong reducing agent, LiAlH$_4$ which had been successfully replaced by a milder reducing agent, NaBH$_4$, for better compatibility in the automated module [13,24]. However, it has been mentioned in the literature that the quenching of a large excess of NaBH$_4$ with acetone is necessary, otherwise, it can further reduce the deprotected ketone produced during the acid deprotection step, resulting in significant decomposition of [^{18}F]FDHT [13]. In this current method, the reduction step was performed on the Sep-Pak by passing (1 mL/min) through an aqueous solution of NaBH$_4$ followed by water wash, which quantitatively reduced intermediate **3** and no quenching of NaBH$_4$ with acetone was needed. The deprotection step was also performed on the Sep-Pak using 6N HCl. HPLC analysis (Figure 4a) of the crude reaction mixture indicated the clean conversion of intermediate **3** to the final product, [^{18}F]FDHT. The final product was purified by HPLC on a semi-preparative column. The overall RCY was 25–32% (decay corrected, $n = 12$) in a 70 min synthesis time with a radiochemical purity of >98% by analytical HPLC (Figure 4b). The identity of the product was confirmed by comparing its HPLC retention time with co-injected, authentic non-radioactive standard, FDHT (Figure 4c). The identity of the product was further confirmed by comparing its mass associated with radiation peak (Figure 4d). The RCY (25–33%, $n = 12$) using the Sep-Pak method was comparable with the literature method in shorter synthesis times but used slightly higher amounts of precursor (5 mg vs. 4 mg). For a direct comparison of the RCY, the literature method was tested with 5 mg of precursor (**2**) [13]. In a typical reaction starting with 5.9 GBq (160 mCi) of [^{18}F]fluoride, 1.8 GBq (49 mCi) (31%, decay corrected) of the product was obtained in 90 min.

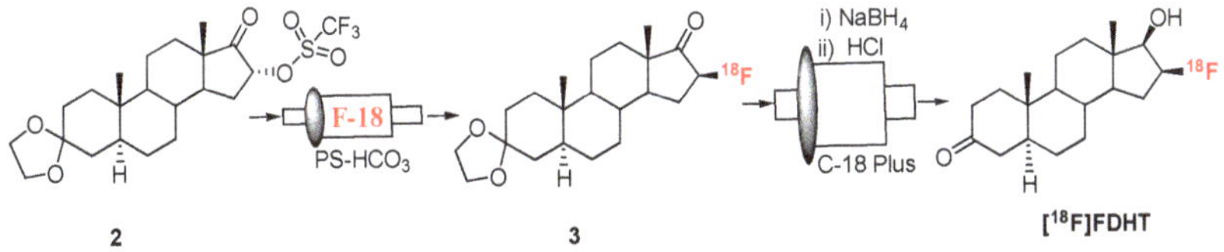

Scheme 2. Synthesis of [^{18}F]FDHT.

Table 2. Elution conditions from the Sep-Pak to prepare compound **3**, the intermediate for [^{18}F]FDHT.

Compound	Amount of Precursor 1 or 2 (mg)	Solvent (0.5 mL)	Eluted from the Sep-Pak (%) [a]
3	1.5	Acetonitrile	14
	3	Acetonitrile	34
	5	Acetonitrile	52
		DMF	13
		DMSO	27
		THF	57 [b]

[a] Radiolabeling was performed with 0.37–0.74 GBq (10–20 mCi) of fluorine-18; [b] $n \geq 6$.

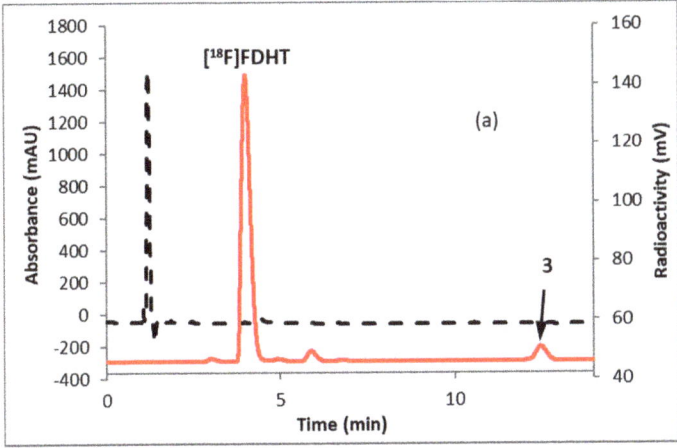

Figure 4. *Cont.*

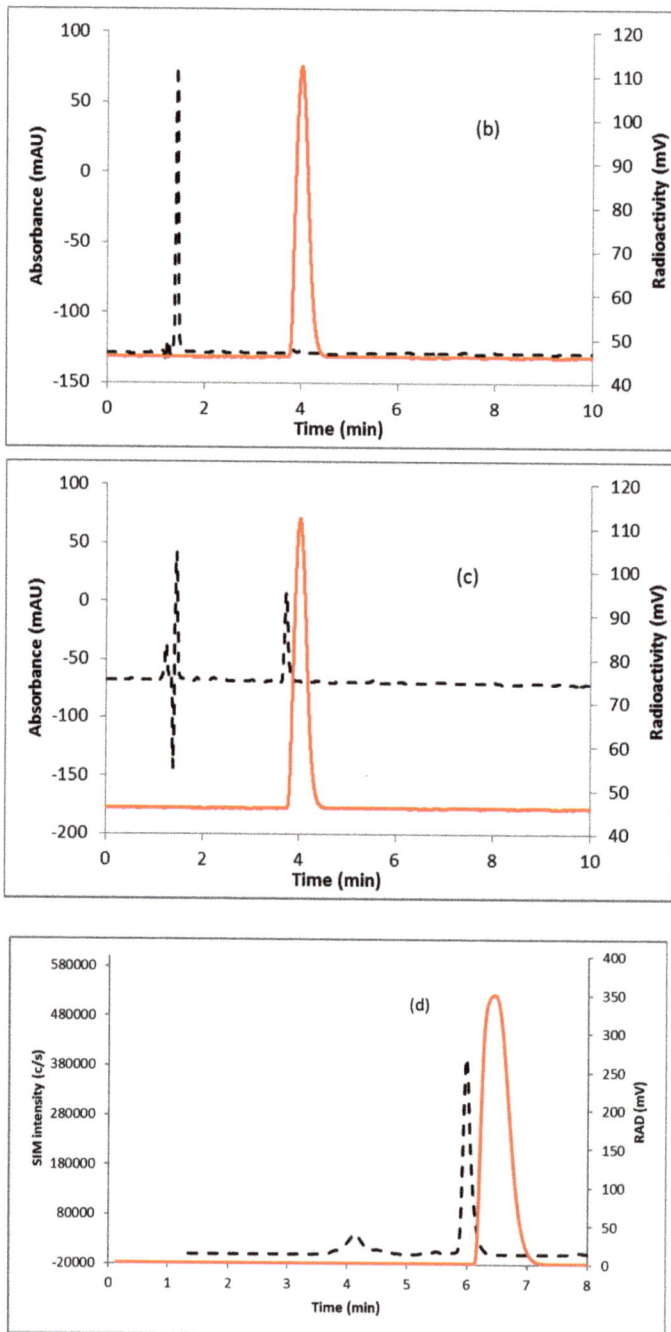

Figure 4. HPLC analysis (method A) of (**a**) crude reaction mixture of [^{18}F]FDHT; (**b**) pure [^{18}F]FDHT; (**c**) [^{18}F]FDHT, co-injected with the non-radioactive standard; (**d**) LC/MS analysis (method E) of [^{18}F]FDHT in SIM mode. Solid line, in-line radiodetector; dotted line, UV detector at 254 nm (for **a**–**c**); for d: dotted line, SIM intensity of *m/z* 350.

To measure the molar activity (the measured radioactivity per mole of the compound), a calibration curve is typically generated by using UV absorption area of known concentration of non-radioactive standard. Using this calibration curve the mass of the labeled tracer is determined. Due to the very poor UV absorption of FDHT (Figure 4c), it was not possible to measure the molar activity by this method. Mass spectrometer connected to liquid chromatography (LC-MS) is a useful technique for quality control analysis of precursors and nonradioactive standards. The mass associated with the radioactive tracer is too low to be detected by a mass spectrometer commonly used for molecular mass determination. Recently our group demonstrated the use of liquid chromatography/tandem mass spectrometry (LC-MS/MS) for the optimization of fluorine-18 labeling reaction with fluorine-19 [25]. Coiller et al. developed an in-line concentration method by using a Sep-Pak to detect the mass of the tracers [26]. We used a similar approach to characterize and determine the molar activity of [^{18}F]FDHT by using LC-MS system. The collected fraction of [^{18}F]FDHT from the HPLC was trapped on the Sep-Pak and eluting with a minimum volume of ethanol (0.2 mL) and water (0.8 mL). A calibration curve was generated using the selected ion monitoring (SIM) of the Advion expression CMS system of a known quantity of nonradioactive standard FDHT. Mass of [^{18}F]FDHT (known amount of radioactivity) was determined using this calibration curve. The molar activity determined by this method was 63–148 GBq/µmol (1700–4000 Ci/mmol, $n = 12$).

2.3. In Vitro Binding Assays

[^{18}F]FDHT exhibited high-affinity binding with a K_d of 2.49 ± 0.52 nM (mean ± SE; $n = 6$) determined from saturation binding studies using either DU145 AR+ (human androgen receptor) transfected cells or tumor cytosols prepared from DU145 AR+ xenograft mouse models (Figure 5). This K_d value compares favorably with the K_d of [^{3}H]FDHT (2.5 nM) and IC$_{50}$ of FDHT indicating the accuracy of the molar activity measurement with this method [22,27]. From the cell assays, the AR concentration (B_{max}) was 2.97 ± 0.66 × 10^5 receptors per cell (mean ± SE; $n = 4$) which was consistent with findings by Pandit-Taskar et al. [22]. The AR concentration in the tumor cytosols (B_{max}) was found to be 0.170 ± 0.048 femtomoles/µg protein (mean ± SE; $n = 2$).

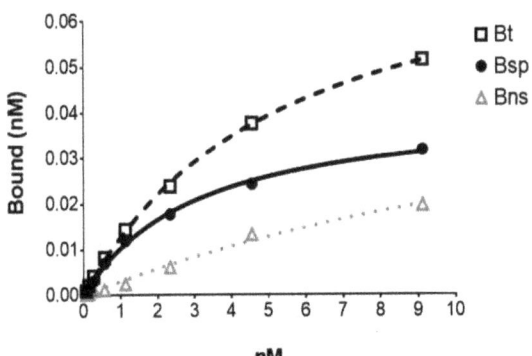

Figure 5. The representative plot from an in vitro [^{18}F]FDHT saturation binding assay using. a DU145 AR+ tumor cytosol with each point representing the average of duplicates. $B_t = B_{total}$; $B_{sp} = B_{specific}$; $B_{ns} = B_{non-specific}$ (determined in the presence of 10^{-6} M unlabeled FDHT).

3. Materials and Methods

3.1. Materials, Chemicals, and Methods

Triflate precursor for the synthesis of [^{18}F]FFNP was purchased from ChemShuttle (Hayward, CA, USA). Other precursors and non-radioactive standards were obtained from ABX GmbH (Radeberg,

Germany). PBS 1X buffer (Gibco) was obtained from Life Technologies (Carlsbad, CA, USA). Normal saline was obtained from Quality Biological (Gaithersburg, MD, USA). PD10 MiniTrap™ columns were obtained from GE Healthcare Bioscience (Pittsburg, PA, USA). All other chemicals and solvents were received from Sigma-Aldrich (St. Louis, MO, USA) and used without further purification. Fluorine-18 was obtained from the National Institutes of Health cyclotron facility (Bethesda, MD, USA). Chromafix 30-PS-HCO$_3$ anion-exchange cartridge was purchased from Macherey-Nagel (Düren, Germany) and the packing material was reduced to half (~20 mg) for better elution efficiency of [^{18}F]fluoride [3]. Phenomenex Luna C18 (2) column (10 × 250 mm, 5 μm) was purchased from Phenomenex (Torrance, CA, USA). All other columns and Sep-Pak cartridges used in this synthesis were obtained from Agilent Technologies (Santa Clara, CA, USA) and Waters (Milford, MA, USA), respectively. Sep-Pak C-18 Plus was conditioned with 5 mL ethanol, 10 mL air, and 10 mL water. Semi-prep HPLC purification and analytical HPLC analyses for radiochemical work were performed on an Agilent 1200 Series instrument equipped with multi-wavelength detectors. Mass spectrometry (MS) was performed on Advion expression compact mass spectrometer (CMS) with an ESI source (Ithaca, NY, USA). The LC inlet was Agilent 1200 series chromatographic system equipped with 1260 quaternary pump, 1260 Infinity autosampler, 1290 thermostatted column compartment, and the radiation detector. Column flow was split (1:4) between the mass spectrometer and the radiation detector. Instrument control and data processing were performed using Advion Mass Express and Quant Express Software.

HPLC conditions: Method A; Column: Agilent XDB C18 column (4.6 × 150 mm, 5 μm). Mobile phase: A: water (0.1% TFA); B: acetonitrile (0.1% TFA). Isocratic: 55% B; flow rate: 1 mL/min. Method B; Column: Phenomenex Luna C18 (2) column (10 × 250 mm, 5 μm). Mobile phase: A: water; B: acetonitrile. Isocratic: 55% B; flow rate: 4 mL/min. Method C; Column: Phenomenex Luna C18 (2) column (10 × 250 mm, 5 μm). Mobile phase: A: water; B: acetonitrile. Isocratic: 40% B; flow rate: 4 mL/min. Method D; Column: Agilent XDB C18 column (4.6 × 150 mm, 5 μm). Mobile phase: A: water (0.1% TFA); B: acetonitrile (0.1% TFA). Isocratic: 55% B; flow rate: 1 mL/min. Method E; Column: Agilent Poroshell 120 EC-C18 column (4.6 × 50 mm, 1.8 μm). Mobile phase: water (0.1% formic acid); B: acetonitrile (0.1% formic acid). Gradient: 5% B for 1 min, 5-95% B within 5.95 min and held at 95% B for 2 min, returned to initial conditions; flow rate: 1 mL/min. The column oven was kept at 40 °C throughout the analysis. The injection volume was 20 μL.

Mass Spectrometric conditions for molar activity determination of [^{18}F]FDHT: Mass spectrometric data were acquired in positive ion mode with the following ESI-MS parameters: Capillary Temperature 300 °C, Capillary Voltage 100 °C, Source Voltage Offset 20V, Source Voltage Span 30V, Source Gas Temperature 200 °C, ESI Voltage 3500V. Nitrogen was used as a desolvation gas. Quantification was done using a scan/SIM switching mode. The parameters for the scan mode was set as follows: scan range 50–800 m/z, scan time 100 ms, scan speed 7500 m/z/sec. In SIM mode dwell time was set at 300 ms for all analytes. Precursor ion (FDHT; calculated 308, found 350 [M + H + ACN], Verapamil; calculated 454, found 455 [M + H]) along with retention time 6.0 min was used to confirm the identity of the analyte in the sample.

3.2. Radiosynthesis of [^{18}F]FFNP

Fluorine-18 labeled target water (7.4 GBq, 200 mCi) was diluted with 2 mL water and passed through an anion-exchange cartridge (Chromafix 30-PS-HCO$_3$). The cartridge was washed with anhydrous acetonitrile (6 mL) and dried for 1 min under vacuum. The trapped [^{18}F]fluoride from the Sep-Pak was manually eluted (0.5 mL/min) with triflate precursor (1, 3 mg) in 0.5 mL acetonitrile. The cartridge was flushed with 1 mL acetonitrile and collected in the same vial. The crude reaction mixture was diluted with 2 mL of water and purified by HPLC using a semi-preparative column (method B). The collected fraction (~15 min) was diluted with 15 mL water and [^{18}F]FFNP was trapped by passing the solution through an activated Sep-Pak plus C-18 cartridge. The product was eluted with ethanol (0.6 mL) and saline (6 mL). The identity and purity of the product were confirmed by analytical HPLC (method A).

3.3. Radiosynthesis of [^{18}F]FDHT

Fluorine-18 labeled target water (7.4 GBq, 200 mCi) was diluted with 2 mL water and passed through an anion-exchange cartridge (Chromafix 30-PS-HCO$_3$). The cartridge was washed with anhydrous acetonitrile (6 mL) and dried for 1 min under vacuum. The trapped [^{18}F]fluoride from the Sep-Pak was manually eluted (0.5 mL/min) with triflate precursor (**2**, 5 mg) in 0.5 mL tetrahydrofuran (THF). The cartridge was flushed with 1 mL acetonitrile, collected in the same vial and diluted with 10 mL of water. The mixture was passed through an activated Sep-Pak plus C-18 cartridge and the cartridge was washed with 6 mL of water. An aqueous solution of sodium borohydride (NaBH$_4$, 40 mg in 1 mL) was slowly passed through the cartridge in 1 min, and the Sep-Pak was washed with 6 mL water followed by slowly passing 6N HCl (2 mL) in 2 min. The cartridge was further washed with 6 mL of water. The crude product was eluted with 2 mL of acetonitrile and diluted with 2 mL of water. The product was purified by HPLC using a semi-preparative column (method C). The collected fraction (~30 min) was diluted with 30 mL water and [^{18}F]FDHT was trapped by passing the solution through an activated HLB light Sep-Pak. The Sep-Pak was first eluted with 0.2 mL ethanol, the eluted fraction was discarded. Next, the Sep-Pak was eluted with ethanol (0.2 mL) and water (0.8 mL) to the product vial. The identity and purity of the product were confirmed by analytical HPLC (method D) and LC/MS (method E).

3.4. Preparation of Calibration Curves for Molar Activity Determination

The molar activity was determined by LC-MS analysis. Briefly, a calibration curve in the range of 1–20 µg/mL (ppm) was prepared for FDHT using verapamil as an internal standard. The curves were fitted least squares linear regression method by measuring the peak area ratio of the analyte to the internal standard. The acceptable criterion for the calibration curve was a correlation coefficient (r^2) of 0.99 or better, and that each back-calculated standard concentration must be within 20% deviation from the nominal value. Signal-to-noise (S/N) of the lowest calibration level was greater than 10.

3.5. In Vitro Studies

Saturation binding studies were performed to determine the K_d of [^{18}F]FDHT and B_{max} with DU-145 AR+ cells [human prostate cancer cells transfected with androgen receptors (AR)] or cytosol preparations of DU-145 AR+ tumors [obtained from xenograft mouse models]. The tumor cytosols were prepared by homogenizing the tumor tissue in 5 to 10 volumes of homogenization buffer [10 mM Tris-HCl (pH 7.5) containing 1.5 mM EDTA] which was followed by ultracentrifugation (200,000× g; 1 h; 4 °C) of the homogenate. After centrifugation, cytosolic supernatants were collected and protein concentrations determined using the Bradford method [28].

To the tumor cytosols in tubes or plated DU-145 AR+ cells (3–4 × 10^5 cells/well; 1 d prior to the assay) increasing concentrations of [^{18}F]FDHT (0.1–40 nM) were added to duplicate tubes; non-specific binding was determined by adding unlabeled FDHT (10^{-6} M) to another set of duplicates. After incubation (2 h, 4 °C), the bound [^{18}F]FDHT was separated from the free as follows: (1) plated cells were washed with phosphate buffered saline (PBS), treated with trypsin, and collected in vials; or (2) to the cytosol in tubes DCC (0.4% dextran-coated charcoal) was added and pelleted by centrifugation from which the supernatants were collected. The bound radioactivity for these samples was determined by gamma counting (Perkin Elmer 2480 Wizard3, Shelton, CT). From the saturation studies, the K_d and B_{max} were determined from 6 to 8 concentrations of [^{18}F]FDHT and analyzed using non-linear regression curve fitting [one-site specific binding, PRISM (version 5.04 Windows), GraphPad Software, San Diego, CA, USA].

4. Conclusions

Two well-known PET tracers [^{18}F]FFNP and [^{18}F]FDHT have successfully been prepared by "fluorination on Sep-Pak" method with moderate to high RCY in a short synthesis time (40–70 min). Both reduction of intermediate 3 and deprotection of reduced intermediate to prepare final [^{18}F]FDHT were performed on a single Sep-Pak cartridge. Due to the poor UV absorption of FDHT the conventional molar activity, determination method was not applicable. We developed a method to determine the mass associated with [^{18}F]FDHT by using LC-MS. The molar activities of [^{18}F]FDHT determined from this newly developed method were found to be accurate and yielded K_d values (~2.5 nM) which were comparable to published values. This "fluorination on Sep-Pak" method might be applicable to other room temperature fluorine-18 labeling reactions.

Author Contributions: Conceptualization, F.B., E.M.J.; Data curation, F.B., X.Z., B.B., M.E.W., E.M.J.; Formal analysis, F.B., X.Z., E.M.J.; Methodology, F.B., X.Z., B.B., M.E.W., E.M.J.; Supervision, R.E.S., P.L.C.; Writing—original draft, F.B.; Writing—review & editing, F.B., X.Z., B.B., M.E.W., E.M.J., P.L.C., R.E.S.

Funding: This project has been funded in whole or in part with federal funds from the National Cancer Institute, National Institutes of Health, under Contract No. HHSN261200800001E.

Conflicts of Interest: The authors declare no conflict of interest.

References

1. Basuli, F.; Zhang, X.; Jagoda, E.M.; Choyke, P.L.; Swenson, R.E. Facile room temperature synthesis of fluorine-18 labeled fluoronicotinic acid-2,3,5,6-tetrafluorophenyl ester without azeotropic drying of fluorine-18. *Nucl. Med. Biol.* **2016**, *43*, 770–772. [CrossRef] [PubMed]
2. Basuli, F.; Zhang, X.; Woodroofe, C.C.; Jagoda, E.M.; Choyke, P.L.; Swenson, R.E. Fast indirect fluorine-18 labeling of protein/peptide using the useful 6-fluoronicotinic acid-2,3,5,6-tetrafluorophenyl prosthetic group: A method comparable to direct fluorination. *J. Labelled Comp. Radiopharm.* **2017**, *60*, 168–175. [CrossRef] [PubMed]
3. Basuli, F.; Zhang, X.; Jagoda, E.M.; Choyke, P.L.; Swenson, R.E. Rapid synthesis of maleimide functionalized fluorine-18 labeled prosthetic group using "radio-fluorination on the Sep-Pak" method. *J. Labelled Comp. Radiopharm.* **2018**, *61*, 599–605. [CrossRef] [PubMed]
4. Zhang, X.; Basuli, F.; Swenson, R.E. An azeotropic drying-free approach for copper-mediated radiofluorination without addition of base. *J. Labelled Comp. Radiopharm.* **2019**, *62*, 139–145. [CrossRef] [PubMed]
5. Siegel, R.L.; Miller, K.D.; Jemal, A. Cancer statistics, 2019. *CA Cancer J. Clin.* **2019**, *69*, 7–34. [CrossRef] [PubMed]
6. Jatoi, I.; Chen, B.E.; Anderson, W.F.; Rosenberg, P.S. Breast cancer mortality trends in the United States according to estrogen receptor status and age at diagnosis. *J. Clin. Oncol.* **2007**, *25*, 1683–1690. [CrossRef] [PubMed]
7. Dasgupta, S.; Lonard, D.M.; O'Malley, B.W. Nuclear receptor coactivators: Master regulators of human health and disease. *Annu. Rev. Med.* **2014**, *65*, 279–292. [CrossRef]
8. Dalm, S.U.; Verzijlbergen, J.F.; De Jong, M. Review: Receptor Targeted Nuclear Imaging of Breast Cancer. *Int. J. Mol. Sci.* **2017**, *18*, 260. [CrossRef]
9. Buckman, B.O.; Bonasera, T.A.; Kirschbaum, K.S.; Welch, M.J.; Katzenellenbogen, J.A. Fluorine-18-labeled progestin 16α, 17α-dioxolanes: Development of high-affinity ligands for the progesterone receptor with high in vivo target site selectivity. *J. Med. Chem.* **1995**, *38*, 328–337. [CrossRef]
10. Salem, K.; Kumar, M.; Yan, Y.; Jeffery, J.J.; Kloepping, K.C.; Michel, C.J.; Powers, G.L.; Mahajan, A.M.; Fowler, A.M. Sensitivity and Isoform Specificity of ^{18}F-Fluorofuranylnorprogesterone for Measuring Progesterone Receptor Protein Response to Estradiol Challenge in Breast Cancer. *J. Nucl. Med.* **2019**, *60*, 220–226. [CrossRef]
11. Poelaert, F.; Van Praet, C.; Beerens, A.-S.; De Meerleer, G.; Fonteyne, V.; Ost, P.; Lumen, N. The Role of Androgen Receptor Expression in the Curative Treatment of Prostate Cancer with Radiotherapy: A Pilot Study. *Biomed. Res. Int.* **2015**, *2015*, 812815. [CrossRef] [PubMed]
12. Dehdashti, F.; Laforest, R.; Gao, F.; Aft, R.L.; Dence, C.S.; Zhou, D.; Shoghi, K.I.; Siegel, B.A.; Katzenellenbogen, J.A.; Welch, M.J. Assessment of progesterone receptors in breast carcinoma by PET with 21-^{18}F-fluoro-16α,17α -[(R)-(1'-α -furylmethylidene)dioxy]-19-norpregn- 4-ene-3,20-dione. *J. Nucl. Med.* **2012**, *53*, 363–370. [CrossRef] [PubMed]

13. Zhou, D.; Lin, M.; Yasui, N.; Al-Qahtani, M.H.; Dence, C.S.; Schwarz, S.; Katzenellenbogen, J.A. Optimization of the preparation of fluorine-18-labeled steroid receptor ligands 16α-[^{18}F]fluoroestradiol (FES), [^{18}F]fluoro furanyl norprogesterone (FFNP), and 16β -[^{18}F]fluoro-5α-dihydrotestosterone (FDHT) as radiopharmaceuticals. *J. Labelled Comp. Radiopharm.* **2014**, *57*, 371–377. [CrossRef] [PubMed]
14. Salem, K.; Kumar, M.; Kloepping, K.C.; Michel, C.J.; Yan, Y.; Fowler, A.M. Determination of binding affinity of molecular imaging agents for steroid hormone receptors in breast cancer. *Am. J. Nucl. Med. Mol. Imaging* **2018**, *8*, 119–126. [PubMed]
15. Joseph, D.R. Structure, function, and regulation of androgen-binding protein/sex hormone-binding globulin. *Vitam. Horm.* **1994**, *49*, 197–280. [PubMed]
16. The Veterans Administration Co-operative Urological Research Group. Treatment and survival of patients with cancer of the prostate. *Surg. Gynecol. Obstet.* **1967**, *124*, 1011–1017.
17. Agus, D.B.; Cordon-Cardo, C.; Fox, W.; Drobnjak, M.; Koff, A.; Golde, D.W.; Scher, H.I. Prostate cancer cell cycle regulators: Response to androgen withdrawal and development of androgen independence. *J. Natl. Cancer. Inst.* **1999**, *91*, 1869–1876. [CrossRef]
18. Isaacs, J.T.; Coffey, D.S. Adaptation versus selection as the mechanism responsible for the relapse of prostatic cancer to androgen ablation therapy as studied in the Dunning R-3327-H adenocarcinoma. *Cancer Res.* **1981**, *41*, 5070–5075.
19. Dehdashti, F.; Picus, J.; Michalski, J.M.; Dence, C.S.; Siegel, B.A.; Katzenellenbogen, J.A.; Welch, M.J. Positron tomographic assessment of androgen receptors in prostatic carcinoma. *Eur. J. Nucl. Med. Mol. Imaging* **2005**, *32*, 344–350. [CrossRef]
20. Larson, S.M.; Morris, M.; Gunther, I.; Beattie, B.; Humm, J.L.; Akhurst, T.A.; Finn, R.D.; Erdi, Y.; Pentlow, K.; Dyke, J.; et al. Tumor localization of 16β-^{18}F-fluoro-5α-dihydrotestosterone versus ^{18}F-FDG in patients with progressive, metastatic prostate cancer. *J. Nucl. Med.* **2004**, *45*, 366–373.
21. Zanzonico, P.B.; Finn, R.; Pentlow, K.S.; Erdi, Y.; Beattie, B.; Akhurst, T.; Squire, O.; Morris, M.; Scher, H.; McCarthy, T.; et al. PET-based radiation dosimetry in man of ^{18}F-fluorodihydrotestosterone, a new radiotracer for imaging prostate cancer. *J. Nucl. Med.* **2004**, *45*, 1966–1971. [PubMed]
22. Pandit-Taskar, N.; Veach, D.R.; Fox, J.J.; Scher, H.I.; Morris, M.J.; Larson, S.M. Evaluation of Castration-Resistant Prostate Cancer with Androgen Receptor–Axis Imaging. *J. Nucl. Med.* **2016**, *57*, 73S–78S. [CrossRef] [PubMed]
23. Liu, A.; Dence, C.S.; Welch, M.J.; Katzenellenbogen, J.A. Fluorine-18-labeled androgens: Radiochemical synthesis and tissue distribution studies on six fluorine-substituted androgens, potential imaging agents for prostatic cancer. *J. Nucl. Med.* **1992**, *33*, 724–734.
24. Ackermann, U.; Lewis, J.S.; Young, K.; Morris, M.J.; Weickhardt, A.; Davis, I.D.; Scott, A.M. Fully automated synthesis of [^{18}F]fluoro-dihydrotestosterone ([^{18}F]FDHT) using the FlexLab module. *J. Labelled Comp. Radiopharm.* **2016**, *59*, 424–428. [CrossRef] [PubMed]
25. Zhang, X.; Dunlow, R.; Blackman, B.N.; Swenson, R.E. Optimization of ^{18}F-syntheses using ^{19}F-reagents at tracer-level concentrations and liquid chromatography/tandem mass spectrometry analysis: Improved synthesis of [^{18}F]MDL100907. *J. Labelled Comp. Radiopharm.* **2018**, *61*, 427–437. [CrossRef] [PubMed]
26. Collier, T.L.; Dahl, K.; Stephenson, N.A.; Holland, J.P.; Riley, A.; Liang, S.H.; Vasdev, N. Recent applications of a single quadrupole mass spectrometer in ^{11}C, ^{18}F and radiometal chemistry. *J. Fluor. Chem.* **2018**, *210*, 46–55. [CrossRef] [PubMed]
27. Bauer, E.R.; Daxenberger, A.; Petri, T.; Sauerwein, H.; Meyer, H.H. Characterisation of the affinity of different anabolics and synthetic hormones to the human androgen receptor, human sex hormone binding globulin and to the bovine progestin receptor. *APMIS* **2000**, *108*, 838–846. [CrossRef] [PubMed]
28. Bradford, M.M. A rapid and sensitive method for the quantitation of microgram quantities of protein utilizing the principle of protein-dye binding. *Anal. Biochem.* **1976**, *72*, 248–254. [CrossRef]

Sample Availability: Samples of the compounds are not available from the authors.

© 2019 by the authors. Licensee MDPI, Basel, Switzerland. This article is an open access article distributed under the terms and conditions of the Creative Commons Attribution (CC BY) license (http://creativecommons.org/licenses/by/4.0/).

MDPI
St. Alban-Anlage 66
4052 Basel
Switzerland
www.mdpi.com

Molecules Editorial Office
E-mail: molecules@mdpi.com
www.mdpi.com/journal/molecules

Disclaimer/Publisher's Note: The statements, opinions and data contained in all publications are solely those of the individual author(s) and contributor(s) and not of MDPI and/or the editor(s). MDPI and/or the editor(s) disclaim responsibility for any injury to people or property resulting from any ideas, methods, instructions or products referred to in the content.